AF905742

Fatigue and Fracture Assessments for Welded Joints

This cutting-edge text covers current experimental approaches to quantify the fracture resistance in non-conventional specimens, assessment of ductile tearing and brittle fracture, and new approaches to estimate welded joints under high and low cycles and combined actions. It also presents a digital twin approach for the fatigue crack sizing and life assessment, along with the example of an offshore structure with applications for welded joints. Using experimental quantification, this bridges the gap between material fracture resistance and cracked engineering structures' fracture resistance such as surface cracks in structural components. Ductile and brittle fracture assessment integrates physics-enhanced models into new engineering assessment approaches applicable to welded joints under combined high and low cycle fatigue loadings.

This book is essential as a textbook for graduate students and a reference for practicing engineers in structural, mechanical, and marine and offshore engineering.

Liuyang Feng is a full professor at the School of Aeronautics, Northwestern Polytechnical University, and key member of the National Key Laboratory of Strength and Structural Integrity, China. He is a member of the Experimental Mechanics Technical Committee, the Chinese Society of Theoretical and Applied Mechanics, specialized in structural integrity assessment.

Xudong Qian is a Dean's chair associate professor in the Department of Civil and Environmental Engineering. He is currently serving as the Director of the Centre for Offshore Research and Engineering, of the Centre for Advanced Materials and Structures, and of the Coastal Protection and Flood Resilience Institute at National University of Singapore. He is also a member of the ASTM Technical Committee on Fatigue and Fracture.

Fatigue and Fracture Assessments for Welded Joints

From physics-based approaches to digital twins

Liuyang Feng and Xudong Qian

CRC Press
Taylor & Francis Group
Boca Raton London New York

CRC Press is an imprint of the
Taylor & Francis Group, an **informa** business

MATLAB® and Simulink® are trademarks of The MathWorks, Inc. and are used with permission. The MathWorks does not warrant the accuracy of the text or exercises in this book. This book's use or discussion of MATLAB® or Simulink® software or related products does not constitute endorsement or sponsorship by The MathWorks of a particular pedagogical approach or particular use of the MATLAB® and Simulink® software.

First edition published 2026
by CRC Press
4 Park Square, Milton Park, Abingdon, Oxon, OX14 4RN

and by CRC Press
2385 NW Executive Center Drive, Suite 320, Boca Raton FL 33431

© 2026 Liuyang Feng and Xudong Qian

CRC Press is an imprint of Informa UK Limited

The right of Liuyang Feng and Xudong Qian to be identified as authors of this work has been asserted in accordance with sections 77 and 78 of the Copyright, Designs and Patents Act 1988.

All rights reserved. No part of this book may be reprinted or reproduced or utilised in any form or by any electronic, mechanical, or other means, now known or hereafter invented, including photocopying and recording, or in any information storage or retrieval system, without permission in writing from the publishers.

For permission to photocopy or use material electronically from this work, access www.copyright.com or contact the Copyright Clearance Center, Inc. (CCC), 222 Rosewood Drive, Danvers, MA 01923, 978-750-8400. For works that are not available on CCC please contact mpkbookspermissions@tandf.co.uk

For Product Safety Concerns and Information please contact our EU representative GPSR@taylorandfrancis.com. Taylor & Francis Verlag GmbH, Kaufingerstraße 24, 80331 München, Germany.

Trademark notice: Product or corporate names may be trademarks or registered trademarks, and are used only for identification and explanation without intent to infringe.

British Library Cataloguing-in-Publication Data
A catalogue record for this book is available from the British Library

ISBN: 978-1-032-83117-6 (hbk)
ISBN: 978-1-032-83537-2 (pbk)
ISBN: 978-1-003-50980-6 (ebk)

DOI: 10.1201/9781003509806

Typeset in Times New Roman
by Apex CoVantage, LLC

Contents

1 Introduction

1.1 BACKGROUND

Fatigue and fracture mechanics address the progressive degradation of structural integrity under static and cyclic loads, which accounts for approximately 80% of mechanical failures [1, 2] across industries such as aerospace and marine structures. These related failures (as shown in Figure 1.1) threaten work safety and cause substantial economic loss. Therefore, the approaches to understand, predict, mitigate, and prevent fatigue failures are critical for designing resilient structures, extending service life, and achieving sustainability goals.

The scientific understanding of fatigue and fracture has evolved significantly since the 19th century. Starting from the documented Paris-Mezieres train axle failure in 1842, the theoretical foundation of the fatigue and fracture mechanics was gradually laid through *S-N* curves by August Wöhler [3, 4], linear elastic fracture mechanics by George Irwin [5], cumulative damage models by Miner-Palmgren [6], Paris' law for crack growth [7] and *J*-integral for elastic-plastic fracture mechanics (EPFM) by J.R. Rice [8]. This progression marks a shift from empirical observations to mechanical understanding, integrating microstructural effects and advancing from safe-life to damage-tolerant approaches [9]. Experimental methodologies have advanced basic mechanical testing to non-destructive and high-resolution methods, enabling precise quantification of fatigue life, crack initiation and growth, and fracture toughness. Conventional test approaches (Figure 1.2) include axial high-cycle fatigue tests [10, 11], fatigue crack growth measurement [11], material fracture toughness tests [12], and strain-controlled low-cycle fatigue tests [13].

Advanced experimental techniques (Table 1.1) incorporate digital image correlation (DIC) [14], acoustic emission (AE) [15], ultrasound [16], and current drop [17] to monitor the strain variation, fracture process zones, and crack evolution. Techniques including scanning electron microscopy (SEM) [18] are adopted to examine the

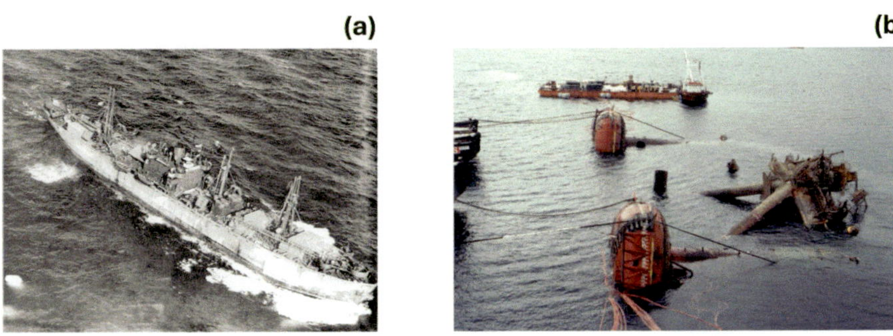

FIGURE 1.1 (a) WWII Liberty ships failure, (b) 1980 Alexander L. Kielland platform capsize.

DOI: 10.1201/9781003509806-1

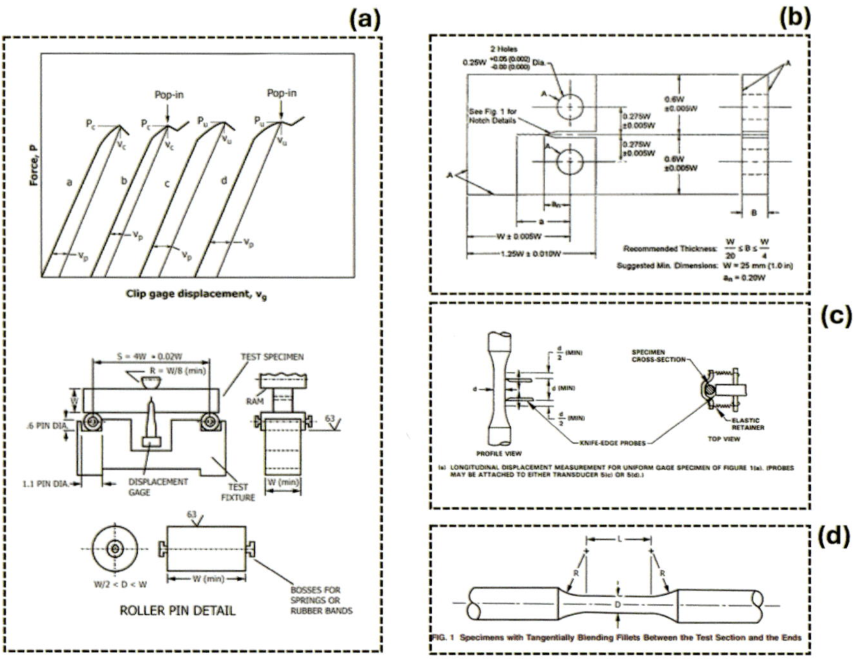

FIGURE 1.2 (a) Fracture toughness test scheme based on E1820; (b) CT specimen for fatigue crack growth based on E647; (c) coupon specimen for strain-controlled fatigue tests based on E606; and (d) coupon specimen for load-controlled fatigue tests based on E466.

TABLE 1.1
Some Examples of Advanced Experimental Techniques for Fatigue and Fracture.

	Instrument examples	Physical illustration	Representative results
Scanning electron microscope		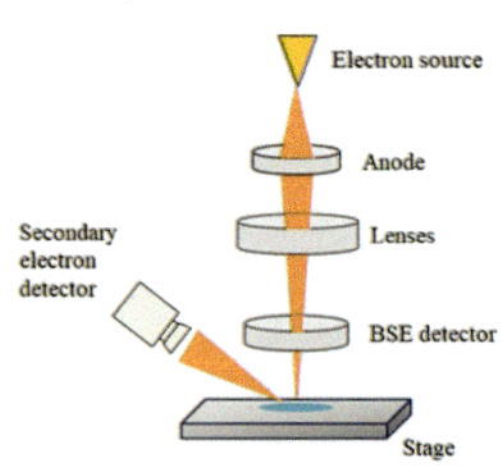 	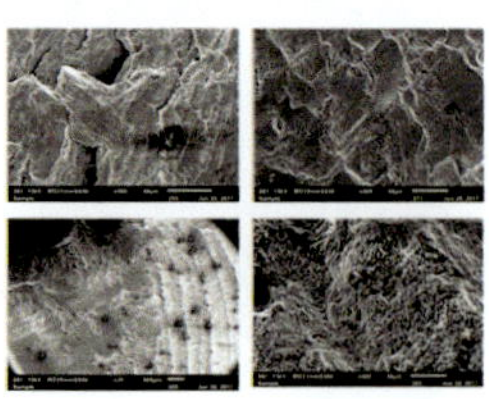
Alternating current potential drop		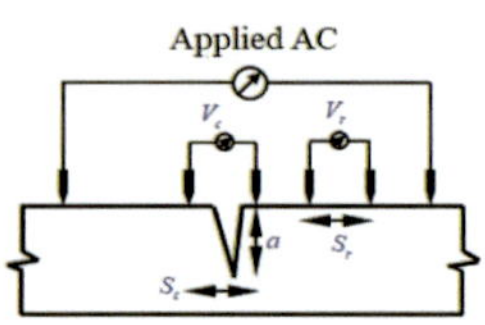 	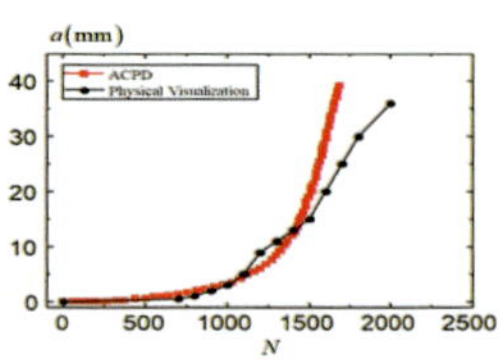

(Continued)

TABLE 1.1 *(Continued)*

Some Examples of Advanced Experimental Techniques for Fatigue and Fracture.

	Instrument examples	Physical illustration	Representative results
Ultrasonic phased array	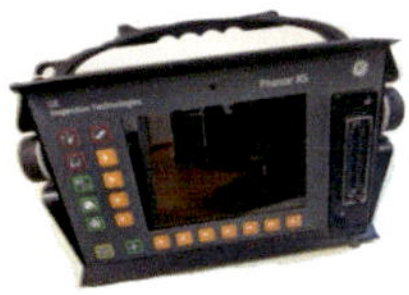	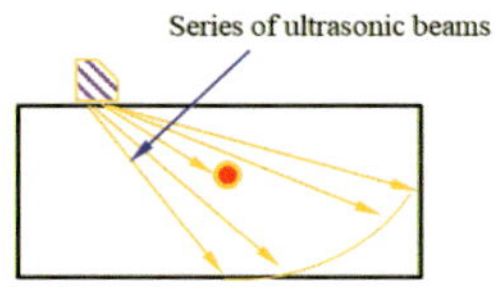	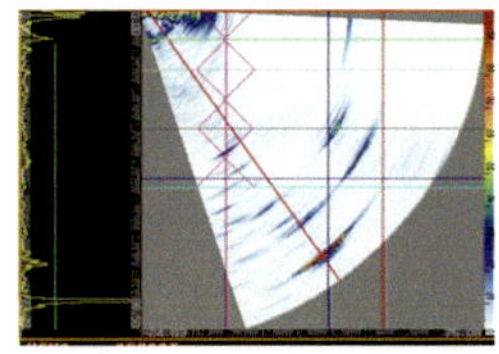
Digital camera correlation		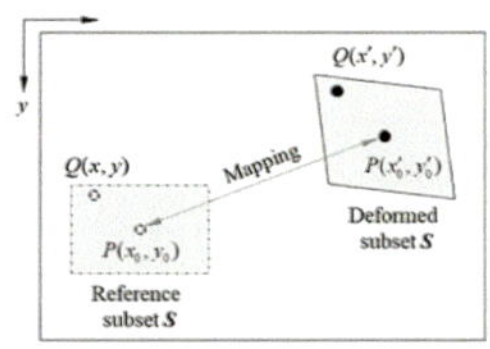	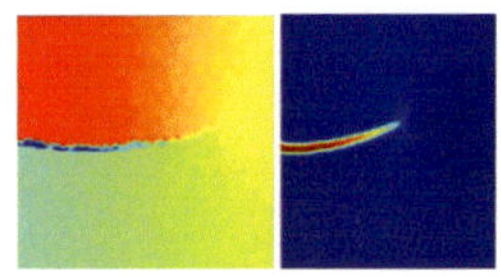
X-ray residual stress measurement		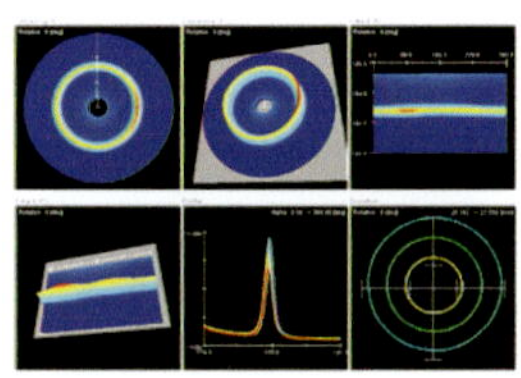	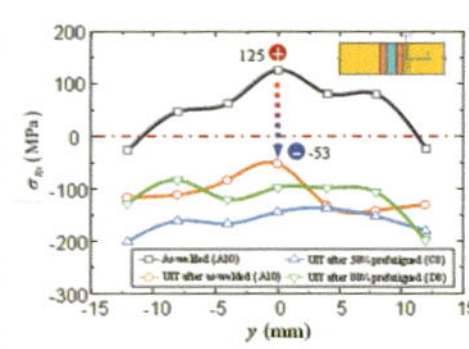

fractography and microstructural effects. Techniques such as X-ray diffraction or hole-drilling are employed to measure the welding residual stress of welded joints [19, 20]. Numerical simulation methodologies (Figure 1.3) have transformed fatigue and fracture analysis from analytical approximations for simple geometries and loading scenarios to predictive and multiscale modeling for complex geometries and loading conditions. The finite element method (FEM) remains a foundational and well-accepted technique for simulating the crack tip stress field, crack initiation, and growth [21, 22]. The extended FEM [23] enhances the effectiveness of simulation for arbitrary crack patterns through the enrichments. Boundary element method (BEM) [24, 25] reduces the dimensionality of the problem by discretizing only the boundaries, leading to fewer degrees of freedom and lower computational costs. Phase-field models [26, 27] are alternative solutions for modeling complex crack propagation through smearing the crack over a diffuse region without explicitly tracking the crack geometry profile.

Welded structures are widely adopted in bridges, ships, offshore platforms, and aerospace components and have been a focal point due to their susceptibility to fatigue and fracture failure. The 1980 Alexander L. Kielland platform capsize [28] and 1989 Exxon Valdez oil spill [29] highlight welding-related vulnerabilities, meanwhile motivating advancements in welding technology and design

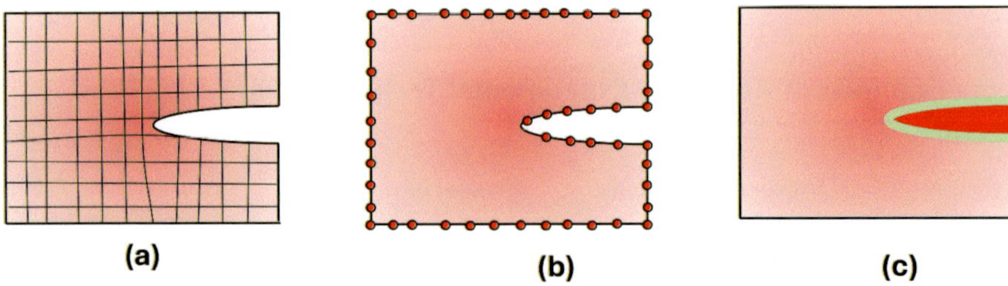

FIGURE 1.3 Illustrative scheme of (a) FEM for crack-related simulation; (b) BEM for crack-related simulation; (c) phase-field for crack-related simulation.

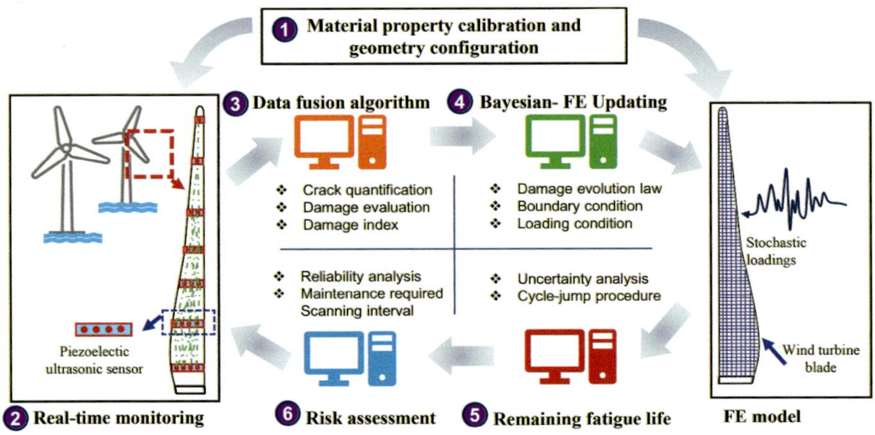

FIGURE 1.4 Illustrative scheme for digital twin application for floating wind turbines.

guidelines. Emerging research has intensified on fatigue and fracture evaluation of welded joints [30, 31], weld profiles optimization [32], and monitoring and repair techniques on welded components [33]. Size effect [34], welding-induced defects [35], complex loadings [36], and environmental factors [17] have been addressed through experimental and numerical simulation. Research studies have endeavored to address the above topics for traditional welded plate joints [37], tubular joints [38, 39] and 3D printing components [40] with different welding processes and geometry configurations.

Originating in the early 2000s for aerospace and manufacturing, digital twins [41] represent a revolutionary leap in monitoring, prediction, and optimization of welded structures, offering a virtual replica that mirrors the physical asset in real-time through integrated data streaming from sensors, simulations, and historical records (as shown in Figure 1.4). Digital twins have evolved to incorporate intelligent algorithms to update predictions dynamically based on operational data. In the context of welded structures, such as bridges, pipelines, offshore structures, and aerospace components, digital twins facilitate the assessment and maintenance of welded components by incorporating multiphysics simulation and intelligent approaches (machine learning and deep learning) under environmental loading conditions. Recent studies highlight their applications in classifying surface cracks [42], optimizing maintenance schedules [43], and predicting residual life [44] for welded structures under variable loads. Physics-informed neural networks (PINN) and digital twin frameworks now enable real-time monitoring of welding residual stresses [45, 46], offering a proactive approach to prevent fatigue failures in critical infrastructure. The virtual replicas of welded joints integrate real-time sensor data – such as strain gauges [47, 48] with high-fidelity finite element (FE) models to simulate fatigue crack initiation and propagation. This convergence of digital and physical domains promises to transform lifecycle management, aligning with Industry 4.0 initiatives. By bridging the gap between empirical data and computational simulations, digital twins of welded structures stand poised to elevate safety standards, sustainability, and efficiency in engineering practices, ultimately contributing to the resilient design of next-generation infrastructures.

1.2 LESSONS LEARNED FROM HISTORICAL FAILURES

The topic of fatigue and fracture has evolved significantly since the 19th century, driven largely by catastrophic failures in engineering applications. Fatigue refers to the progressive weakening of materials or structures under cyclic loading, which results in fatigue crack initiation and short-long crack propagation. Fracture mechanics quantifies the driving forces of crack growth-induced ductile or brittle failure. The development of fatigue and fracture mechanics starts with empirical testing and matures into a rigorous science that incorporates mechanics, materials, and artificial intelligence. As listed in Table 1.2, the field of fatigue and fracture began with failures like the 1842 Paris-Mezieres train axle, where stress concentrations at geometric discontinuities caused brittle failures in iron under cyclic loads. Figure 1.5 presents the typical brittle fracture and ductile fracture failure surface in steel. Wilhelm Albert's first published article on fatigue in chains marked its formal recognition [4]. Subsequent railway axle reports by William Rankine, Joseph Glynn, and August Wöhler [4] led to systematic studies of fatigue and fracture failure. By the 1850s–1860s, Wöhler [3, 4] developed the stress-fatigue life (*S-N*) curve and the endurance limit concept via rotating-bend tests, showing materials could withstand infinite cycles below a stress threshold (fatigue strength). These events and research work expand engineering design from static load to cyclic loads, laying the groundwork for protocols to predict fatigue and fracture in metals.

The 1919 Boston Molasses Disaster implies that the brittle fracture stems from flaws due to poor ductility and neglected maintenance. The unstable energy release with increasing crack growth promotes the final failure modes (Figure 1.6a). James Alfred Ewing [49] demonstrated fatigue originating from microscopic cracks, while A.A. Griffith [50] introduced fracture mechanics via the potential energy required to create new surfaces (crack propagation). WWII Liberty ships and LNG tanks failed brittlely at welded joints under low temperatures. Poorly welded components are

TABLE 1.2
A Timeline of Key Milestones from Historical Failures.

Year/Period	Historical failures	Lessons	Milestone	Key contributors	Significance
1830s–1840s	Paris Mezieres Train Axle Failure (1842)	Stress concentration at geometric discontinuities	First published article on fatigue, noting failure in iron chains under repeated loading	Wilhelm Albert	Marked the formal recognition of fatigue as a distinct failure mode, beyond static overload
			Reports on railway axle brittle failures due to cyclic stress	William Rankine, Joseph Glynn, Wohler	Highlighted the role of repeated loading in brittle-like failures, sparking systematic investigations
1850s–1860s			Development of *S-N* curves (stress versus number of cycles) and endurance limit concepts through rotating-bend tests.	August Wöhler	Established fatigue life prediction for metals, showing materials could endure infinite cycles below a threshold stress
1900s–1920s	Boston Molasses Disaster (1919)	Brittle fracture of the tank due to poor ductility and higher loads	Demonstration of fatigue failures originates from microscopic cracks	James Alfred Ewing	Shifted focus on fatigue crack initiation
		Neglected maintenance and inadequate pressure testing contributed to the failure	Introduction of fracture mechanics with the Griffith criterion, explaining brittle fracture via energy balance for crack propagation.	A.A. Griffith	Shifted focus from bulk strength to flaw tolerance, foundations for understanding why materials fail below theoretical strengths
1940s	World War II Liberty ships (1940s)	Brittle failure at the welded steel joints	Miner Palmgren linear damage hypothesis for cumulative fatigue damage under variable loading	Miner and Palmgren	Improved fatigue life estimation for complex loads
	Liquefied natural gas storage tank (1944)				
1950s	De Havilland Comet aircraft failures (1954)	Fatigue crack initiation in pressurized skins due to stress concentration, development of the fatigue "safe-life" approach	Recognition of brittle fractures in welded structures (e.g., Liberty Ships); early LEFM with stress intensity factor (K).	George Irwin	Integrated elasticity theory to model crack-tip stresses, enabling quantitative failure predictions
			Fatigue crack growth via plastic strain for low-cycle fatigue (Coffin-Manson relationship)	L.F. Coffin and S.S. Manson	Enhanced models for high-stress fatigue

(Continued)

TABLE 1.2 *(Continued)*

A Timeline of Key Milestones from Historical Failures.

Year/Period	Historical failures	Lessons	Milestone	Key contributors	Significance
1960s	F-111 aircraft No.94 wing pivot fitting (1969)	Fatigue failure due to material defects in high-strength steel	Paris' law ($da/dN = C(\Delta K)^m$) for fatigue crack growth rate under cyclic loading	Paul Paris and Erdogan	Allowed prediction of crack propagation life, crucial for damage-tolerant design in aerospace
			Extension to elastic-plastic fracture mechanics (EPFM) with *J*-integral for ductile materials	J.R. Rice	Addressed limitations of LEFM in plastics and metals with yielding, improving models for real-world failures
			Rainflow-counting for analyzing random loadings	Tatsuo Endo and M. Matsuishi	Enhanced fatigue life predictions for random loads
1970s			Crack closure mechanism improving fatigue life prediction	Wolf Elber	Improved accuracy in crack growth predictions
			Multiaxial fatigue life is governed by the plane with maximum damage	M.W. Brown and K.J. Miller	Complex stress state
1980s	Kielland oil Platform capsize (1980)	Poor welding in structural bracing amplifies fatigue under cyclic wave loads	Probabilistic fracture mechanics and environmental effects (e.g., corrosion fatigue, thermal fatigue)	Various standards and codes	Incorporated randomness in loads/ flaws and factors like corrosion, leading to safer designs via standards
	Japan Airlines Flight (1985)	Faulty repairs can initiate fatigue cracks			
	Aloha incident Boeing 737 (1988)	Accelerated corrosion and multiple fatigue crack initiation			
1990s-Present			Advanced simulations technologies and integration with composites/ biomaterials/additive manufacturing process		Emphasizes microstructural effects, predictive modeling, and interdisciplinary applications

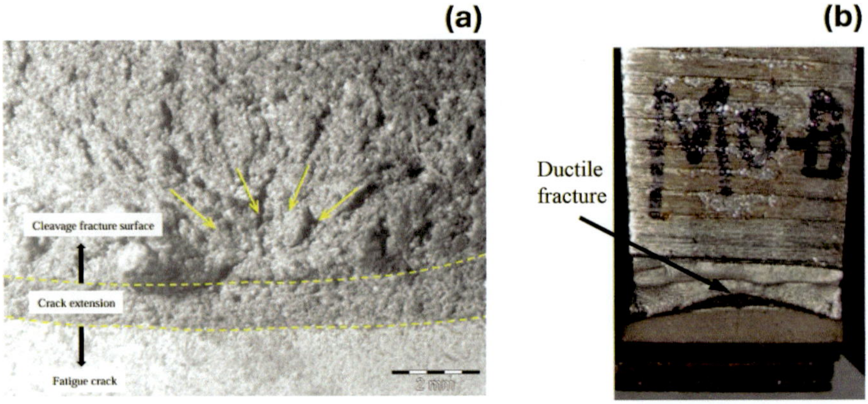

FIGURE 1.5 (a) Cleavage fracture of SE(B) specimens of S550 under low temperatures; (b) ductile fracture of SE(B) specimens of X65 under room temperatures.

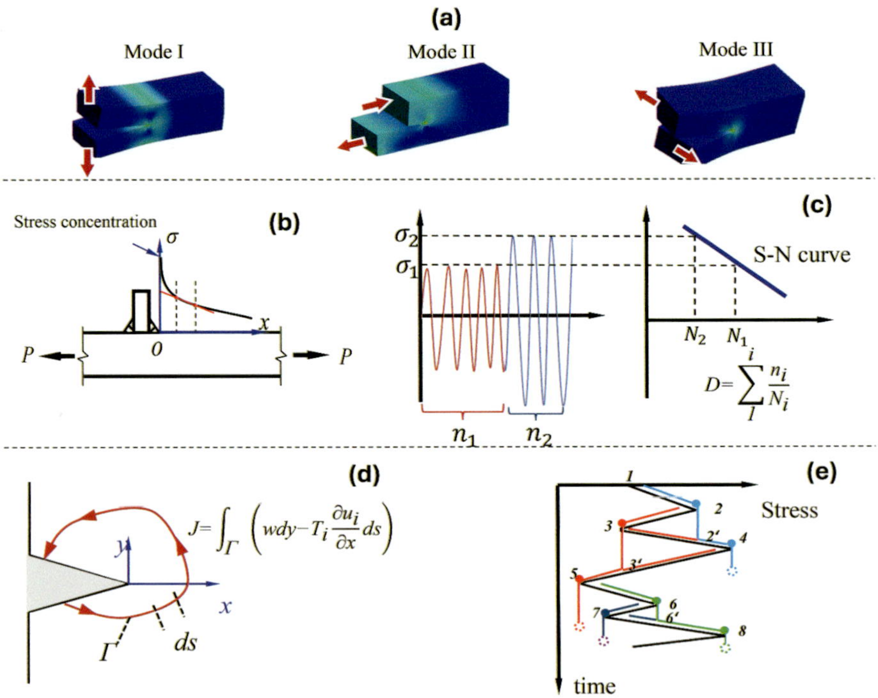

FIGURE 1.6 (a) Three modes of fracture failure; (b) stress concentration at the weld toe; (c) linear damage accumulation law [6]; (d) *J* integral calculations [8]; and (e) rainflow fatigue cycle counting method [53].

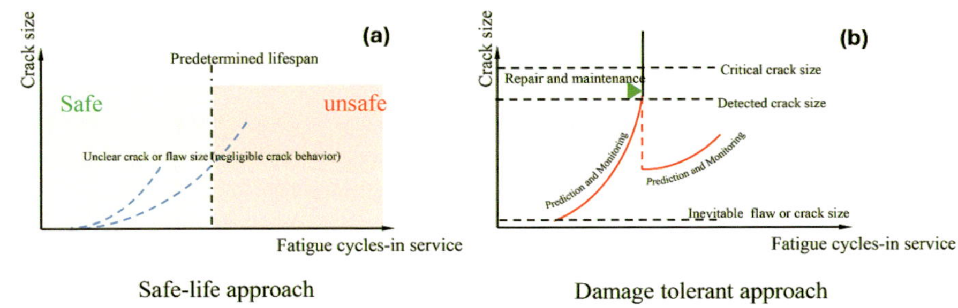

FIGURE 1.7 Illustrative scheme of (a) "Safe-life" approach and (b) "Damage-tolerant" approach.

prone to brittle fractures, especially for steels under low temperatures. Continuous cracks easily occur in welded joints due to stress concentration (Figure 1.6b) and finally lead to brittle fracture. Therefore, the selection of tougher structural steel is critical for welded joining in the shipbuilding industry. The Miner-Palmgren [6] linear damage hypothesis emerged for cumulative fatigue under variable loads in Figure 1.6c. The 1954 De Havilland Comet crashes revealed fatigue cracks from stress concentrations in pressurized skins, motivating the development of a "safe-life" design approach [9] in Figure 1.7. Theoretically, George Irwin [5] advanced linear elastic fracture mechanics with the stress intensity factor, while Coffin Manson [51, 52] related plastic strain to low-cycle fatigue. Enhanced life estimation methods have been developed for complex loads through energy release rate and stress-strain related parameters.

During the 1960s to 1970s, Paul Paris and Erdogan [7] proposed Paris' law to predict fatigue crack growth rate. Rice [8] proposed J integral (Figure 1.6d) to characterize the elastic-plastic crack tip stress for ductile materials. More detailed descriptions about the crack initiation and growth transfer the "safe-life" design approach to the "damage-tolerant" design approach, as shown in Figure 1.7. The "safe-life" approach neglects crack behavior in structural components during service. The structures are designed to remain safe and functional for a predetermined lifespan, after which it is retired. The "damage-tolerant" approach assumes the structure is designed to tolerate small cracks or flaws, with prediction and monitoring of the cracks or defects under loadings. The designed structures can continue operating safely until the damage is detected and repaired regularly. Matsuishi and Endo [53] proposed the rainflow counting method (Figure 1.6e) to analyze the fatigue life under random loadings. Wolf Elber [54] revealed the crack closure to refine fatigue life predictions. Brown and Miller [55] addressed multiaxial fatigue on planes of maximum damage. With increasing acknowledged details in crack nucleation and propagation, the design philosophy relating to fatigue and fracture failure gradually changes. Recent failures like the 1980 Kielland oil platform capsize, 1985 Japan Airlines Flight, and 1988 Aloha Boeing 737 incident (Figure 1.8) repeatedly highlighted that amplification effects due to complicated environmental loads, corrosion, temperatures, multiple cracks, etc. on the crack nucleation and growth further increase the probability

FIGURE 1.8 Structural failure of Aloha Boeing 737 in 1988 from *Independent News.*

of failures. Existing standards and codes [56–58] tend to provide conservative results to incorporate the randomness from the load, environment, and flaws. Nowadays, progressive advancement in numerical and experimental development facilitates the understanding of fatigue and fracture on multiple scales for emerging materials and structures. To sum it up, some common lessons from historical failures are: 1) Selection of material is critical and carefully determined according to application purposes, environments, and fabrication process. 2) Fatigue is an insidious failure under repeated stresses, even below yield strength; it consists of crack nucleation, short crack propagation, long crack propagation, and final unstable failure. 3) Cracks initiate from flaws microscopically and propagate according to energy/stress intensity, affected by crack closure, local plasticity state, stress concentration and environmental effects. 4) Fracture occurs when the crack reaches a critical size (material toughness) and can be significantly rapid for brittle fracture. 5) Design philosophy evolved from "safe-life" approach to "damage-tolerant" approach, accounting for probabilities and randomness from load, material, and manufacturing process, etc. 6) Periodic inspection and timely maintenance are essential to prevent brittle catastrophic failures.

Looking ahead, the multidisciplinary nature of fatigue and fracture research evolves from traditional mechanics to advanced materials and applications in energy, aerospace, and biomedical science. Through analyzing 500 highly cited papers from Web of Science from 2021 to 2025 on the topic "fatigue and fracture of material and structures," Figure 1.9 presents the keywords and network built

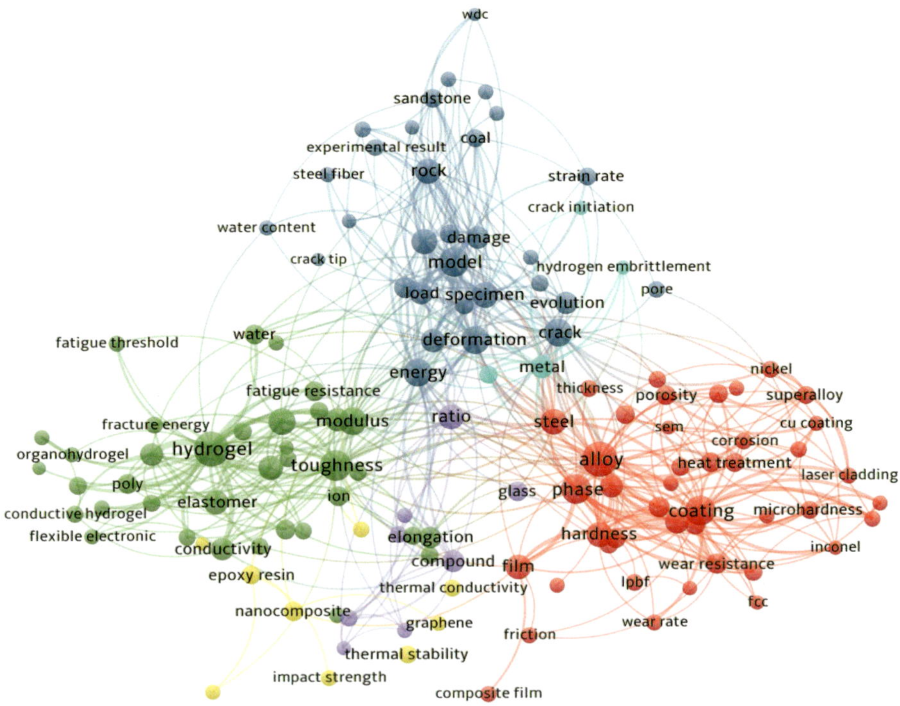

FIGURE 1.9 Network visualization of 500 highly cited papers from 2021 to 2025 relating to "fatigue and fracture of material and structures."

through VOS viewer. It shows a dense, interconnected network with three main clusters (cluster on the left, cluster in the center top, and cluster on the right) and a smaller cluster at the bottom. "Fatigue threshold," "fracture energy," "organic hydrogel," "polyhydrogel," "flexible electronic," "epoxy resin," and "nanocomposite" frequently occur in the left cluster. Biomimetic and sustainable materials for biomedical devices, wearables, and electronics have surged in the last 5 years. Soft and advanced materials in this cluster emphasize enhancing toughness, modulus, and energy absorption for fatigue resistance in flexible or bio-inspired applications. "Water content," "crack tip," "damage model," "load spectrum," "deformation," "crack initiation," "hydrogen embrittlement," and "strain rate" have dominated in the center top cluster. Aligning with global energy transitions, research work focusing on brittle materials like rocks and porous structures, accounting for dynamic loadings and environmental factors, has raised research attention significantly. The cluster on the right deals with high-performance metals, focusing on new modification approaches to combat fatigue, corrosion, and wear, indicated by the following keywords: "superalloy," "semi corrosion," "cu coating," "heat treatment," "laser cladding," "wear resistance," and simulation techniques "phase." This is due to an explosion in aerospace and energy, with emphasis on additive manufacturing and coatings for extreme conditions. "Impact strength," "composite film," and "thermal stability" in the bottom cluster bridge polymers and metals, focusing on thin films and nanomaterials for impact resistance and thermal properties. Overall, increasing interdisciplinary shifts are observed, from classical fracture mechanics of metals to failure mechanisms of advanced materials and simulation techniques.

1.3 ORGANIZATION OF THE BOOK

This book integrates these advancements, offering a comprehensive examination of fatigue and fracture in non-welded and welded structures, with an emphasis on experimental methodologies, FE simulation, welding stress management, and digital twin applications to deliver practical engineering solutions.

Chapter 2 delivers experimental approaches for quantifying fracture resistance, covering conventional test methods, proposed hybrid methods, and the DIC method used for surface-cracked structural components.

Chapter 3 examines in detail the ductile tearing phenomenon and engineering assessment diagrams.

Chapter 4 addresses the brittle fracture failure via ASTM standards and the Weibull stress method. The effect of prior ductility and mixed-mode loadings on the brittle fracture failures is investigated.

Chapter 5 focuses on high-cycle fatigue assessment of welded joints, focusing on the welded tubular and plate joints, using the effective notch stress (ENS) approach and hot-spot methods.

Chapter 6 investigates low-temperature and size effects on the low-cycle fatigue of welded joints, develops a unified approach for high- and low-cycle fatigue, and describes the residual stress evolution experimentally and numerically.

Chapter 7 reviews the digital twin studies and introduces digital twin methodologies for crack classification, updating the scheme via sensor data and application to the welded joints in lab-scale.

Chapter 8 synthesizes conclusions and future directions.

This book synthesizes advancements into actionable insights through the following:

Filling Knowledge Gaps: This book bridges theory and practice, covering experimental techniques (Chapter 2) beyond conventional tests for real-world welded pipes, addressing limitations in traditional assessments.

Practical Engineering Focus: This book focuses on practical evaluation of ductile/brittle fracture (Chapters 3–4), high/low-cycle fatigue (Chapters 5–6), and welding stresses (Chapter 6) and provides tools for engineers to manage residual stresses and extend structure life.

Innovative Digital Integration: This book emphasizes applying digital twin concepts (Chapter 7) for real-time crack classification, residual life assessment through FE analysis, machine learning, updating algorithms, and enabling predictive maintenance in welded structural components.

REFERENCES

[1] K.-H. Chang, e-Design: Computer-aided engineering design, Academic Press, 2015.
[2] A. Fajri, A.R. Prabowo, N. Muhayat, D.F. Smaradhana, A. Bahatmaka, Fatigue analysis of engineering structures: State of development and achievement, Procedia Structural Integrity, 33 (2021) 19–26.
[3] A. Wöhler, Über die festigkeitsversuche mit eisen und stahl, Ernst & Korn, 1870.
[4] A.F. Ralph, I. Stephens, Robert R. Stephens, Henry O. Fuchs, Metal fatigue in engineering, John Wiley & Sons, 2001.
[5] G.R. Irwin, Analysis of stresses and strains near the end of a crack traversing a plate, Journal of Applied Mechanics, 24 (1957) 361–364.
[6] M.A. Miner, Cumulative damage in fatigue, ASME. Journal of Applied Mechanics, 12(3) (1945) A159–A164.
[7] P. Paris, F. Erdogan, A critical analysis of crack propagation laws, Journal of Basic Engineering, 85 (1963) 528–533.
[8] J.R. Rice, A path independent integral and the approximate analysis of strain concentration by notches and cracks, Journal of Applied Mechanics, 35 (1968) 379-386.
[9] F.C. Campbell, Fatigue and fracture: Understanding the basics, ASM International, 2012.
[10] ASTM, ASTM E466–21, Standard practice for conducting force controlled constant amplitude axial fatigue tests of metallic materials, in, ASTM International, West Conshohocken, PA, 2021.
[11] ASTM, ASTM E647–24, Standard test method for measurement of fatigue crack growth rates, in, ASTM International, West Conshohocken, PA, 2024.
[12] ASTM, ASTM E1820–22, Standard test method for measurement of fracture toughness, in, ASTM International, West Conshohocken, PA, 2022.
[13] ASTM, ASTM E606/E606M-21, Standard test method for strain-controlled fatigue testing, in, ASTM International, West Conshohocken, PA, 2021.
[14] C. Chen, L. Feng, Effect of geometrical configuration and strain rate on aluminum alloy 5083 and S550 steel characterized by digital image correlation, Sensors, 25 (2025) 3607.
[15] T.B. Quy, J.-M. Kim, Crack detection and localization in a fluid pipeline based on acoustic emission signals, Mechanical Systems and Signal Processing, 150 (2021) 107254.
[16] L. Feng, X. Qian, Enhanced crack sizing and life estimation for welded tubular joints under low cycle actions, International Journal of Fatigue, 137 (2020) 105670.
[17] L. Feng, X. Qian, Size effect and life estimation for welded plate joints under low cycle actions at room and low ambient temperatures, Thin-Walled Structures, 132 (2018) 195–207.
[18] L. Feng, X. Qian, Low cycle fatigue test and enhanced lifetime estimation of high-strength steel S550 under different strain ratios, Marine Structures, 61 (2018) 343–360.
[19] T. Li, Y. Wang, L. Feng, J. Gan, Y. Shao, Fatigue life extension of the prefatigued welded joints by ultrasonic impact treatment, Journal of Constructional Steel Research, 229 (2025) 109508.
[20] T. Li, C. Zhang, G. Huang, L. Feng, Fatigue life prediction of prefatigued welded Q235 steel butt-joints treated by high frequency mechanical impact, International Journal of Steel Structures, 24 (2024) 839–848.
[21] L. Feng, X. Qian, R. Jiang, A uniform volume-based fatigue indicator for high-and low-cycle assessment of notched components, International Journal of Fatigue, 144 (2021) 106048.

[22] L. Feng, T. Liu, X. Qian, C. Chen, A complete integrity assessment of welded connections under high and low cycle fatigue followed by fracture failure, Steel and Composite Structures, An International Journal, 43 (2022) 465–481.
[23] N. Moës, T. Belytschko, Extended finite element method for cohesive crack growth, Engineering Fracture Mechanics, 69 (2002) 813–833.
[24] A. Yan, H. Nguyen-Dang, Multiple-cracked fatigue crack growth by BEM, Computational Mechanics, 16 (1995) 273–280.
[25] E.D. Leonel, W.S. Venturini, A. Chateauneuf, A BEM model applied to failure analysis of multi-fractured structures, Engineering Failure Analysis, 18 (2011) 1538–1549.
[26] P. Li, W. Li, B. Li, S. Yang, Y. Shen, Q. Wang, K. Zhou, A review on phase field models for fracture and fatigue, Engineering Fracture Mechanics, 289 (2023) 109419.
[27] K. Seleš, F. Aldakheel, Z. Tonković, J. Sorić, P. Wriggers, A general phase-field model for fatigue failure in brittle and ductile solids, Computational Mechanics, 67 (2021) 1431–1452.
[28] H. Petroski, Design paradigms: Case histories of error and judgment in engineering, Cambridge University Press, 1994.
[29] J.W. Barnard, Exxon Collides with the "Valdez Principles", Business and Society Review 74 (32–35) (1990).
[30] X. Liao, Y. Wang, L. Feng, H. Ban, Y. Chen, Fatigue crack initiation and energy-based life analysis for Q345qD bridge steel at low temperatures, Journal of Constructional Steel Research, 180 (2021) 106571.
[31] L. Feng, L. Zhang, X. Liao, W. Zhang, Probabilistic fatigue life of welded plate joints under uncertainty in Arctic areas, Journal of Constructional Steel Research, 176 (2021) 106412.
[32] L. Wang, X. Qian, Optimization-improved thermal–mechanical simulation of welding residual stresses in welded connections, Computer-Aided Civil and Infrastructure Engineering, 39 (2024) 1275–1293.
[33] M. Lukić, C. Cremona, Probabilistic optimization of welded joints maintenance versus fatigue and fracture, Reliability Engineering & System Safety, 72 (2001) 253–264.
[34] L. Feng, X. Qian, A hot-spot energy indicator for welded plate connections under cyclic axial loading and bending, Engineering Structures, 147 (2017) 598–612.
[35] A.S. Madhvacharyula, A.V.S. Pavan, S. Gorthi, S. Chitral, N. Venkaiah, D.V. Kiran, In situ detection of welding defects: A review, Welding in the World, 66 (2022) 611–628.
[36] C. Sonsino, M. Kueppers, Multiaxial fatigue of welded joints under constant and variable amplitude loadings, Fatigue & Fracture of Engineering Materials & Structures, 24 (2001) 309–327.
[37] S. Maddox, Hot-spot stress design curves for fatigue assessment of welded structures, International Journal of Offshore and Polar Engineering, 12 (2002).
[38] Y. Choo, X. Qian, J. Wardenier, Effects of boundary conditions and chord stresses on static strength of thick-walled CHS K-joints, Journal of Constructional Steel Research, 62 (2006) 316–328.
[39] X. Qian, K. Jitpairod, P. Marshall, S. Swaddiwudhipong, Z. Ou, Y. Zhang, M.R. Pradana, Fatigue and residual strength of concrete-filled tubular X-joints with full capacity welds, Journal of Constructional Steel Research, 100 (2014) 21–35.
[40] Ö. Karakaş, F.B. Kardeş, P. Foti, F. Berto, An overview of factors affecting high-cycle fatigue of additive manufacturing metals, Fatigue & Fracture of Engineering Materials & Structures, 46 (2023) 1649–1668.
[41] M. Singh, E. Fuenmayor, E.P. Hinchy, Y. Qiao, N. Murray, D. Devine, Digital twin: Origin to future, Applied System Innovation, 4 (2021) 36.

[42] X. Zhou, S. He, L. Dong, S.N. Atluri, Real-time prediction of probabilistic crack growth with a helicopter component digital twin, Aiaa Journal, 60 (2022) 2555–2567.
[43] Q. Yan, H. Wang, F. Wu, Digital twin-enabled dynamic scheduling with preventive maintenance using a double-layer Q-learning algorithm, Computers & Operations Research, 144 (2022) 105823.
[44] L. Feng, X. Qian, W. Zhang, Adaptive fatigue assessment of welded plate joints based on crack measurements, in: Proceedings of the 17th East Asian-Pacific conference on structural engineering and construction, 2022: EASEC-17, Singapore, Springer, 2023, pp. 223–235.
[45] F. Tao, M. Zhang, A.Y.C. Nee, Digital twin driven smart manufacturing, Academic Press, 2019.
[46] T. Zhou, S. Jiang, T. Han, S.-P. Zhu, Y. Cai, A physically consistent framework for fatigue life prediction using probabilistic physics-informed neural network, International Journal of Fatigue, 166 (2023) 107234.
[47] E.W.W. Cheok, X. Qian, S.T. Quek, M.B.I. Si, A digital twin approach toward unlocking elasto-plastic fatigue crack growth rates under negative load ratios, Engineering Fracture Mechanics, 301 (2024) 110034.
[48] L. Feng, X. Qian, Enhanced real-time crack monitoring and updating in welded structural components with limited measurement data, Mechanical Systems and Signal Processing, 198 (2023) 110403.
[49] J.A. Ewing, X. Experimental researches in magnetism, Philosophical Transactions of the Royal Society of London (1885) 523–640.
[50] A.A. Griffith, VI. The phenomena of rupture and flow in solids, Philosophical Transactions of the Royal Society of London. Series A, Containing Papers of a Mathematical or Physical Character, 221 (1921) 163–198.
[51] L.F. Coffin Jr, A study of the effects of cyclic thermal stresses on a ductile metal, Transactions of the American Society of Mechanical Engineers, 76 (1954) 931–949.
[52] S.S. Manson, Behavior of materials under conditions of thermal stress, National Advisory Committee for Aeronautics, 1953.
[53] M. Matsuichi, T. Endo, Fatigue of metals subjected to varying stress, Japan Society of Mechanical Engineering, 1968.
[54] E. Wolf, Fatigue crack closure under cyclic tension, Engineering Fracture Mechanics, 2 (1970) 37–45.
[55] M.W. Brown, K. Miller, A theory for fatigue failure under multiaxial stress-strain conditions, Proceedings of the Institution of Mechanical Engineers, 187 (1973) 745–755.
[56] A. Plaza, Fatigue assessment of offshore structures, Northchase Drive Houston, TX, 77060 (2003).
[57] A. Hobbacher, J. Baumgartner, Recommendations for fatigue design of welded joints and components, Springer, 2024.
[58] API, Recommended practice for planning, designing and constructing fixed offshore platforms–working stress design–, API Recommended Practice 2A-WSD, 21th ed., 2000.

2 Experimental Approaches to Quantify Fracture Resistance

2.1 INTRODUCTION

The experimental fracture mechanics aims to quantify the fracture resistance through measurable fracture parameters in the experimental procedure. Analogous to a stress-strain curve (Figure 2.1a), which characterizes the material resistance against yielding, typically measured in a uniaxial setting, the fracture test quantifies the material fracture resistance through the fracture resistance curve, also known as the *J*-*R* curve (Figure 2.1b), measured from a fracture specimen, for example, the compact tension C(T) specimen in Figure 2.1b. The vertical axis in Figure 2.1b denotes the energy release rate, quantified by the *J*-integral,

$$J = -\frac{1}{B}\frac{dU}{da} \tag{2.1}$$

where U represents the strain energy absorbed by the specimen, or the area under the load-deformation curve of the specimen, and B denotes the thickness of the fracture specimen. The denominator in Eq. (2.1) thus denotes the change in crack area, assuming that the through-thickness crack front experiences an approximately uniform crack extension. The energy release rate, J, hence quantifies the energy released per unit crack extension. Anderson [1] provides a detailed discussion on the theoretical background of the *J*-integral.

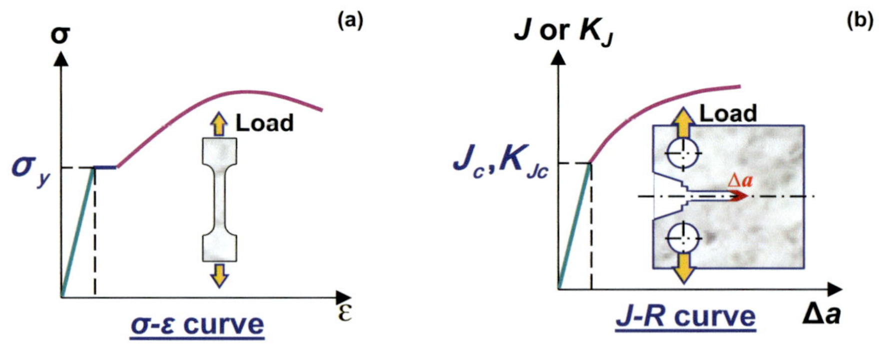

FIGURE 2.1 (a) Stress-strain curve measured from uniaxial tension test; (b) fracture resistance curve (often known as *J*-*R* curve) measured from fracture test.

DOI: 10.1201/9781003509806-2

The experimental determination of the *J-R* curve in Figure 2.1b thus requires the measurement of both the load-deformation curves and the amount of crack extension to characterize the material's fracture resistance. To ensure the validity of the *J*-integral measurement, the experimental procedure often imposes a small-scale yielding (SSY) condition on the fracture specimens, measured by the remaining ligament size, thickness, or crack size in the material testing standards [2]. Meanwhile, to reflect the realistic crack-front constraint conditions, the fracture specimen typically entails the same thickness as the structure of interest.

2.2 CONVENTIONAL FRACTURE TEST METHODS

2.2.1 Multiple Specimen Approach

The pioneering effort to measure *J*-integral values, as reported by Begley and Landes [3], originates from the desire to quantify the partial derivative in Eq. (2.1) in an experimental setup. To achieve this, they have utilized multiple fracture specimens fabricated from the same material along the same orientation, with the same dimensions but different crack depths, say from a_1 to a_n. The experimental procedure will first record the load-displacement curve of the specimen. The area under the load-displacement curves should ideally represent the energy absorbed by the specimen primarily due to the crack advancement. Wherever possible, this area should exclude energy absorbed via other sources, for example, local indentation in the specimen near the loading point. The selected displacement parameter is often the load-line displacement (LLD) or the crack-mouth opening displacement (*CMOD*), with the latter preferred due to its high sensitivity to the crack advancement.

Figure 2.2a illustrates schematically the load-displacement curves measured from the same series of specimens with different crack sizes, which allows calculation of the strain energy for different crack depths at different displacement levels. Figure 2.2b plots the variation of the strain energy with respect to the crack depth. The slope of each curve at a given crack depth provides the basis to calculate the material fracture resistance given in Eq. (2.1).

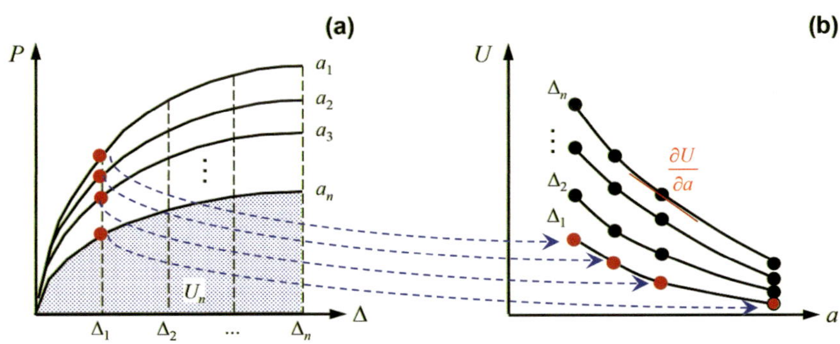

FIGURE 2.2 (a) Load-displacement curves for a series of fracture specimens with the same dimensions but different crack sizes; (b) strain energy variation with respect to the crack size measured from specimens with different crack sizes.

The multiple specimen method assumes that the crack extension in each of the fracture specimens remains negligible during the loading process, which can be a challenging requirement for low-toughness materials. If the amount of crack extension is substantial in the specimens, the physical correspondence of the measured slope to the energy release rate becomes questionable. Additionally, testing multiple specimens to measure the J value may not deliver an economical solution in practice.

2.2.2 Single Specimen η-Approach with Unloading Compliance

Rice *et al.* [4] have proposed an approach to measure the J-integral from a single specimen, known as the η-approach, which has been widely implemented in many engineering testing standards, for example, in ASTM E1820 [2], ISO 12135 [5]. The single specimen allows measurement of both the J value and the amount of crack extension from a single specimen with a growing crack front.

The η-approach originates from the relationship between the load resistance of a fracture specimen with the extending crack size. For a single-edge notched bend, SE(B), specimen shown in Figure 2.3, the load applied follows the following relationship with the plastic moment resistance (M_P) at the mid-span of a beam under a three-point bending, or,

$$P = \frac{4M_P}{L} \tag{2.2}$$

where L defines the span of the SE(B) specimen. For an elastic-plastic material with strain hardening characterized by a hardening exponent, n, for example, in the context of Ramberg-Osgood hardening model, the moment resistance becomes,

$$P = \frac{Bb^2}{4}\sigma_y h\left(\frac{2\Delta_{Pl}}{L}, \frac{E}{\sigma_y}, n\right) \tag{2.3}$$

where B refers to the thickness of the SE(B) specimen and b denotes the remaining ligament, as shown in Figure 2.3. The non-dimensional function h equals 1.0 for

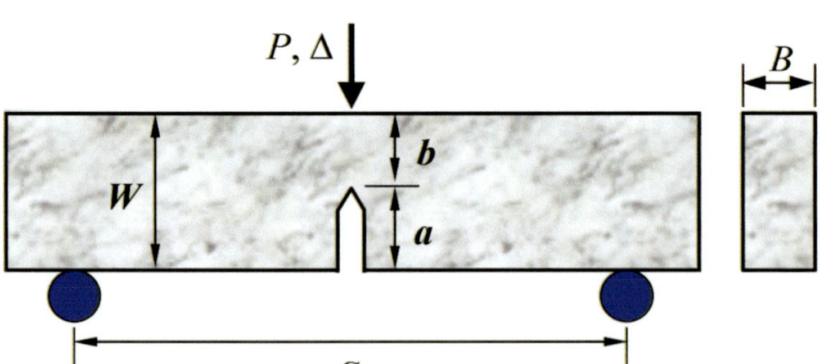

FIGURE 2.3 A single-edge notched bend [SE(B)] specimen.

elastic perfectly plastic materials. For a hardening material, the function h depends on the non-dimensional vertical displacement at the mid-span ($2\Delta_{Pl} / L$), the elastic modulus to the yield strength ratio (E / σ_y), and the hardening exponent (n).

For a SE(B) specimen under displacement-controlled loading, Eq. (2.1) becomes,

$$J = -\frac{1}{B}\int_0^{\Delta} \frac{\partial P(\Delta,a)}{\partial a} d\Delta \tag{2.4}$$

The remaining ligament has a unique relationship with the crack depth, as seen from Figure 2.3,

$$b = W - a \tag{2.5}$$

The derivative in Eq. (2.4) thus becomes,

$$\frac{\partial P(\Delta,a)}{\partial a} = -\frac{\partial P(\Delta,b)}{\partial b} = -\frac{Bb}{2}\sigma_y h\left(\frac{2\Delta_{Pl}}{L}, \frac{E}{\sigma_y}, n\right) = -\frac{2P}{b} \tag{2.6}$$

By inserting Eq. (2.6) into Eq. (2.4), we have,

$$J = \frac{1}{B}\int_0^{\Delta} \frac{2P}{b} d\Delta = \frac{2}{Bb}\int_0^{\Delta} P d\Delta \tag{2.7}$$

Since the integral $\int_0^{\Delta} P d\Delta$ represents the strain energy U stored in the specimen.

For general fracture specimens, the energy release rate, J, follows,

$$J = \frac{\eta U}{Bb} \tag{2.8}$$

The application of Eq. (2.8) to standard fracture specimens often separates the J-value into two components, an elastic component J_{le} and a plastic component J_{pl},

$$J = J_{le} + J_{pl} \tag{2.9}$$

The elastic component J_{le} derives from the stress-intensity factor solutions documented for the fracture specimens,

$$J_{le} = K_I^2 \frac{1-\upsilon^2}{E} \tag{2.10}$$

where υ is the Poisson's ratio of the material. The plastic component J_{pl} utilizes the area under the load versus plastic deformation curve, U_{pl},

$$J_{pl} = \frac{\eta U_{pl}}{Bb} \tag{2.11}$$

The non-dimensional parameter η depends on the geometry, loading, and boundary conditions of the specimens, as well as on the hardening properties of the material. For the same specimen, η assumes different values depending on whether the plastic strain energy U_{pl} is evaluated using the LLD or the *CMOD*.

To relate the measured energy release rate to the extent of crack extension, the single-specimen approach integrates unloading-reloading events in the testing procedure to quantify unloading compliance, which has a unique relationship with the crack size.

$$\frac{a}{W} = g(C) \tag{2.12}$$

where C refers to the compliance of the specimen defined by,

$$C = \frac{\Delta u}{\Delta P} \tag{2.13}$$

Equation (2.12) often derives from a polynomial regression of a calibrated linear-elastic FE database. The displacement u in Eq. (2.13) refers to either the LLD or the *CMOD*, with the latter more sensitive to the crack size and often preferred for higher accuracy.

The J-value measured from Eq. (2.9), coupled with the crack length inferred from Eq. (2.12) prior to the same unloading events in the test, allows a direct evaluation of the fracture resistance curve for standard fracture specimens.

2.3 A HYBRID NUMERICAL AND EXPERIMENTAL APPROACH

2.3.1 Fundamental Methodology

Inspired by the multiple specimen approach, Qian and Yang [6] proposed a hybrid approach to determine the fracture resistance curve for both mode I and mixed-mode I and II specimens. The fundamental principle behind the hybrid numerical and experimental approach is similar to that of the conventional multiple-specimen experimental approach described above. Instead of using multiple experimental specimens with different crack sizes and negligible crack growth in each, the hybrid approach combines a single experimental specimen with an extending crack and replaces the remaining experimental specimens with FE models with different crack sizes, as shown in Figure 2.4. Large-deformation elastic-plastic analyses subsequently generate a series of P-Δ curves from these FE fracture specimens with non-growing stationary cracks. In parallel, the experimental test records the load deformation curves of the single fracture specimen, which experiences ductile

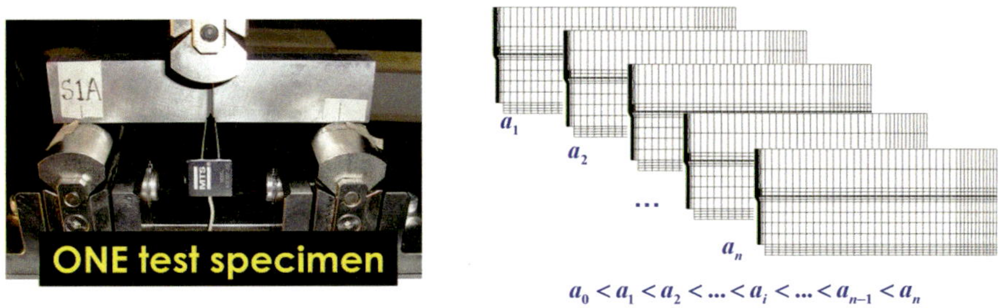

FIGURE 2.4 The hybrid approach combining one experimental specimen with multiple finite element (FE) fracture models.

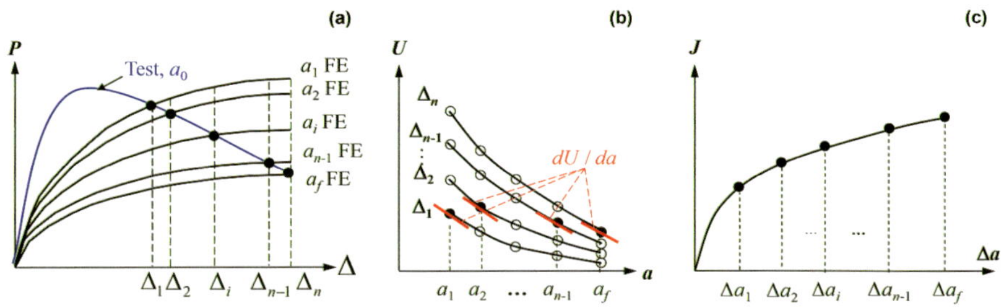

FIGURE 2.5 The hybrid approach combining one experimental specimen with multiple finite element (FE) fracture models.

crack extension. The fracture specimen in the experiment shall conform to the size requirement and the specimen preparation described in the testing standards.

Figure 2.5a illustrates the load-displacement (P-Δ) curve obtained from the experimental specimen with the crack size a_0 and those obtained from the FE models with various crack sizes a_1 to a_n. The intersection point between the experimental P-Δ curve and the numerical P-Δ curve represents a common loading and displacement level in the FE specimen with a stationary crack, and the experimental specimen experiencing crack growth. Assuming no experimental uncertainties, the hybrid approach theoretically equates the extended crack size in the experimental specimen to the crack size in the corresponding FE specimen at this intersection point. As the crack size in the FE model matches the current crack size in the experimental specimen, the energy release rate, calculated from the multiple FE specimens in a manner similar to the conventional multiple-specimen experimental method, provides the J-value in the experimental specimen with the corresponding crack size.

The displacements at the intersection points between the experimental and numerical curves, that is, Δ_1 to Δ_n in Figure 2.5a, define the displacement levels used to compute the strain energy U for each crack depth. Figure 2.5b presents a schematic representation of the variation in strain energy with crack depth, derived from multiple FE models. To facilitate the calculation of the energy release rate from the FE models, the hybrid approach employs regression analyses to approximate polynomial functions in terms of the crack size (a), effectively describing the strain energy curves depicted in Figure 2.5b.

The solid circles in Figure 2.5b represent the displacement levels where the energy release rate calculated from multiple FE models matches, theoretically, the energy release rate of the experimental fracture specimen with a growing crack. The energy release rate at these points is computed using Eq. (2.1). Figure 2.5c illustrates the J-values determined at these solid circles as a function of the corresponding crack extensions. The fundamental energy principle of the hybrid approach allows for its application to both mode I and mixed-mode (I and II) specimens, by measuring the mode I strain energy U_I and the mode II strain energy U_{II}, respectively.

The following outlines the detailed steps for the hybrid approach.

Step 1: Obtain the experimental load-displacement curve from a fatigue pre-cracked fracture specimen, such as a C(T) or SE(B) specimen, as per ASTM E-1820 [2]. The displacement parameter here can be either the LLD or the *CMOD*. Use duplicate specimens to minimize experimental scatter. Measure both the initial crack size, a_0, and the final crack size, a_f, according to ASTM E-1820 [2]. Record the uniaxial stress-strain relationship from the standard coupon test, following guidelines such as ASTM E-8 [7].

Step 2: Create a series of FE models based on the geometry of the experimental fracture specimen used in Step 1, with the crack size ranging from a_0 to a_f. The smallest crack size, a_0, in the FE models should match the initial crack size in the experimental fracture specimen, while the largest crack size a_f should match the final crack size in the experiment. Use the uniaxial stress-strain curve obtained in Step 1 to compute the P-Δ curve for each FE model. The computed P-Δ curve for the smallest crack size a_0 should replicate the initial compliance of the experimental specimen.

Step 3: Plot the experimental P-Δ curve alongside the numerical P-Δ curves from the FE models, as shown in Figure 2.5a. Determine the intersection points between the experimental curve and each numerical curve for crack sizes ranging from a_1 to a_f. Denote the displacement at each intersection point as $\Delta_i(a = a_i)$.

Step 4: Calculate the strain energy (U) from the area under each P-Δ curve at each $\Delta_i(a = a_i)$ point for the FE models, with crack sizes varying from a_1 to a_f, as illustrated in Figure 2.5b.

Step 5: Perform a regression analysis to derive approximate polynomial functions representing the strain energy (U) at each $\Delta_i(a = a_i)$ as a function of the crack depth.

Step 6: For the strain energy function U at $\Delta_i(a = a_i)$, calculate the energy release rate J at the corresponding crack size, a_i, using Eq. (2.1). Since $\Delta_i(a = a_i)$ corresponds to the intersection point between the experimental P-Δ curve and the numerical P-Δ curve for the FE model with the crack size a_i, the calculated energy release rate represents the fracture resistance at the crack extension, $\Delta a_i = a_i - a_0$.

Step 7: Plot the J-values obtained in Step 6 against the crack extension $\Delta a_i = a_i - a_0$.

2.3.2 Validation on Fracture Specimens

Qian and Yang [8] have presented the validation of the hybrid approach using single-edge-notched specimens subjected to two types of loading conditions: mode I (three-point bending) and mixed-mode I/II (four-point loading). The verification for mode I loading utilizes the experimental results reported by Joyce [9] and Zhu and Joyce [10] for HY80 steel, while the validation for mixed-mode I/II loading makes use of the test data from Tohgo and Ishii [11] for the aluminum alloy 6061-T651.

Figure 2.6 shows the true stress-true strain relationships for the two materials under consideration. The HY80 steel has a Young's modulus of 207 GPa and a Poisson's ratio of 0.3, with a yield strength of 630 MPa and an ultimate strength of 735 MPa. In contrast, the aluminum alloy 6061-T651 used for the mixed-mode I/II test has a Young's modulus of 68.9 GPa and a Poisson's ratio of 0.35, with a yield strength of 288 MPa and an ultimate strength of 317 MPa.

The validation study reported by Qian and Yang [8] uses the FE research code WARP3D [12] for numerical computation, leveraging its open-source capabilities to solve large-scale 3D problems. WARP3D employs the B-bar formulation to mitigate hourglass modes in 8-node brick elements under plastic, incompressible deformation. The element formulation used in this study follows a large-deformation framework.

2.3.2.1 Mode I SE(B) Specimens

Zhu and Joyce [10] conducted an extensive experimental study on the *J*-resistance curves for HY80 steel, using side-grooved SE(B) specimens and the normalized

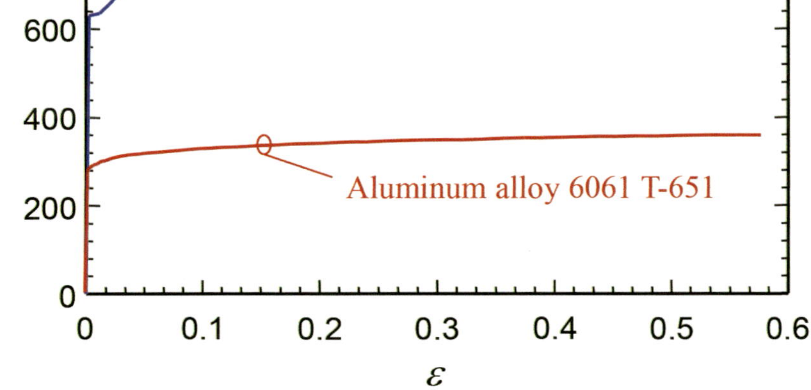

FIGURE 2.6 Uniaxial true stress – true strain curve for two metallic materials HY 80 steel and aluminum alloy 6061 T-651.

method [10, 13]. The SE(B) specimen they used had a total thickness of $B = 25.4$ mm, with the net-thickness after side-grooving equal to 80% of the total thickness, that is, $B_N = 0.8B$. The width of the SE(B) specimen, W, as previously illustrated in Figure 2.3, equals 50.8 mm, while the span over width, S / W, has a fixed value of 4 for all specimens. The initial crack depth over the width ratio, a_0 / W, ranges from 0.135 to 0.83 for all specimens reported by Zhu and Joyce (2007) [10]. In the validation study, Qian and Yang (2011) [8] selected two SE(B) specimens with the a_0 / W ratios equal to 0.186 and 0.549, respectively. The SE(B) specimen with a relatively shallow crack depth ($a_0 / W = 0.186$) represents a fracture specimen with low crack-front constraints, while the deep crack ($a_0 / W = 0.549$) corresponds to a high crack-front constraint condition, complying with the requirements in ASTM E-1820.

Figure 2.7 presents a typical half FE model for the mode I SE(B) specimens, constructed using 3D 8-node brick elements. The FE model features a single layer of elements along the thickness direction, with all nodes constrained against out-of-plane displacements to simulate plane-strain conditions. Due to the presence of a plane of symmetry, only half of the specimen needs to be modeled, with displacement degrees of freedom for all nodes on the symmetry plane constrained in the direction normal to that plane. To enhance numerical convergence under large deformation, the crack tip has a focused mesh incorporating an initial root radius of 25.4 μm, as depicted in Figure 2.7. The total number of nodes in the FE models varies between 2400 and 3400, depending on the crack depth, with corresponding element counts ranging from 1100 to 1600.

Figure 2.8 demonstrates the close agreement between the experimentally measured J-Δa curve and the corresponding J-Δa curve obtained using the hybrid approach, thereby confirming the validity of the hybrid method for fracture specimens under mode I loading for both low-constraint and high-constraint conditions. The measurement of the experimental data in Figure 2.8 follows the η-approach with unloading compliance, as described in Section 2.2.2.

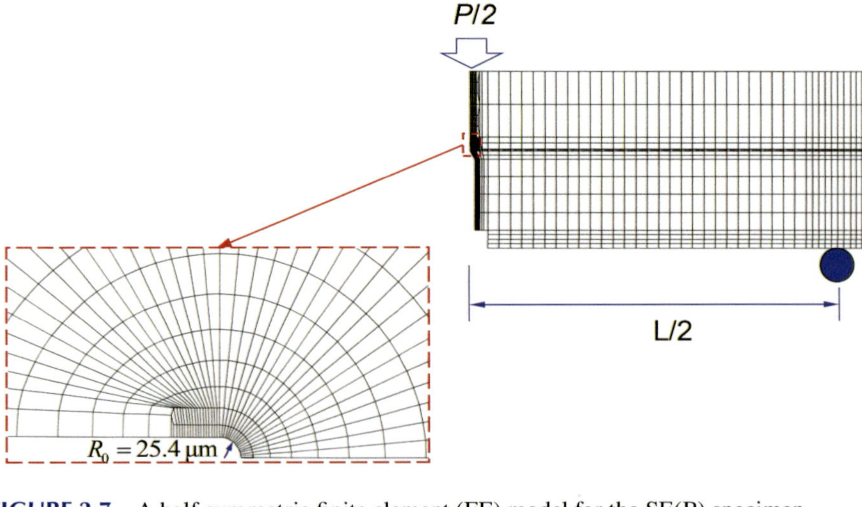

FIGURE 2.7 A half-symmetric finite element (FE) model for the SE(B) specimen.

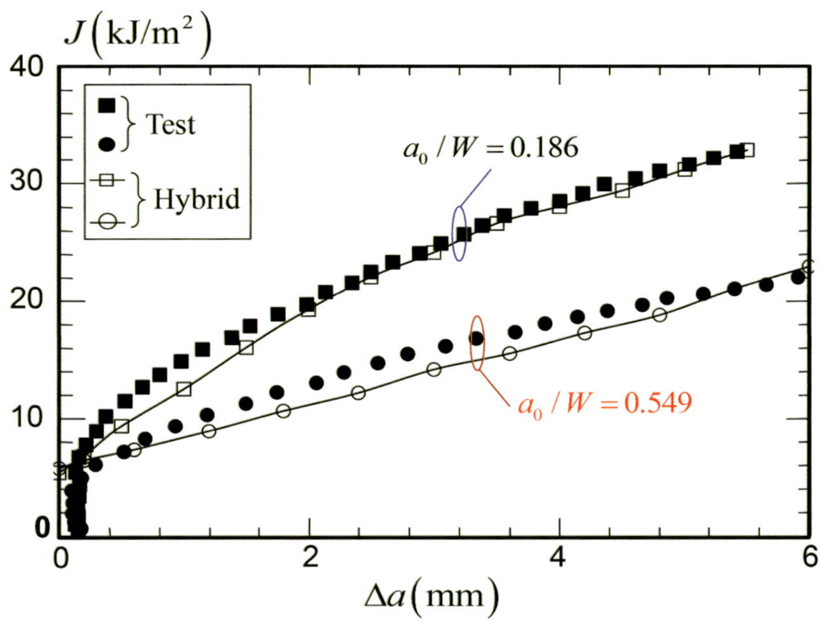

FIGURE 2.8 Comparison of the *J-R* curve determined from the experiment and the hybrid method for pure mode I specimens.

2.3.2.2 Mixed-Mode I and II Specimens

Yang and Qian [14] conducted an experimental investigation into the fracture resistance of aluminum alloy 5083 specimens under mixed-mode I and II loading. Figure 2.9 illustrates the configuration of the test specimen. Varying the distance S_0 between the load line and the crack plane leads to different mode mixity in the specimen. At $S_0 = S$, the specimen experiences a pure mode II condition. Hallback and Nilsson [15] define the linear-elastic mode-mixity via an equivalent angle, β_{eq},

$$\beta_{eq} = \tan^{-1}\left(K_I / K_{II}\right) \tag{2.14}$$

The mode I and mode II stress-intensity factors, K_I and K_{II}, typically derives from the linear-elastic interaction integral approach, implemented in the FE codes [16, 17].

In their mixed-mode test, Yang and Qian [14] measured the *CMOD* by mounting a crack opening displacement (COD) gauge at the mouth of a crack. Measuring the *CMOD* enables the calculation of the crack-plane rotation, assuming a fixed center of rotation for the crack plane,

$$\theta = \frac{CMOD}{a + r_p(W - a)} \tag{2.15}$$

where r_p represents the plastic rotation factor. ASTM E1820 [2] recommends a value of 0.44 for mode I SE(B) specimens. For the surface plate, the rotation center derives

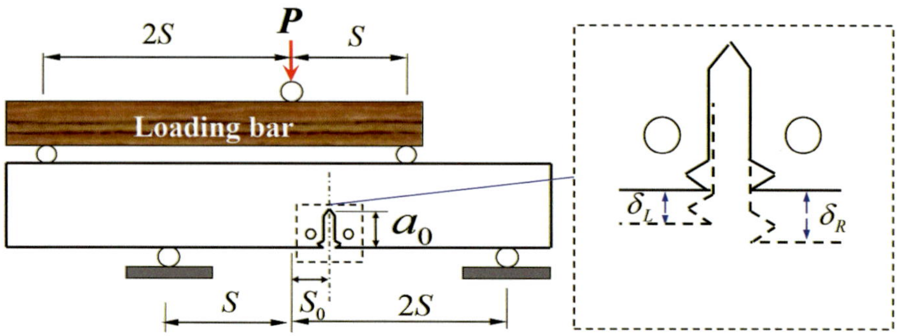

FIGURE 2.9 The experimental setup and measurement of displacement.

from a linear fitting of the numerically computed deformation of the crack plane. For the plates considered in the current study, the r value equals 0.18.

To measure the shear displacement across the crack plane, they machined four holes with inner threads near the edges of the crack plane and attached four stiff cantilever bars on both sides of the crack plane. This setup enables the measurement of the displacement parallel to the crack plane on the left side, δ_L, and that on the right side, δ_R, as shown in Figure 2.9. The shear displacement of the crack plane thus equals,

$$\delta_V = \delta_R - \delta_L \tag{2.16}$$

The shear force and bending moment acting on the crack plane derive from the equilibrium equations, with the shear force equal to one third of the applied load and the bending moment depending on S_0. The experimental measurements provide the record for the shear force (F_V) versus the relative shear displacement (δ_V) and the bending moment (M) versus the rotation angle (θ) at the crack plane. The total energy release during the mixed-mode crack extension, J_T, follows,

$$J_T = J_I + J_{II} \tag{2.17}$$

where J_I is the mode I energy release rate,

$$J_I = \frac{K_I^2(1-\upsilon^2)}{E} + \frac{2}{bB}\int_0^{\theta} M d\theta \tag{2.18}$$

where b refers to the length of the remaining ligament and J_{II} corresponds to the shear-mode energy release rate,

$$J_{II} = \frac{K_{II}^2(1-\upsilon^2)}{E} + \frac{2}{bB}\left[\int_0^{\delta_V} F_V d\delta - \frac{1}{2}F\delta_V\right] \tag{2.19}$$

The above procedure estimates the J-values for the mixed-mode specimens. The J-R curve requires the measurement of crack extension during the test. For mixed-mode specimens, the crack size measurement employs different approaches for the mode I dominant specimens and the mode II dominant specimens. The variation of *CMOD* compliance exhibits a unique relationship with the crack size for mode I dominant specimens, while the striations imprinted on the fracture surfaces of mode II dominant specimens enable a physical measurement of the crack extension.

Similar to pure mode I specimens, the compliance versus crack size relationship for mode I dominant specimens derives from linear-elastic finite element analysis (FEA) for the mixed-mode specimen with the specific mode mixity angle. However, for mode I dominant specimens, the compliance relies on the direction of the crack extension, which remains unknown prior to testing. Consequently, the relationship between compliance and crack extension is established using a fitted compliance function obtained from post-test FEA, based on the crack direction measured near the mid-thickness of each specimen.

For mode II dominant specimens, however, the specimen compliance does not indicate a clear and sensitive variation with respect to the crack size. For the aluminum alloy specimens investigated by Yang and Qian [14], the fracture surface for the mode II dominant specimens exhibits distinctive striations created by the unloading and reloading cycles during the test. The number of striations equals the number of unloading-reloading cycles after the detection of the crack initiation, as revealed in Figure 2.10. However, the presence of the striations depends on the materials. For materials where striations are not observable, an alternative approach should be explored.

FIGURE 2.10 Striations on the fracture surface of mode II dominant aluminum alloy specimen caused by unloading-reloading cycles.

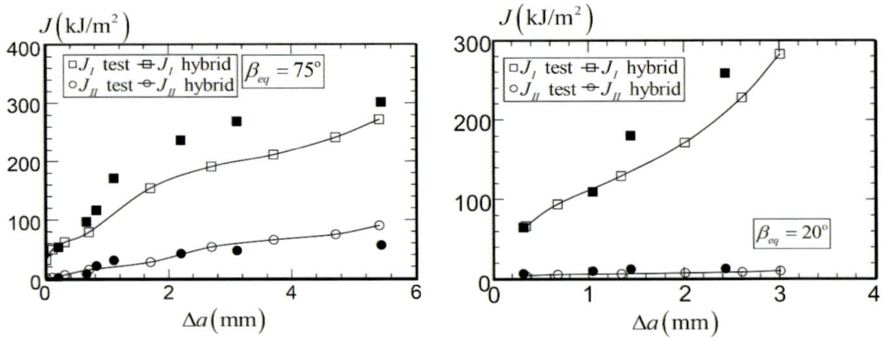

FIGURE 2.11 Comparison of the *J-R* curve determined from the experiment and the hybrid method for mixed-mode I-II specimens.

Figure 2.11 compares the evolution of mode I and mode II *J* values with respect to the crack extension, measured from the experiments and estimated from the hybrid approach, for a mode I dominant specimen with $\beta_{eq} = 75°$ and a mode II dominant specimen with $\beta_{eq} = 25°$. Both specimens are made of the same material, that is, aluminum alloy 5083 [6, 14]. The comparison indicates a close agreement in both the mode I and mode II *J* values when the crack extension is small, with increasing deviations as the crack grows further.

2.4 TEST METHOD FOR SURFACE-CRACKED PLATES

Existing testing guidelines [2, 5, 18] prescribe the standard protocol to determine the material fracture resistance using planar specimens with a straight through-thickness crack. The energy release rate per unit area of the crack extension, derived often from the area under the load versus the LLDs using the single or multiple-specimen approach [8, 19], represents the material's resistance against crack extension averaged over the crack front. The amount of crack extension, on the other hand, derives from the relationship between the crack length and the compliance of the specimen developed from extensive FE database. If the crack extension remains approximately uniform along the crack front, the energy release rate (measured by the *J*-integral) coupled with the amount of crack extension (Δa), forms the *J-R* curve and characterizes the material's resistance against ductile crack extension.

The fundamental principles outlined above, often named as the unloading-compliance η-approach, provide the basic framework to measure the fracture resistance curve directly from the structural components. This section presents the details to extend these fundamental principles to measure the fracture resistance curve for surface cracks in two types of structural components, namely the surface-cracked plate and a welded pipe with a circumferential crack as the weld toe. The approaches presented here remain extensible for direct fracture resistance measurement in other types of structural components.

This section presents an approach to extend the unloading-compliance η-approach to the high-strength steel plate specimens with a surface crack subjected to four-point bending. The experimental approach determines the average energy release rate along the curved crack front from the area under the moment-rotation curve measured for the cracked section during the experiment, using an η-approach. The energy release rate at different crack-front locations depends on the experimentally measured average J-value and a numerically derived relationship for the J-value variation along the crack front. The compliance measurement during the experiment determines the amount of crack extension, using the compliance calculated from linear-elastic analysis of plates containing different crack sizes.

The extension of the unloading-compliance η-approach outlined for surface cracked specimens faces two critical challenges: 1) the determination of the varying energy release rate along the curved front of a surface crack from experimentally measurable quantities; and 2) the evaluation of the extending curved crack front in the experiment. The energy release rate determined from the area under the load-displacement curves provides a single J-value at a fixed applied displacement for numerous locations along the entire curved crack front. This J-value estimates the average energy release rate along the curved crack front. In contrast to the through-thickness fracture specimens, for which the average energy release rate provides a close estimation of the J-values over a predominant portion of the crack front, the curved front for surface cracks can experience a significant variation in J-values along the crack front. The determination of the energy release rate at discrete crack-front locations requires a pre-determined relationship between the J-values and the crack-front location parameters. In addition, surface cracks often experience differential extensions along the crack front. The compliance approach to determine the crack size for surface cracked specimens faces the uniqueness problem, as the compliance value calculated for two specimens with different crack depths and aspect ratios can be equal. To overcome this uniqueness problem, the compliance versus the crack size relationship calculated should correspond to the specific crack-front profiles in each individual specimen based on the post-test sectioning measurement.

The solution to the above two issues requires experimental investigations. Figure 2.12a shows the geometric configuration of the surface-cracked specimen under four-point bending. The plate specimen contains a machined semi-elliptical surface notch at the center of the plate, introduced by a wire-cutting machine. The experimental procedure includes a fatigue pre-cracking process to generate a sharp crack front ahead of the machined notch. Figure 2.12b shows the four-point bend test set-up for the plate specimens. The testing procedure involves multiple unloading approaches to measure the compliance of the specimen, similar to the standard fracture specimens with a through-thickness crack.

Figure 2.13 illustrates the scheme to determine the J-R curves for the surface cracked specimens. Similar to the unloading-compliance approach for through-thickness fracture specimens, the experimental procedure measures the load, the corresponding $CMOD$ and the compliances of the specimen in multiple unloading-reloading cycles incurred at different load levels. The measured load allows the calculation of the applied moment M on the cracked section,

$$M = \frac{1}{2} PS \tag{2.20}$$

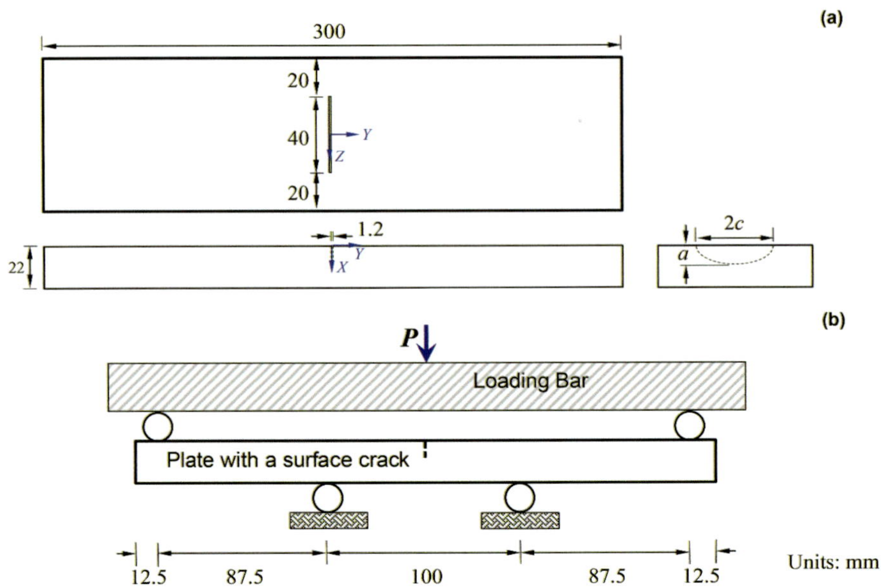

FIGURE 2.12 A plate specimen with a surface crack: (a) geometric dimensions and (b) four-point bending setup.

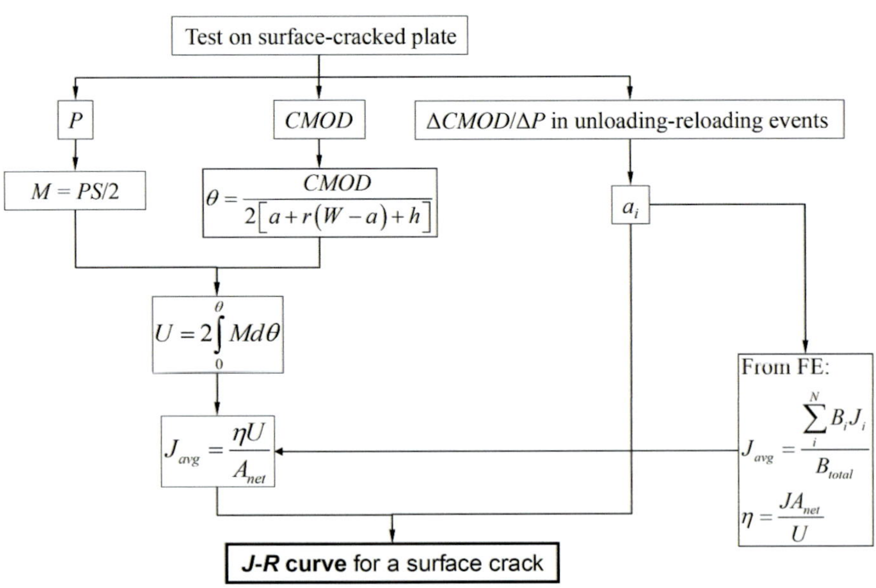

FIGURE 2.13 The scheme to measure *J-R* curve for the surface cracked plate.

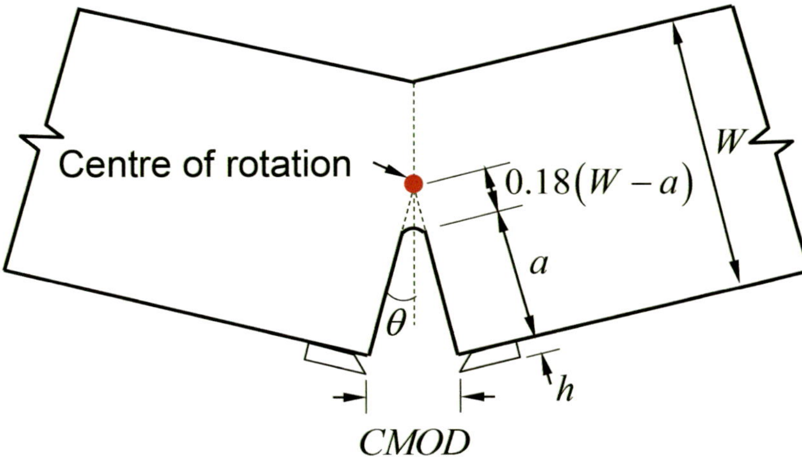

FIGURE 2.14 Rotation of the crack plane.

where S denotes the horizontal distance between the upper roller and its adjacent lower roller support and is equal to 87.5 mm for the plate shown in Figure 2.12b. The measured *CMOD* facilitates the calculation of the crack plane rotation (θ), under the assumption that the crack plane pivots around a point within the remaining ligament (see Figure 2.14),

$$\theta = \frac{CMOD}{2\left[a + r(W - a) + h\right]} \tag{2.21}$$

where W refers to the width of the specimen and r denotes a dimensionless coefficient, which calculates the distance from the center of rotation to the crack tip when multiplied by (W-a). The crack depth a in Eq. (2.21) refers to the crack depth at the deepest crack-front location. Equation (2.21) calculates the rotation of the crack plane based on the current crack depth determined from the unloading compliance method, similar to Eq. (2.13). The material testing standard [2] suggests a r-value of 0.44 for SE(B) specimens. For a surface-cracked specimen, FEA assist in determining the value of r. The parameter h in Eq. (2.21) corresponds to the thickness of the knife-edge attachment mounted on the surface of the plate and equals 1.8 mm for the current test (see Figure 2.14).

The strain energy, U, stored in the specimen thus equals,

$$U = 2\int_0^{\theta} M d\theta \tag{2.22}$$

The multiple unloading-reloading cycles during the experiment allow the measurement of the compliance of the specimen as the crack extends. Together with a regressed compliance formulation based on the compliance computed from multiple FE models for the plate specimens with varying crack sizes, the measured compliance allows the determination of the current crack size in the specimen. For each test specimen, the crack sizes simulated in FE models derive from a linear interpolation of the post-test measurement of the fatigue pre-crack profile and the final fracture profile, to avoid a non-unique compliance-crack size relationship for surface cracks. This crack size provides the estimation of the crack area (A_{crack}) in the specimen. The average energy release rate (J_{avg}) along the crack front thus becomes,

$$J_{avg} = -\frac{\Delta U}{\Delta A_{crack}} \tag{2.23}$$

Similar to the η-approach developed for standard fracture specimens, the current study simplifies Eq. (2.23) into,

$$J_{avg} = \frac{\eta U}{A_{net}} \tag{2.24}$$

where A_{net} refers to the net intact area of the cracked section. The η value derives from the FEA for the surface-cracked plates under four-point bend loading. By dividing the entire crack-front into multiple segments (say, a total of N segments), the average energy release rate along the crack-front also derives from the energy release rate along different segments of the crack front.

$$J_{avg} = \frac{\sum_{i}^{N} B_i J_i}{B_{total}} \tag{2.25}$$

where B_i denotes the length of an individual segment along the crack front and B_{total} refers to the total crack-front length. Coupling the J_{avg} calculated from the area under the M-θ curve from a FE model and the domain-integral values computed using the same FE model allows the determination of the η-value for the plate specimens.

Equation (2.23), together with the calculated crack size, leads to an average fracture resistance curve for the surface crack in a plate specimen. A precise calculation of the fracture resistance for a material point along the curved front of a surface crack in a plate specimen requires a closed-form expression for the J-integral variation along the crack front, or,

$$J_i(\phi) = J(\pi/2) f(\phi) \tag{2.26}$$

where $J(\pi/2)$ defines the energy release rate at the deepest crack-front point ($2\phi/\pi = 1$) and $f(\phi)$ corresponds to a dimensionless function prescribing the variation of the energy release rate along the crack front. The angle ϕ denotes the position

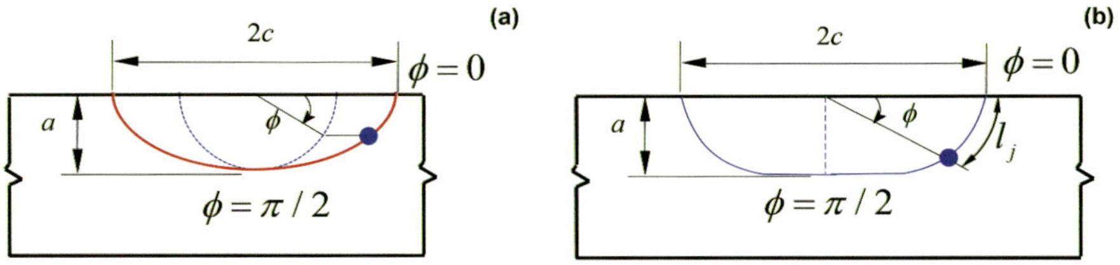

FIGURE 2.15 Two different definitions of the crack-front position: (a) based on angular position and (b) based on crack-front length.

along the crack front. The crack-front angle, ϕ, for elliptical surface cracks follows the definition in Figure 2.15,

$$\phi = \pi \frac{l_i}{B_{total}} \tag{2.27}$$

where $l_j = \sum_{i=1}^{j} B_i$ sums the length of all crack-front segments from the surface point ($\phi = 0$ and $i = 1$) to the crack-front segment of interest (segment j).

Substituting Eq. (2.26) into Eq. (2.25) leads to the energy release rate at the deepest crack-front point, which can often be represented by a polynomial fit along the crack front. Coupling the extended crack front profiles derived from the compliance measured in the unloading-reloading cycles and the *J*-value calculated at different crack-front locations leads to the *J-R* curves at different crack-front locations for surface cracks in plate specimens.

The above procedure requires the variation of the *J*-integral values along the curved surface crack front [in Eqs. (2.25) and (2.26)] determined from numerical analysis. This study computes the energy release rate at each individual segment along the crack front, that is, the J_i value. The energy release rate for realistic structural details often requires a large-deformation analysis which describes more accurately the stress distributions in structural details under large, remote loadings [20, 21] than does the small displacement theory. As demonstrated by previous researchers [22, 23], the far field *J*-integral values, which retain the path independence, characterize closely the crack driving forces in large-deformation analysis. As such, the computation of J_i values in Eq. (2.25) should adopt large-deformation, elastic-plastic analysis, with the domain-integral value extracted from the furthest domain away from the crack tip.

Table 2.1 illustrates the dimensions of the four plate specimens in the experimental program. All specimens contain a machined crack depth of 10 mm with a crack aspect ratio (a/c) of 0.5. The fatigue pre-cracking employs a three-point bending set-up with a small eccentricity to force fatigue cracks on a single side of the machined notch. The maximum load in the fatigue pre-cracking generates a crack driving force less than 30% of the critical fracture resistance measured from a C(T) specimen made of the same material, that is, S690 steel. Figure 2.16 illustrates the uni-axial stress-strain relationships for the S690 steel. The pre-cracking cycles have a fixed

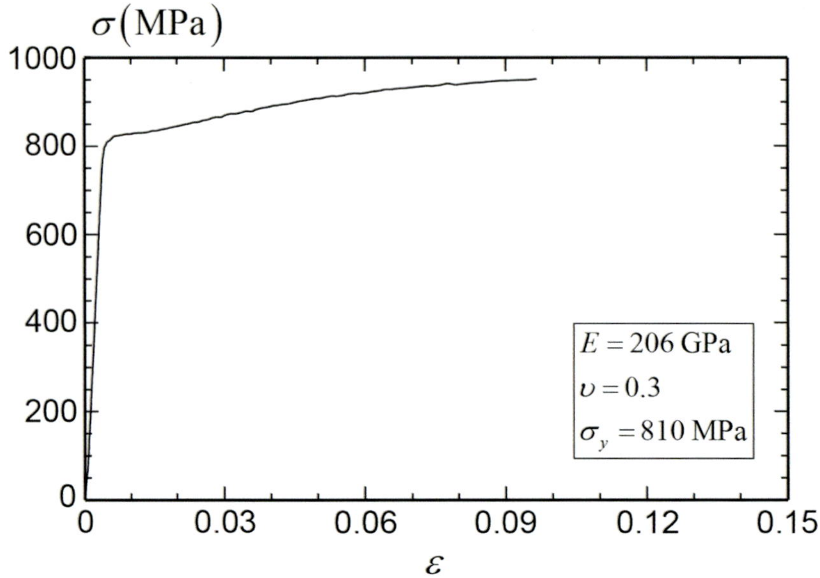

FIGURE 2.16 Uniaxial true stress and true strain curve for S690 steel materials.

load ratio of 0.1 (with the load range of 5 kN to 50 kN and the corresponding ΔK_I in range from 20 MPa$\sqrt{\text{m}}$ to 50 MPa$\sqrt{\text{m}}$ for different crack-front locations) and a frequency of 5 Hz. A crack opening displacement (COD) gauge monitors the compliance of the specimen at the center of the specimen ($Y = Z = 0$ in Figure 2.12) during the pre-cracking cycles. The fatigue pre-cracking terminates as the monitored COD compliance increases by about 20% compared to the original compliance of the specimen for SP1 and SP2, and by about 30% for SP3 and SP4.

The fracture test on these surface-cracked plates utilizes the four-point-bend set-up, instead of the three-point-bend configuration as in the fatigue pre-cracking process. The four-point-bend loading avoids the large plastic deformation near the loading point in the three-point-bend set-up, which may interfere with the plastic deformations near the crack tip.

This section illustrates the determination of the *J-R* curve using the first plate specimen, SP1, in Table 2.1. Figure 2.17a presents the load versus *CMOD* evolutions for the plate specimens SP1, measured during the experiment. The test on each specimen involves multiple unloading events throughout the testing procedure. Each unloading event includes three unloading-reloading cycles to measure the compliance of the specimen. Similar to the practice in ASTM E1820 [2], the ΔP used excludes the upper 30% and lower 30% of the load range in each unloading-reloading cycle. The experimental procedure calculates the compliance of the specimen by fitting a straight line for the screened data measured in each of the three unloading-reloading cycles in an unloading event. The measured compliance of the specimen at the unloading event thus equals the average of the three measured compliance in the three unloading-reloading cycles.

TABLE 2.1
Plate Specimens with Surface Cracks.

	Fatigue pre-crack		Final crack	
Specimen	a_0 **(mm)**	a_0/c_0	a_f **(mm)**	a_f/c_f
SP1	12.0	0.497	13.6	0.532
SP2	12.0	0.496	13.9	0.552
SP3	12.4	0.484	14.3	0.537
SP4	12.3	0.483	14.4	0.541

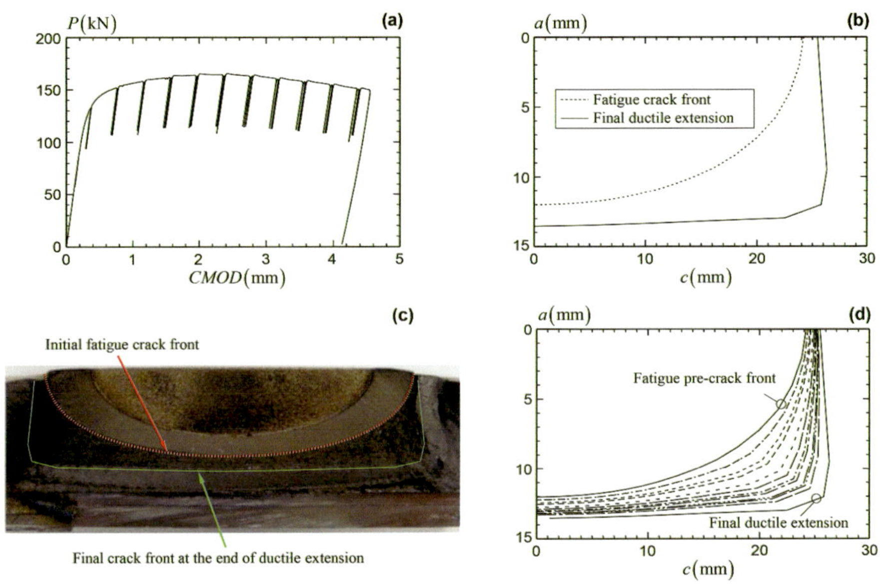

FIGURE 2.17 For SP1, (a) measured load versus the crack mouth opening displacement (*CMOD*); (b) measured crack-front profiles at the end of the fatigue pre-cracking and at the end of the ductile crack extensions; (c) a typical fracture surface revealed by post-test sectioning; (d) crack-front profiles corresponding to the measured compliances.

The final stage of the test entails a cyclic loading, which aims to generate a smooth fatigue surface beyond the ductile fracture surface produced during the fracture test, similar to the approach adopted in previous experimental investigations [14]. The post-test sectioning reveals the amount of the fatigue crack and ductile crack extension along the crack front, as illustrated in Figures 2.17b and 2.17c for SP1.

The crack-size versus the compliance relationship derives from linear-elastic FEA for multiple FE models with various crack sizes. To ensure uniqueness in the compliance-crack size relationship for surface cracks, each specimen requires a separate set of FE models with the crack sizes linearly interpolated based on the measured fatigue pre-cracked front profile and the crack-front profile at the end of the ductile crack extension. The cracks simulated in the FE mesh include seven different crack sizes. The smallest crack size corresponds to the measured fatigue crack-front, and the largest crack size refers to the measured crack-front profile at the end of the ductile crack extension. The other five crack sizes follow a linear interpolation between the smallest and the largest crack sizes. This study utilizes a second-order polynomial to express the maximum crack depth along the crack front as a function of the compliance. For SP1, the relationship follows,

$$\frac{a}{t} = 4.83 \times 10^{-3} \left(ECt\right)^2 + 0.214\left(ECt\right) + 3.72 \tag{2.28}$$

The linearly interpolated crack-front profiles assume proportional extensions at different locations along the entire crack front and remain a reasonable representation of the evolving crack-front profile during the ductile crack extension of the plate specimens, in the absence of detailed instruments to monitor the growing crack-front during the test. This study thus assumes a linear interpolation in determining the crack-front profile corresponding to each compliance value measured in the unloading-reloading cycles. This implies that the location for each front point depends only on the maximum crack size determined using Eq. (2.28). Figure 2.17d shows the twelve crack-front profiles determined from the 12 measured compliances for specimen SP1. The calculation of the average energy release rate from Eq. (2.24) requires a non-dimensional η value. This study determines the η value through a series of large-deformation, elastic-plastic FEA. The FEA compute the domain integral values along the stationary crack front. The average energy release rate for the entire crack front derives from Eq. (2.25), using the calculated J_i values along the crack front. The M-θ relationship computed from the FE analysis allows the evaluation of U for the surface cracked specimen using Eq. (2.24). The average value of η equals 1.1 for the surface-cracked plates considered in this study.

The evaluation of the energy release rate at different crack-front positions requires a relationship of the J-value variation along the crack front. The current study performs a polynomial regression of the domain-integral values computed using large-deformation, elastic-plastic analysis along the curved crack front. The polynomial expression follows,

$$\begin{aligned} f(\phi) = \frac{J_i(\phi)}{J(\pi/2)} &= 1 + \left(8.66\left(\frac{a}{t}\right) - 5.51\right)\left(1 - \frac{2\phi}{\pi}\right) \\ &+ \left(-86.1\left(\frac{a}{t}\right) + 53.8\right)\left(1 - \frac{2\phi}{\pi}\right)^2 + \left(91.2\left(\frac{a}{t}\right) - 54.4\right)\left(1 - \frac{2\phi}{\pi}\right)^3 \end{aligned} \tag{2.29}$$

Figure 2.18 presents the J-R curves determined based on the linearly interpolated crack-front profiles at different crack-front locations for the four surface-cracked plates, together with the fracture resistance curve obtained from the standard C(T) specimens. The J-R curves measured at three different crack-front positions demonstrate similar levels of fracture resistance, which agrees with the fracture resistance obtained from the C(T) specimens, at short crack extensions. This indicates that the

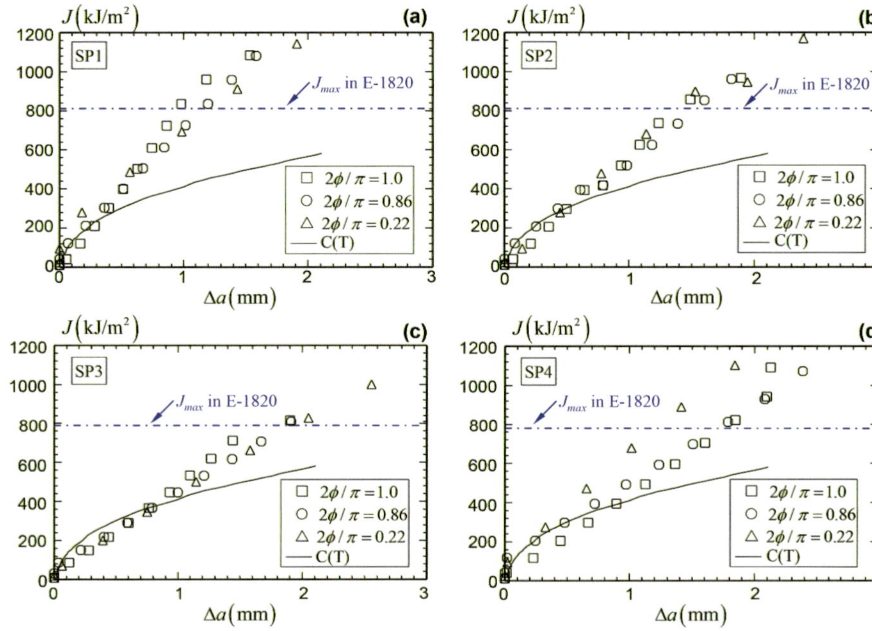

FIGURE 2.18 Comparison of the *J-R* curves measured at different crack-front locations for the plate specimens with the *J-R* curve obtained from a side-grooved C(T) specimen for (a) SP1, (b) SP2, (c) SP3, and (d) SP4.

linearly interpolated crack-front profiles assumed in deriving the crack-size versus the compliance relationship are a reasonable assumption, in the absence of detailed instrumentation to monitor the crack growth along the curved surface crack front. As the load increases, the crack-front near the free surface ($2\phi/\pi = 0.22$ in Figure 2.18) experiences more significant crack extensions than the crack-front material near the deepest crack-front point, due to the higher crack driving forces near the free surface. The *J-R* curves measured from the four specimens reflect similar fracture resistance for all plates. The comparison in Figure 2.18 also implies that the *J-R* curves derived by assuming that the crack extends only in the thickness direction (*i.e.*, the crack-size versus the compliance relationship derives from the assumed crack-front profiles with fixed $c = c_0$) remain much lower than the plane-strain *J-R* curves determined from the C(T) specimen. The set of crack-front profiles with a fixed crack length $c = c_0$ does not describe the realistic crack extensions along the curved crack front.

2.5 TEST METHOD FOR WELDED PIPES WITH CIRCUMFERENTIAL SURFACE CRACK

The above approach to measure *J-R* curve in surface cracked plates applies to surface cracks in other types of structural details, for example, in pipes with a circumferential surface crack along the girth welds. The following discussions illustrate the detailed steps for this.

2.5.1 Unloading Compliance η-Approach

Similar to the surface cracked plates, the measurement of J-R curve in pipes with a circumferential crack requires the η method to quantify the J value and the unloading compliance technique to size the crack extension. As the load-deformation curve of each cracked pipe yields an evolution of a single U value in Eq. (2.24), the η method thus provides the measurement of a single J history, corresponding to the average J-value over the surface crack front. This J derives from the area below the experimentally recorded P-$CMOD$ curve and a numerically determined non-dimensional parameter η. The relationship between the average J and the point-wise domain-integral value along the crack front, determined from detailed numerical analysis, leads to the location-specific J-Δa curve. The measured compliance during the test allows estimation of the crack extension using the numerically determined compliance versus the crack size relationship.

The procedure to measure the J-Δa curve for pipes with a circumferentially oriented surface crack includes the J measurement based on the load and $CMOD$ records, and the Δa measurement from the crack-front profiles and the specimen compliance. The testing procedure includes: 1) a cyclic fatigue pre-cracking on each machine-notched pipe specimen, and 2) a monotonically increasing load with numerous intermediate unloading-reloading cycles to monitor the change in compliance caused by crack extension. Figure 2.19 displays the three-point bend setup of a pipe with the cracked cross-section. The experimentally measured load versus δ_{LLD} and load versus $CMOD$ curves with intermediate unloading and reloading cycles at multiple displacement values allow the calculation of the stored strain energy in the pipe. Previous researchers [24–27] have suggested $CMOD$ to be a more reliable measurement compared to δ_{LLD} for low-constraint specimens. The strain energy thus derives from the area below the P-$CMOD$ curve as,

$$U = U_{le} + \int_0^{CMOD_{pl}} P dCMOD_{pl} \tag{2.30}$$

where U_{le} denotes the elastic component of the strain energy and the integral in Eq. (2.30) refers to the plastic component of the strain energy.

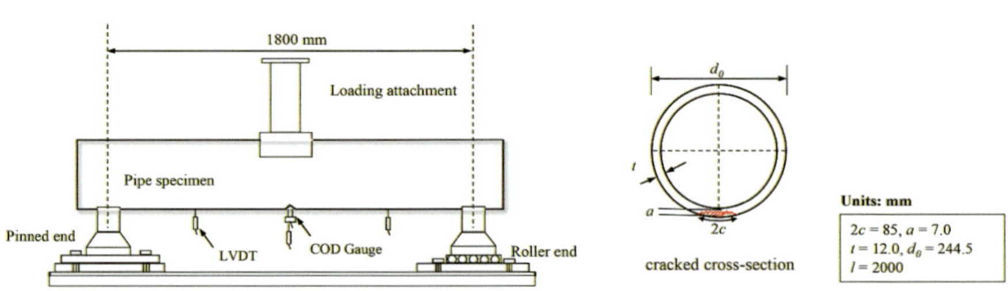

FIGURE 2.19 Experimental setup for the pipe with a circumferential surface crack.

The measurement of the exact amount of crack extension at each unloading point remains crucial in evaluating the J-Δa curve. The realistic crack-front profile in structural components may deviate significantly from that of an often assumed semi-elliptical profile, as evidenced in previous experimental investigations [28–30]. To resolve this, the compliance-crack depth relationship should be derived from the actual crack profile in the pipe specimens. The crack-front profiles assumed in deriving the compliance function follow the crack profiles interpolated linearly between the profile after the cyclic pre-cracking and that at the final ductile tearing. The compliance function then derives from linear-elastic analyses on a series of pipe models incorporating the assumed crack profiles. The specimen compliance measured at the kth unloading-reloading cycle, C_k, leads to the corresponding crack size a_k. The amount of crack extension thus follows,

$$\Delta a = a_k - a_0 \tag{2.31}$$

The value of a_k and the corresponding linear interpolations of the crack-front profile facilitates the calculation of the current net intact area $A_{k,net}$ at the kth unloading-reloading cycle.

Unlike the through-thickness cracked specimen, the energy release rate in a pipe varies substantially over the crack front. The J-integral determined from the experimental P-$CMOD$ curves using Eqs. (2.9)–(2.11) [31] for surface cracks yields a single J value at a given displacement. This J represents the average J-value (J_{avg}) along the entire crack front. Based on the η method, Eq. (2.32) calculates J_{avg} as,

$$J_{avg} = J_{le-av} + \frac{\eta_{pl-avg} U_{pl}}{A_{net}} \tag{2.32}$$

Equating the average J determined from Eq. (2.32) with the average J estimated from the domain-integral approach in an elastic-plastic analysis leads to an estimated η_{pl-avg} value. For the domain-integral approach, the average energy release rate, J_{avg} employs the same relationship in Eq. (2.25) as the surface cracked plates. Calculation of the J-Δa curve at a specific crack-front location needs a pre-determined expression for the J-variations, that is, Eq. (2.26), similar to the surface cracked plate.

2.5.2 Experimental Validation

The experimental program consists of tension specimens to measure the mechanical properties of the material, standard fracture specimens to measure the material J-Δa curve, and pipes with a circumferential surface crack made of S355 steel to measure the J-Δa curves in the pipes.

The fracture tests consist of six full-scale, surface-cracked pipe specimens. Each specimen comprised two 1-meter-long straight pipe segments joined by circumferential girth welding. A machined circumferential surface notch was introduced near the weld toe using wire cutting. The notch had a central depth, a, of 7 mm and an outer surface length, $2c$, of 85 mm (Figure 2.19). Each pipe had an outer diameter d_0

TABLE 2.2
Geometric Dimensions for Pipe Specimens.

				Machined notch		Fatigue pre-crack	
Specimen no.	d_0 (mm)	t (mm)	Span (mm)	Crack depth a (mm)	Crack length $2c$ (mm)	Crack depth a_0 (mm)	Crack length $2c_0$ (mm)
Pipe_01	244.5	12.0	1800	7.0	85.0	9.14	85.0
Pipe_02	244.5	12.0	1800	7.0	85.0	9.27	85.0
Pipe_03	244.5	12.0	1800	7.0	85.0	9.42	85.0
Pipe_04	244.5	12.0	1800	7.0	85.0	8.98	85.0
Pipe_05	244.5	12.0	1800	7.0	85.0	9.43	85.0
Pipe_06	244.5	12.0	1800	7.0	85.0	9.30	85.0

of 244.5 mm and a wall thickness, t, of 12 mm. Table 2.2 details the dimensions of all pipe specimens.

The experimental procedure consists of two main steps: (1) fatigue pre-cracking and (2) fracture testing, both conducted using a three-point bend setup with a span length of 1.8 m, as shown in Figure 2.19. The load applies via a semi-circular loading attachment that grips the pipe over a finite area, minimizing local plastic deformation during the test.

For fatigue pre-cracking, a sinusoidal cyclic load is applied with a maximum load set at approximately 70% of the yield load of the cracked section. The loading frequency remains fixed at 2 Hz, with a constant load ratio R of 0.1. The crack opening displacement (COD) is monitored using a COD gauge attached to two aluminum plates with knife edges positioned on either side of the notch (Figure 2.19). The gauge tracks the initial compliance of the pipe specimen and the subsequent changes in compliance caused by fatigue crack initiation and growth.

Fatigue pre-cracking continues until the compliance change reaches approximately 20%–30%. The number of cycles required to achieve this compliance change varies between 300,000 and 450,000 cycles [31]. After fatigue pre-cracking, a displacement-controlled load is applied to the pipe specimens at room temperature, using the same experimental setup. The setup, as shown in Figure 2.19, includes linear variable displacement transducers (LVDTs) positioned at the mid-span and quarter spans to monitor displacement. The loading rate ranges from 0.2 mm/min to 0.5 mm/min.

Figures 2.20a and b illustrate the measured load (P) versus crack mouth opening displacement ($CMOD$) responses for two selected pipe specimens, which include multiple unloading-reloading cycles. The specimen compliance is determined from a linear fit of the P-$CMOD$ data within each unloading-reloading cycle, excluding the upper 30% and lower 30% of the data points to improve accuracy.

The estimation of the J–Δa curve requires post-test measurements of the crack profiles. These measurements are obtained using quick-setting silica putty, which is mixed and cured at room temperature to create replicas of the cracked surfaces. The hardened putty captures detailed surface profiles, including the notch size, fatigue pre-cracking length, and ductile tearing.

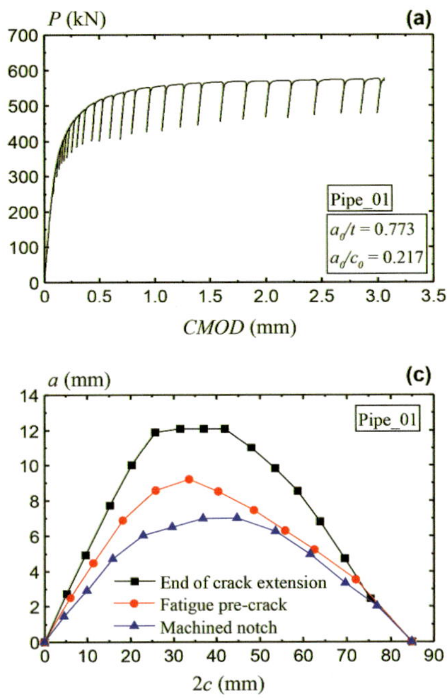

FIGURE 2.20 Measured load versus *CMOD* for (a) Pipe_01 and (b) Pipe_02; and measured crack-front profiles for (c) Pipe_01 and (d) Pipe_02.

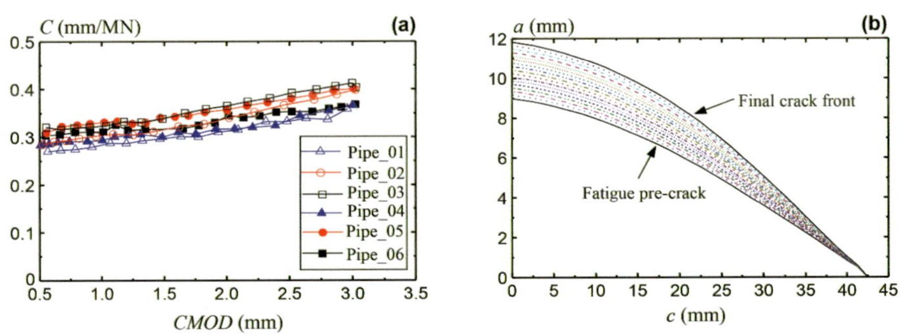

FIGURE 2.21 (a) Measured compliance versus *CMOD* for: six pipe specimens and (b) linearly interpolated crack-front profiles.

Figures 2.20c and d show the measured crack profiles for four selected pipe specimens, revealing that crack propagation primarily occurs through the pipe thickness, while the crack length ($2c$) remains constant.

Figure 2.21a illustrates the variation in pipe compliance ($\Delta\delta / \Delta P$) at different unloading *CMOD* levels for all six specimens. The relationship between specimen compliance and crack size for pipes with circumferential surface cracks is derived from extensive linear-elastic FEA. These analyses model 15 distinct crack profiles,

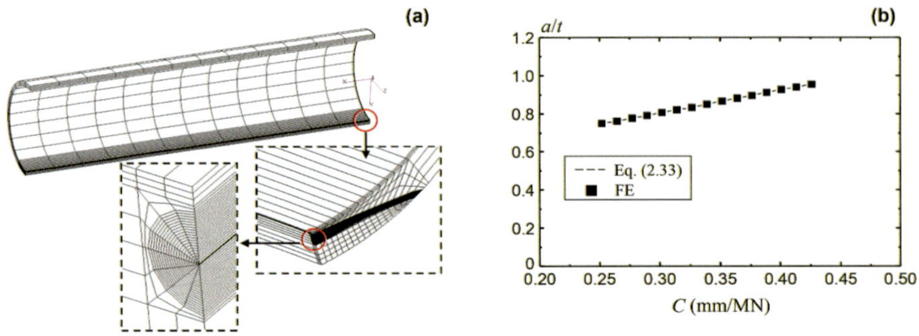

FIGURE 2.22 (a) A typical FE mesh of the pipe specimen; (b) crack depth versus compliance relationship.

with the first representing the cyclic pre-cracked profile and the last corresponding to the final ductile tearing. The intermediate 13 profiles are linearly interpolated between the initial and final profiles (Figure 2.21b). Figure 2.22a presents a typical FE model of a surface-cracked pipe, which utilizes 20-node hexahedral elements with reduced integration for enhanced computational efficiency. Figure 2.22b depicts the decreasing crack size in relation to increasing pipe compliance, as determined from the FEA. A second-order polynomial [Eq. (2.33)] is applied to quantify the crack depth as a function of the non-dimensional pipe compliance, providing a robust method for characterizing the surface crack behavior.

$$\frac{a}{t} = 0.0364\left(ECt\right)^2 + 0.4116\left(ECt\right) + 0.4723 \tag{2.33}$$

The calculation of J values employs η values derived from a series of elastic-plastic FEA for pipes with circumferential surface cracks. The crack profiles used in the FEA are interpolated from experimental measurements (Figure 2.21b). To ensure numerical convergence at large displacement levels, the FE models feature a minimum initial root radius of 15 μm at the crack tip (Figure 2.22a). The crack-front mesh is divided into 20 domains, each containing eight elements around the crack tip in the circumferential direction.

A typical FE model consists of approximately 50,000–60,000 nodes and 11,000–12,000 elements. The numerical procedure, implemented in the open-source FE code WARP3D, uses the domain-integral approach developed by Shih *et al.* [32] to compute the energy release rate. Additionally, the FEA generate the *P-CMOD* curve, which is used to determine the η value.

The *CMOD*-dependent η expression derived from multivariate regression analyses, ensuring a more accurate representation of the crack-tip driving forces.

$$\begin{aligned}\eta_{pl-avg} = {} & 1.514\left(CMOD_{pl}/t\right)^2 - 35.98\left(a/t\right)^2 - 47.66\left(CMOD_{pl}/t\right)\left(a/t\right) \\ & + 28.93\left(CMOD_{pl}/t\right) + 60.65\left(a/t\right) - 20.50\end{aligned} \tag{2.34}$$

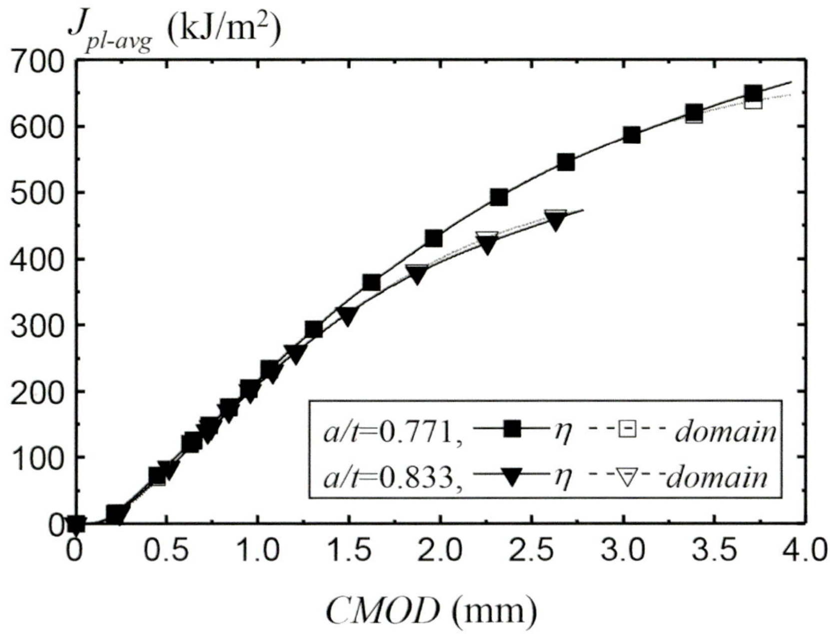

FIGURE 2.23 Comparison between J-value based on the η approach and the domain-integral method.

Figure 2.23 shows the good agreement between the J_{pl-avg} computed using Eq. (2.34) and the average J-integral value.

The above procedure calculates the average J-integral value using the η-method. To evaluate the location-specific energy release rate (J_i) along the crack profile, further computations are performed using Eq. (2.35).

$$J_i(\phi) = J(\pi/2) f(\phi) \tag{2.35}$$

The variation of J-integral values along the crack front shows an increasing trend from the free surface toward the deepest point. These variations reflect the combined influence of geometry, plasticity-induced constraint differences at various crack-front locations, and three-dimensional effects. Previous three-dimensional FEA results [33] have demonstrated similar behavior, particularly for stress intensity factors (SIFs) under mode I loading. Rapid decreases in mode I SIF values near the free surface and sharp increases in mode II SIF values have been attributed to the vertex or corner singularity at the crack-front intersection with a free surface. Reviews by [34] have highlighted that the stress field near a corner point results from the combined effects of two singularities.

To capture the variation in the normalized J-integral values along the crack front, a fourth-order polynomial is fitted to the data presented in Figure 2.24,

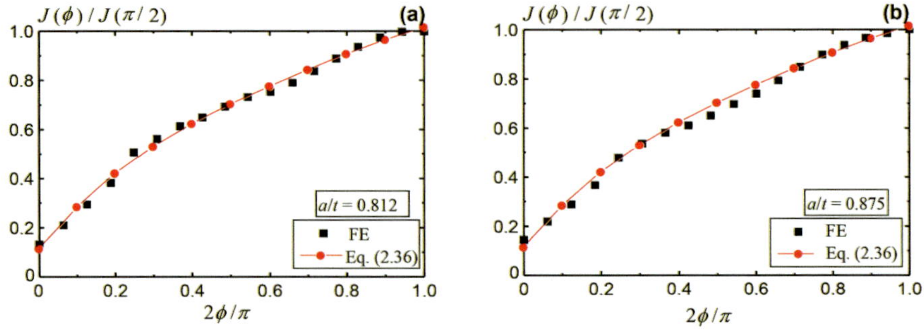

FIGURE 2.24 Comparison between Eq. (2.36) and FE data.

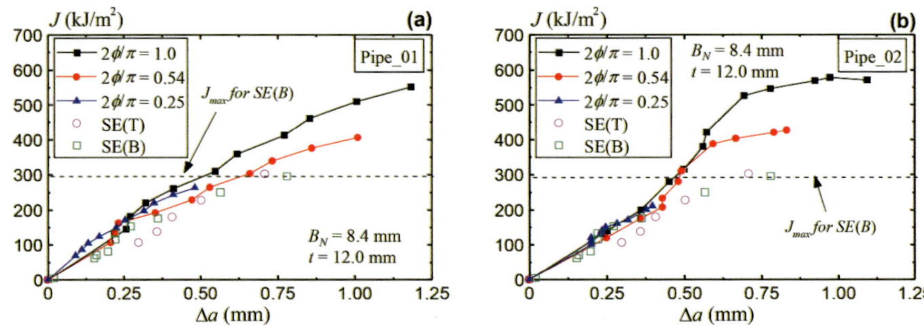

FIGURE 2.25 Comparison of *J-R* curve measured from pipe specimen and standard fracture specimens: (a) Pipe_01; and (b) Pipe_02.

providing a functional representation of the observed trends, as represented by Eq. (2.36).

$$f(\phi) = -0.069\left(\frac{2\phi}{\pi}\right)^4 + 1.952\left(\frac{2\phi}{\pi}\right)^3 - 2.263\left(\frac{2\phi}{\pi}\right)^2 + 1.908\left(\frac{2\phi}{\pi}\right) + 0.113 \quad (2.36)$$

Figure 2.24 compares Eq. (2.36) and the variations of computed *J*-integrals for two different a/t ratios. Figure 2.25 compares the *J*–Δ*a* curves for all selected pipes at three distinct crack-front locations: near the free surface ($2\phi/\pi = 0.25$), midway along the crack front ($2\phi/\pi = 0.54$), and at the deepest point ($2\phi/\pi = 1$). Due to the lower crack driving force near the free surface, this location experiences the least crack extension compared to the deepest point. In contrast, the *J*–Δ*a* curve at the deepest point shows the greatest crack extension, driven by the highest crack driving force (Figure 2.24).

The figure also includes *J*–Δ*a* curves from standard SE(B) and SE(T) specimens. At small crack extensions, the pipe specimens exhibit fracture resistance similar to the SE(B) and SE(T) specimens. However, as the crack extends further, the fracture resistance of the pipe specimens surpasses that of the SE(B) and SE(T) specimens. The difference in fracture resistance between the SE(B) and SE(T) specimens remains minimal, particularly for small crack extensions.

2.5.3 Hybrid Approach

An alternative to the unloading compliance η-method, the hybrid approach discussed in Section 2.3, provides a practical method for estimating the *J–R* curve in pipes with circumferential cracks.

Similar to its application in standard fracture specimens, the hybrid method for pipe specimens requires conducting a fracture test on a single fatigue pre-cracked pipe using the unloading compliance method and performing FEA on multiple models with varying crack sizes. The unloading compliance test supplies experimental data to validate the hybrid approach, while uniaxial stress-strain curves obtained from tension tests serve as essential inputs for the FEA.

The crack sizes in the FE models are derived through linear interpolation between the fatigue pre-cracking profile and the final crack-front profile obtained at the conclusion of the fracture test. The final crack-front profile can be determined either via the compliance method or through post-test measurements of the fracture surface. Large-deformation, elastic-plastic analyses performed on these FE pipe models generate a series of *P–CMOD* curves.

Unlike the approach used for standard fracture specimens with straight crack fronts, which relies on the intersection of experimental and numerical *P–CMOD* curves, the hybrid method for cracked pipes calculates the strain energy (U) corresponding to the *CMOD* levels recorded prior to unloading and reloading cycles in the pipe test. By coupling the strain energy (U) with crack sizes determined from the unloading compliance test, the derivative $\partial U/\partial a$ can be computed. This enables the construction of the *J–R* curve for the cracked pipes.

The validation of the modified hybrid method entails the same six 2 m long pipe specimens discussed above, each with a circumferential surface notch near the toe of the girth welds at its mid-length. The post-test examination reveals the fatigue pre-crack front profile a_0 and the final crack-front profile a_f from replica-reproduced fracture surfaces. Linear-elastic numerical analyses on pipes with linearly interpolated crack-front profiles between a_0 and a_f lead to a relationship defining the *CMOD* compliance dependence on the crack size. Figure 2.26a shows the experimentally measured *CMOD* prior to each unloading and reloading cycle against the crack depth, a, estimated using the compliance-crack size relationship in [31]. The modified hybrid method makes use of this specimen-specific compliance-crack size relationship to determine the local displacement level, at which the strain energy is evaluated.

The strain energy, U, from the *P-CMOD* curve and the crack area for a semi-elliptical surface crack, leads to the average J value,

$$J = -\frac{dU}{dA_{crack}} = -\frac{1}{0.5\pi c}\frac{dU}{da} \tag{2.37}$$

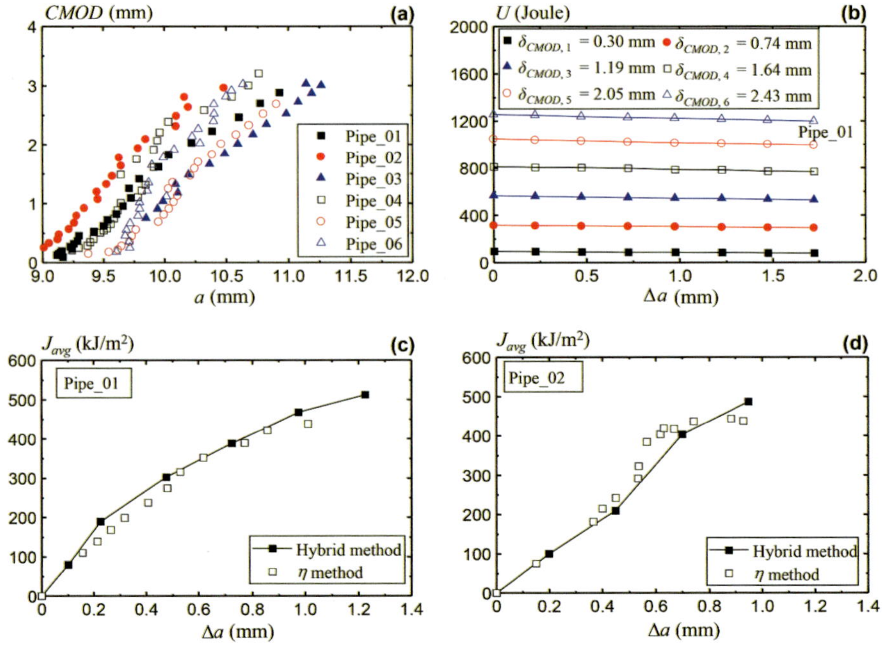

FIGURE 2.26 (a) Variation of the experimental *CMOD* versus crack depth for pipe specimens; (b) strain energy versus crack extension for Pipe_01; Comparison of the *J-R* curve measured using the η method and modified hybrid approach for (c) Pipe_01; and (d) Pipe_02.

The strain energy, U, computed at each $\delta_{CMOD,i}$ level, as shown in Figure 2.26b against the crack extension, Δa, leads to the average J-value following Eq. (2.37) for specimen Pipe_01. Figures 2.26c and d exemplify the good agreement in the $J - \Delta a$ curve estimated from the hybrid method and that measured using the η-compliance method based on *P-CMOD* curve for two selected pipes with a circumferential surface crack.

2.6 DIGITAL IMAGE CORRELATION METHOD

Although the traditional methods have found successful applications in the laboratory *J-R* curve measurement, they rely mainly on the in-contact instruments to record the load, the deformation, or the crack profile. Recent development in the DIC technology [35, 36] provides a more convenient means to quantify the *J*-integral from the free-surface measurement. The DIC utilizes non-contact cameras to trace the specimen deformation and derives the *J*-integral from the local field analyses. Chen and Qian [37] extended the DIC method to plane-strain specimens and investigate the thickness effects on the deviation of free-surface and through-thickness average *J* results. While the DIC technology has integrated the *J*-integral assessment successfully, the measurement of crack extension under elastic-plastic conditions is not satisfactory. The crack tunneling leads to a shorter crack length on the free-surface than that on the mid-thickness plane [38]. The surface measurement thus underestimates the average crack size for an accurate *J-R* curve quantification. Besides, the shear lips induce non-straight crack shapes on the free surface and cause discontinuity issues on the local crack-tip field [39]. To avoid the disruptions of the crack tunneling and shear lips, the ASTM E1820 [2] recommends side grooves for the *J-R* curve measurement, as discussed in Section 2.3. The V-shape and machine-slit grooves lead to more uniform plastic zone shapes along the crack front. Considering the popularity of side-grooved specimens in the laboratory tests [32, 40, 41], the research gap between the *J*-integral measurement and the DIC application needs to be addressed. The non-slant crack profile from the side-grooved specimens also requires more convenient methods to determine the crack size from the standard specimens, rather than using the complicated image processing algorithms that aim at random cracks [42].

2.6.1 DIC-Based Contour Integral

Figure 2.27a illustrates the discretization of the image for the following correlation process. The region of interest (ROI) is selected as the desired region, covered by the randomly distributed speckles. ROI is further divided by several subsets with finite grids. Every subset has the same dimension but contains unique speckle patterns from the assembly of different grids.

Figure 2.27b demonstrates the deformation of an arbitrary subset from the reference to the deformed configuration. $P(x_0, y_0)$ is the center of the subset $\boldsymbol{S}$. In order to track $P(x_0', y_0')$ in the deformed configuration, the pixel intensities inside the subset $\boldsymbol{S}$ are assembled into a matrix. This matrix is compared with matrices from any random subset $\boldsymbol{R}$ (with the same dimension) to locate the subset $\boldsymbol{S}$ in the deformed configuration. A correlation coefficient (denoted as C) is defined using the cross-correlation criterion.

$$C = \sum\nolimits_{x,y \in \boldsymbol{S}; x',y' \in \boldsymbol{R}} \left[\frac{f(x,y) - f_m}{\Delta f} - \frac{g(x',y') - g_m}{\Delta g} \right]^2 \tag{2.38}$$

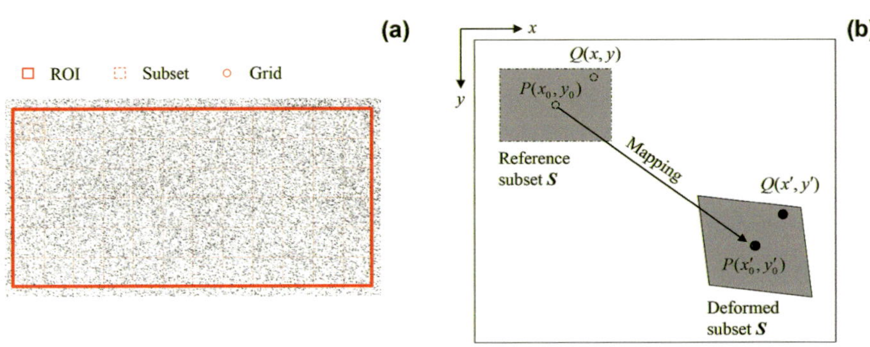

FIGURE 2.27 (a) Definition of ROI, subset and grid; and (b) the displacement mapping.

$$\Delta f = \sqrt{\sum\nolimits_{x,y\in \boldsymbol{S}} \left[f(x,y) - f_m\right]^2} \tag{2.39}$$

$$\Delta g = \sqrt{\sum\nolimits_{x',y'\in \boldsymbol{R}} \left[g(x',y') - g_m\right]^2} \tag{2.40}$$

where f and g represent the pixel intensity under reference and deformed configurations, f_m and g_m are the mean intensity value, Δf and Δg are the corresponding variations over the subset $\boldsymbol{S}$ and $\boldsymbol{R}$. An extremum value of the correlation coefficient C indicates the same subset ($\boldsymbol{R} = \boldsymbol{S}$) in different configurations [43].

Based on the assumption of deformation continuity, all the grids inside a reference subset can be mapped into the same subset in every configuration. Point $Q(x, y)$ is an arbitrary point inside the reference subset $\boldsymbol{S}$. The coordinates of $Q(x', y')$ under the deformed configuration derive from the following expressions.

$$x' = x + u_x + \frac{\partial u_x}{\partial x}(x - x_0) + \frac{\partial u_x}{\partial y}(y - y_0) \tag{2.41}$$

$$y' = y + u_y + \frac{\partial u_y}{\partial x}(x - x_0) + \frac{\partial u_y}{\partial y}(y - y_0) \tag{2.42}$$

where u_x and u_y denote the horizontal and vertical displacement at the center point $P(x_0, y_0)$, $i, j \in \mathbf{S}$. A Newton-Raphson optimization method helps to enhance the displacement mapping efficiency significantly [44].

Rice [45] demonstrates the path-independent J values over contours that enclose the crack tip in 2D bodies. The contour integral has the following expression,

$$J = \int W_e dy - T_i \frac{\partial u_i}{\partial x} ds \tag{2.43}$$

where W_e is the strain energy density, T_i indicates the traction vector, and u_i describes the displacement along the contour ds. The subscript i entails the x and y directions in 2D conditions. With the displacement measured from DIC, the strains are derived from strain-displacement relationship, and the stresses are derived from the constitutive models. This study employs the Green-Lagrange strain [46],

$$\varepsilon_{ij} = \frac{1}{2}(F_{ki}F_{kj} - \delta_{ij}) \tag{2.44}$$

where δ_{ij} represents the Kronecker sign. The deformation gradient F derives from a pointwise least-square (PLS) method [47]. The PLS method calculates the deformation gradients from the least square method, using displacement data approximated by first-order polynomials. Figure 2.28 illustrates the PLS-based displacement filtering using a low-pass window. The PLS method helps to reduce the displacement noise effectively and derive more accurate strain results from the DIC measurement [48–50].

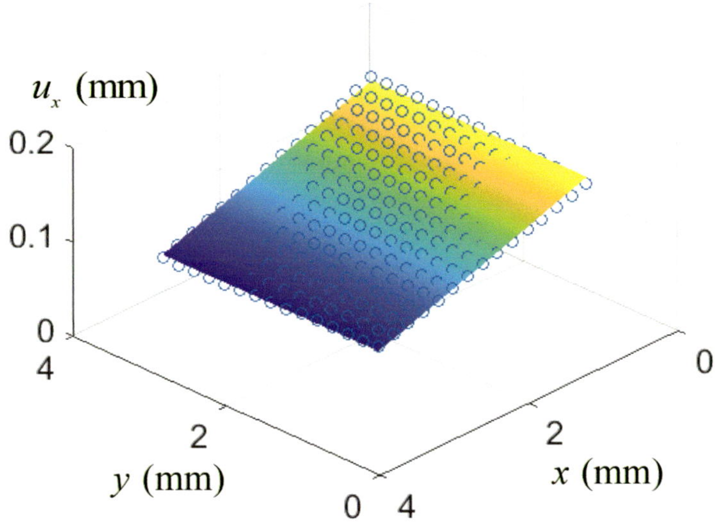

FIGURE 2.28 The displacement filtering.

The stress field derives from the strain solutions through the constitutive model. Under linear-elastic conditions, the stress-strain relationship [51] reads,

$$\sigma_{ij} = \lambda \varepsilon_{kk} \delta_{ij} + 2\mu \varepsilon_{ij} \tag{2.45}$$

where σ_{ij} indicates the stress components, λ describes the Lame coefficient, and μ refers to the shear modulus. Under plastic conditions, the Prandtl-Reuss equation [52] follows,

$$d\varepsilon_{ij}^{p} = s_{ij} d\lambda \tag{2.46}$$

where $d\varepsilon_{ij}^{p}$ indicates the plastic strain increment, s_{ij} represents the deviatoric stress components, and $d\lambda$ denotes the plastic incremental scalar. In this study, the von Mises yielding criterion is adopted for the power-law hardening material. The equivalent stress σ_{eq} reads,

$$\sigma_{eq} = \sqrt{\frac{3}{2} s_{ij} s_{ij}} \tag{2.47}$$

The strain energy density is integrated by the stress and strain as,

$$W_e = \int \sigma_{ij} d\varepsilon_{ij} \tag{2.48}$$

This chapter constructs the contours from the nodal points instead of the Gauss points due to their convenience for numerical and experimental implementation. Although the Gauss points provide more accurate stress solutions than the nodal points, the difference in J-integral becomes insignificant after the integration process.

2.6.2 Experimental Setup

Figure 2.29a presents the geometry dimensions of the SE(B) specimen. There are three specimens in total with a thickness of 26 mm and a width of 76 mm. The span length between the two supports is 304 mm. The V-shape side groove has an angle of 60° (Figure 2.29b) on both sides and contributes to a total reduction of $0.2B$ in the thickness direction. The side grooves follow the same geometry (Figure 2.29c), with $B_N / B = 0.8$ under a 60° opening angle. A cyclic fatigue pre-cracking process precedes the fracture test to create the sharp crack front. The initial a / W ratio, including the fatigue pre-crack size, is around 0.5 for the side-grooved SE(B) specimen. Figure 2.29c illustrates the experimental setup for the fracture test of the SE(B) specimen, made of aluminum alloy 5083.

Table 2.3 summarizes the dimensions and initial crack size of the three SE(B) specimens. The specimen surface is sprayed with speckles required for DIC analysis (details shown in Figure 2.29c). The DIC camera has a resolution of 6016×4016 pixels (0.03 mm/pixel), with an image recording frequency of 0.1 Hz. The fracture test, performed under an Instron universal testing machine, is subjected to a displacement-controlled load at a rate of 0.3 mm/min. A crack opening displacement (COD) gauge records the $CMOD$ during the test. Figure 2.30 records the P-$CMOD$

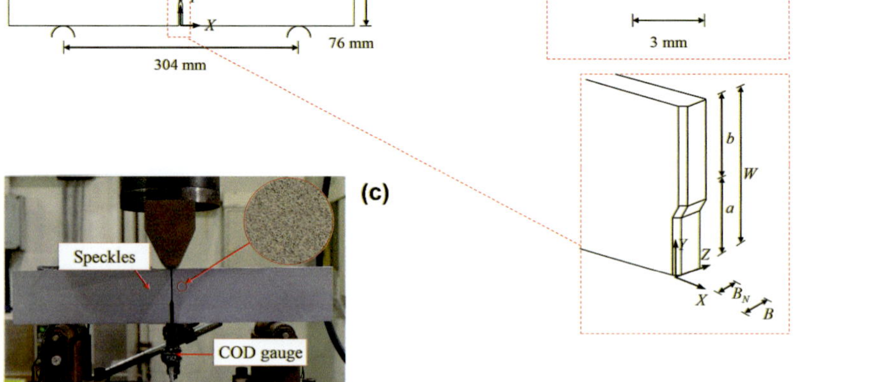

FIGURE 2.29 (a) Geometry configuration of the SE(B) specimen; (b) details of local regions; and (c) experimental setup of the SE(B) specimen.

TABLE 2.3
Summary Information of SE(B) Specimens.

Specimen No.	Dimension (mm)	B_N / B	Initial crack size (mm)
SEB-1	$445 \times 76 \times 26$	0.8	37.4
SEB-2	$445 \times 76 \times 26$	0.8	38.0
SEB-3	$445 \times 76 \times 26$	0.8	37.7

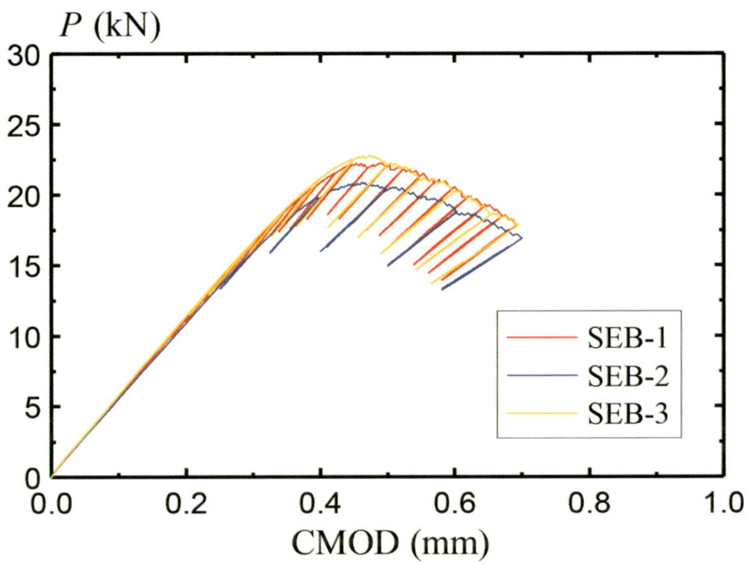

FIGURE 2.30 Experimental *P-CMOD* curves for three specimens.

curves of the SE(B) specimens from the fracture test. Several unloading-reloading sequences are applied to measure the elastic compliance during the test. The fracture test ends at a *CMOD* of 0.7 mm.

2.6.3 Validation of *J-R* Curve by DIC

Figure 2.31a compares the *CMOD* values from the DIC and the COD gauge measurement of the specimen SEB-1. The close agreement demonstrates the accuracy of DIC in the displacement measurement. The unloading and reloading process does not affect the DIC's accuracy. Figure 2.31b displays the DIC-measured horizontal displacement on the free surface of the specimen SEB-1 at the end of the test (*CMOD*=0.7 mm). The DIC algorithm selects a region of interest (ROI) $W \times W$ centered around the crack plane. Regions near the crack plane ($|X / W| < 0.02$) are excluded from the DIC analysis due to the presence of side grooves. The subset size is

4×4 mm for the displacement mapping, and the grid distance is 0.2 mm. The experimental measurement contains noise in the DIC-based displacement fields. Although the noise is trivial to the displacement magnitude, the differentiation from the displacement results in amplifying the noise in the strain fields. Figure 2.31c illustrates the escalating noise level in the original ε_{XX} field derived from the displacement in Figure 2.31b. Therefore, this study applies the PLS method (Figure 2.28) to filter the displacement noise and obtain more accurate strain fields. The noise filtering deploys a 3×3 mm low-pass window. Figure 2.31d presents the strain field after the displacement filtering. The improved accuracy in the strain results facilitates the following J-integral calculation.

The accurate strain and stress estimation from DIC allows the calculation of the J-integral using the contour [Eq. (2.43)] integration methods. The contour definition consists of three paths, that is, the upper path ahead of the crack tip, the lower path behind the crack tip, and the lateral path parallel to the crack plane, as shown in Figure 2.32. This study specifies the upper path as passing through the rotation center for two reasons. First, the stress components across the thickness demonstrate the closest agreement ahead of the crack tip around the rotation center. Second, the near-zero stress level minimizes the side-grooving effects on the free-surface integration. The contribution from the side-grooved materials to the J-integral is thus negligible near the rotation center. The definition of the lower path is not so strict as the upper path, since the stress level on the free surface remains close to that in

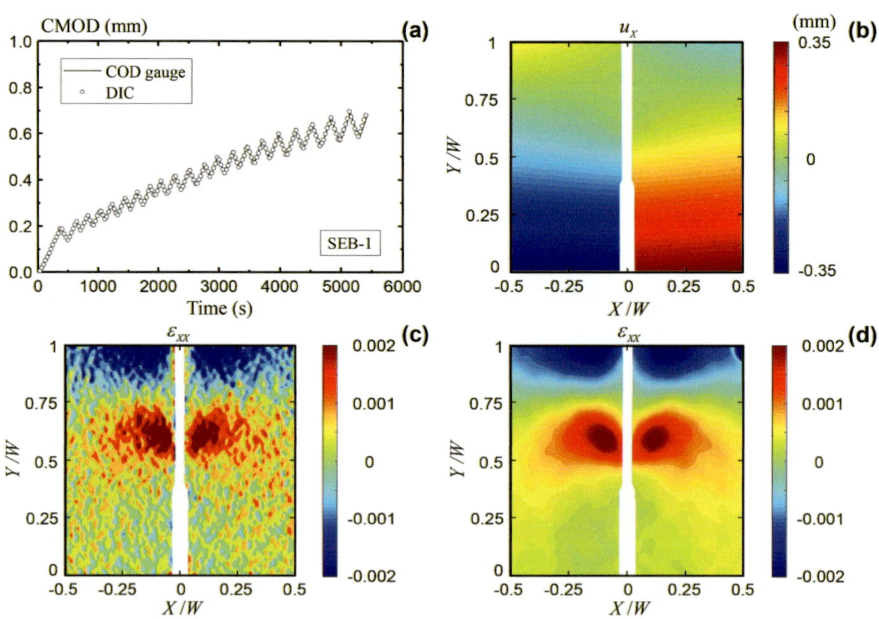

FIGURE 2.31 (a) *CMOD* results by DIC and COD gauge; (b) the DIC-measured u_X field; (c) the DIC-measured ε_{XX} field without displacement filtering; and (d) the DIC-measured ε_{XX} field with displacement filtering.

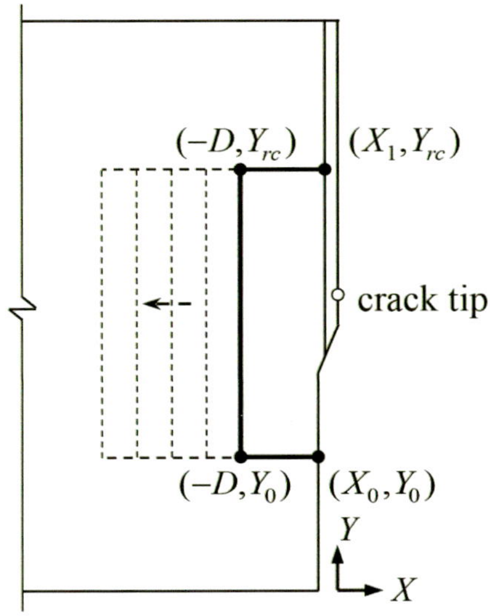

FIGURE 2.32 The contour integration definition.

the mid-thickness over a wide range of regions behind the crack tip. This section defines the lower path at $Y/W = 0.2$. To investigate the path independence of the contour integral, the lateral path varies from the near-crack region to the far field. Figure 2.32 illustrates the contour path definition at the free surface. Y_{rc} denotes the vertical coordinate of the rotation center, Y_0 equals $0.2W$, X_0 and X_1 represent the horizontal coordinates at the edge of the notch and the side groove, and the variable D indicates the distance from the lateral contour to the crack plane.

The through-thickness averaged J-integral, denoted as J_{avg}, represents the average crack driving force over the crack front,

$$J_{avg} = \frac{\sum_{i}^{m} B_i^N J_i}{B_N} \tag{2.49}$$

where B_i^N denotes each element thickness across the crack front, J_i represents the corresponding J value at the crack-front location, and m defines the number of elements across the net thickness $\left(\sum_{i}^{m} B_i^N = B_N\right)$. The path independence is achieved when D reaches $0.25W$ for the contour integral [53]. The total energy release rate can be represented by the free-surface J values (J_{fs}), calculated by contour integral method from DIC displacement maps as follows:

$$J_{avg} = \frac{J_{fs} B}{B_N} \tag{2.50}$$

The validity of Eq. (2.50) originates from the definition of the *J*-integral [54],

$$J = -\frac{1}{B}\frac{\partial U}{\partial a} \tag{2.51}$$

where U represents the strain energy stored in the cracked body. Applying the net thickness B_N yields the J_{avg}, while adopting the total thickness B leads to an approximate J_{fs}. Equation (2.50) reveals the relationship between the J_{fs} and J_{avg}, which remains valid for both the side-grooved and plane-sided specimens under SSY conditions [37].

The DIC method determines the location center by using the horizontal displacement of vertical lines parallel to the crack plane. Figure 2.33a plots the DIC-measured u_X along $X/W = -0.05$. A linear regression, applied to u_X around the rotation region, reduces the noise of experimental measurement and determines the rotation center by $u_X = 0$. Figure 2.33b summarizes the rotation center evolution during the test from $u_X = 0$ in vertical lines with different X/W ratios. The position of the rotation center, Y_{rc}/W, converges when determined from vertical lines sufficiently close to the crack plane ($|X/W| \le 0.05$). The rotation center moves after around 0.4 mm *CMOD* with the extending crack front.

The fracture resistance curve requires the measurement of crack size at every step. This study correlates the crack depth to the rotation center using the rotation factor r.

$$r = \frac{Y_{rc} - a}{W - a} \tag{2.52}$$

The r value depends on the initial a/W ratio [54]. According to the numerical results in Section 3, this study sets r as 0.4 for the tested SE(B) specimens ($a/W = 0.5$) and assumes the r factor as a constant within the maximum crack extension ($\Delta a < 0.25\, b_0$). Substituting the Y_{rc} results from Figure 2.33b into Eq. (2.52) derives the DIC-based crack extension in Figure 2.34a. The ASTM E1820 calculates the crack size based on the compliance method. The reasonable agreement

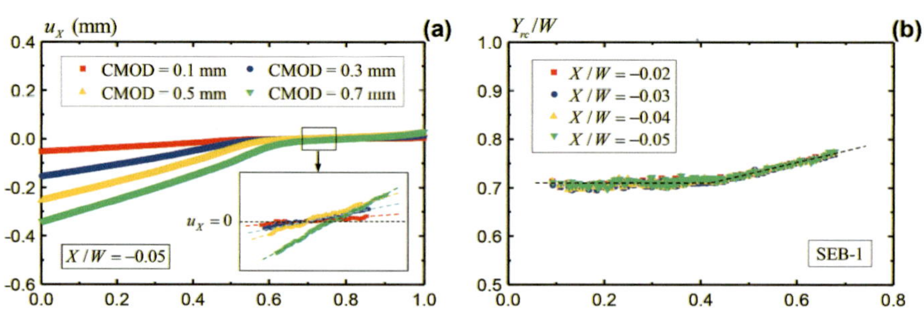

FIGURE 2.33 (a) The horizontal displacement along $X = -0.05W$ and (b) the location of the rotation center.

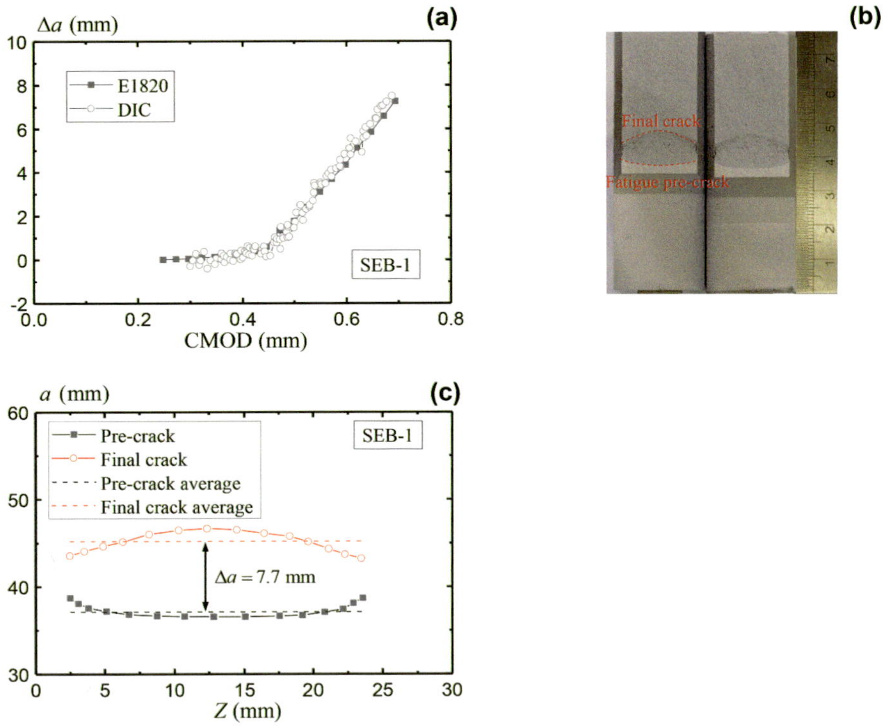

FIGURE 2.34 (a) the DIC-measured crack size; (b) the fracture surface of specimen SEB-1; and (c) the in-situ crack size measurement.

validates the accuracy of the current DIC estimation, which provides a convenient and effective approach to determine the crack size in a side-grooved SE(B) specimen. Figure 2.34b shows the fracture surface of the specimen SEB-1 after the fracture test. The presence of side grooves enforces the Mode-I fracture in the SE(B) specimen, demonstrated by the straight crack profile in Figure 2.34c. Both the pre-crack and final crack illustrate a symmetric shape across the net thickness. Hence, the crack sizing from DIC and E1820 method achieves consistent results in Figure 2.34a. Figure 2.34c quantifies the initial and final crack size through in-situ measurement and estimates the average amount of crack extension around 7.7 mm, which agrees with the results in Figure 2.34a. The measurement from the fracture surface further demonstrates the accuracy of the DIC-based crack sizing method in this study.

Figures 2.35a and b verify the DIC measurement of the J-integral and crack size for the specimen SEB-1, respectively. Figure 2.35a combines J and Δa into the J-R curve. Since the DIC data involves some oscillations during the test, Figure 2.35a includes an exponential fit of the DIC data to represent the J-R curve ($J = 20.8\Delta a^{0.26}$). In general, the DIC measured J-R curve agrees well with the J-R curve measured based on ASTM E1820 [2]. A J_{ini} value of 14 kJ/m^2 is observed by an intersection between the J-R curve and the 0.2 mm offset line. Following the same procedures, Figure 2.35b presents the J-R curves for the specimens in Table 2.3 to evaluate the repeatability of

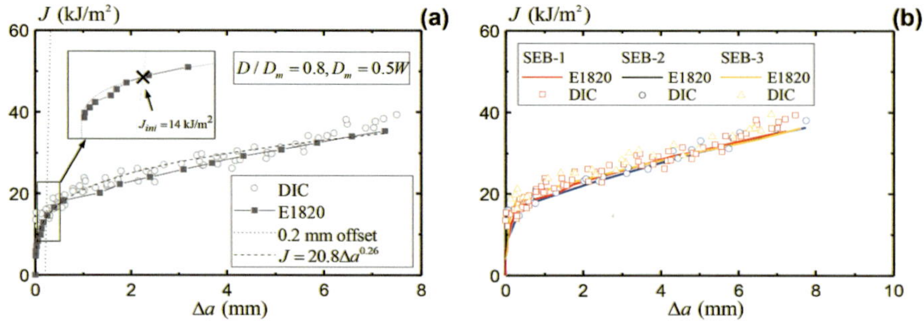

FIGURE 2.35 (a) The *J-R* curve of SEB-1; and (b) the DIC-measured *J-R* curves.

the DIC approach. The results demonstrate the robustness and accuracy of the proposed DIC method for the *J-R* curve measurement of side-grooved specimens.

REFERENCES

[1] T.L. Anderson, Fracture mechanics: Fundamentals and applications, 4th ed., Taylor and Francis CRC Press, 2017.

[2] ASTME1820, ASTM E1820–22, Standard test method for measurement of fracture toughness, in, ASTM International, West Conshohocken, PA, 2022.

[3] J. Begley, J. Landes, The J integral as a fracture criterion, in: Fracture toughhhness proceedings of the 1971 national symposium on fracture mechanics, ASTM International, 1972, pp. 1.

[4] J. Rice, P. Paris, J. Merkle, Some further results of J-integral analysis and estimates, in: Progress in flaw growth and fracture toughness testing, ASTM International, 1973, pp. 231–245.

[5] ISO12135, Metallic materials-Unified method of test for the determination of quasistatic fracture toughness, in, International Standard Organization, 2016.

[6] X. Qian, W. Yang, Initiation of ductile fracture in mixed-mode I and II aluminum alloy specimens, Engineering Fracture Mechanics, 93 (2012) 189–203.

[7] ASTME8, ASTM E8–22, Standard test methods for tension testing of metallic materials, in, ASTM International, West Conshohocken, PA, 2022.

[8] X. Qian, W. Yang, A hybrid approach to determine fracture resistance for mode I and mixed-mode I and II fracture specimens, Fatigue & Fracture of Engineering Materials & Structures, 34 (2011) 305–320.

[9] J.A. Joyce, J-resistance curve testing of short crack bend specimens using unloading compliance, in: Fracture mechanics: Proceedings of the 22nd national symposium, Volume 1, 1992, pp. 904–924.

[10] X.-K. Zhu, J.A. Joyce, J–Resistance curve testing of HY80 steel using SE (B) specimens and normalization method, Engineering Fracture Mechanics, 74 (2007) 2263–2281.

[11] T. Keiichiro, I. Hitoshi, Elastic-plastic fracture toughness test under mixed mode I-II loading, Engineering Fracture Mechanics, 41 (1992) 529–540.

[12] B. Healy, A. Gullerud, K. Koppenhoefer, A. Roy, S. RoyChowdhury, M. Walters, B. Bichon, K. Cochran, A. Carlyle, J. Sobo, WARP3D-release 17.2: 3-D dynamic nonlinear fracture analysis of solids using parallel computers, Structural Research Series (SRS), 607 (2012).

[13] R. Herrera, J. Landes, A direct JR curve analysis of fracture toughness tests, Journal of Testing and Evaluation, 16 (1988) 427–449.
[14] W. Yang, X. Qian, Fracture resistance curve over the complete mixed-mode I and II range for 5083 aluminum alloy, Engineering Fracture Mechanics, 96 (2012) 209–225.
[15] N. Hallbäck, F. Nilsson, Mixed-mode I/II fracture behaviour of an aluminium alloy, Journal of the Mechanics and Physics of Solids, 42 (1994) 1345–1374.
[16] X. Qian, Stress-intensity factors for circular hollow section V-joints with a rack-plate chord, Fatigue & Fracture of Engineering Materials & Structures, 32 (2009) 61–79.
[17] X. Qian, R.H. Dodds Jr, Y. Choo, Mode mixity for tubular K-joints with weld toe cracks, Engineering Fracture Mechanics, 73 (2006) 1321–1342.
[18] ISO12737, Metallic materials-unified method of test for the determination of quasi-static fracture toughness, in, International Standard Organization, 2010.
[19] ISO22889, Metallic materials-method of test for the determination of resistance to stable crack extension using specimens of low constraint, in, International Standard Organization, 2013.
[20] X. Qian, R.H. Dodds, Y. Choo, Elastic–plastic crack driving force for tubular X-joints with mismatched welds, Engineering Structures, 27 (2005) 1419–1434.
[21] X. Qian, R.H. Dodds, Y. Choo, Elastic–plastic crack driving force for tubular K-joints with mismatched welds, Engineering Structures, 29 (2007) 865–879.
[22] R. McMeeking, D. Parks, Crack-tip fields in large-scale, Elastic-Plastic Fracture, 668 (1979) 175.
[23] D. Carka, C.M. Landis, On the path-dependence of the J-integral near a stationary crack in an elastic-plastic material, Journal of Applied Mechanics, 78 (2011) 011006.
[24] Y.-J. Kim, J.-S. Kim, S.-M. Cho, Y.-J. Kim, 3-D constraint effects on J testing and crack tip constraint in M (T), SE (B), SE (T) and C (T) specimens: Numerical study, Engineering Fracture Mechanics, 71 (2004) 1203–1218.
[25] Y.J. Kim, P. Budden, Plastic η factor solutions of homogeneous and bi-material SE (T) specimens for toughness and creep crack growth testing, Fatigue & Fracture of Engineering Materials & Structures, 24 (2001) 751–760.
[26] Y.J. Kim, K. Schwalbe, On experimental J estimation equations based on CMOD for SE (B) specimens, Journal of Testing and Evaluation, 29 (2001) 67–71.
[27] Y.-Y. Wang, The limits of applicability of J and CTOD estimation procedures for shallow-cracked SENB specimens, in: Proc. Int. Conf. Shallow Crack Fracture Mechanics, Toughness Tests and Applications, Cambridge, 1992.
[28] S.-P. Chiew, S.-T. Lie, C.-K. Lee, Z.-W. Huang, Fatigue performance of cracked tubular T joints under combined loads. I: Experimental, Journal of Structural Engineering, 130 (2004) 562–571.
[29] X. Qian, K. Jitpairod, P. Marshall, S. Swaddiwudhipong, Z. Ou, Y. Zhang, M.R. Pradana, Fatigue and residual strength of concrete-filled tubular X-joints with full capacity welds, Journal of Constructional Steel Research, 100 (2014) 21–35.
[30] X. Qian, Y. Li, A compliance-based approach to measure fracture resistance curve for surface cracked steel plates, International Journal of Fracture, 182 (2013) 1–19.
[31] N. Parool, X. Qian, C. Koh, An η-compliance method to estimate the J-Δa curve for pipes with a circumferential surface crack, Fatigue & Fracture of Engineering Materials & Structures, 40 (2017) 1624–1639.
[32] C. Shih, H. De Lorenzi, W. Andrews, Elastic compliances and stress-intensity factors for side-grooved compact specimens, International Journal of Fracture, 13 (1977) 544–548.
[33] Z. He, A. Kotousov, F. Berto, Effect of vertex singularities on stress intensities near plate free surfaces, Fatigue & Fracture of Engineering Materials & Structures, 38 (2015) 860–869.

[34] L.P. Pook, F. Berto, A. Campagnolo, State of the art of corner point singularities under in-plane and out-of-plane loading, Engineering Fracture Mechanics, 174 (2017) 2–9.
[35] B. Pan, Recent progress in digital image correlation, Experimental Mechanics, 51 (2011) 1223–1235.
[36] B. Pan, Digital image correlation for surface deformation measurement: Historical developments, recent advances and future goals, Measurement Science and Technology, 29 (2018) 082001.
[37] C. Chen, X. Qian, Quantifying through-thickness J for plane strain specimens using digital image correlation considering constraint effects, Engineering Fracture Mechanics, 267 (2022) 108430.
[38] M. James, J. Newman Jr, The effect of crack tunneling on crack growth: Experiments and CTOA analyses, Engineering Fracture Mechanics, 70 (2003) 457–468.
[39] E. Fagerholt, E. Østby, T. Børvik, O.S. Hopperstad, Investigation of fracture in small-scale SENT tests of a welded X80 pipeline steel using digital image correlation with node splitting, Engineering Fracture Mechanics, 96 (2012) 276–293.
[40] Y. Imai, T. Matake, Effect of side grooves for three-point bending fracture toughness specimens: Three-dimensional elastic-plastic finite element analysis, Bulletin of JSME, 27 (1984) 909–916.
[41] K. Guan, D. Wang, J. Dobrovská, K. Matocha, Evaluation of the ductile-brittle transition temperature of anisotropic materials by small punch test with un-notched and U-notched specimens, Theoretical and Applied Fracture Mechanics, 102 (2019) 98–102.
[42] A. Mohan, S. Poobal, Crack detection using image processing: A critical review and analysis, Alexandria Engineering Journal, 57 (2018) 787–798.
[43] M.A. Sutton, W. Wolters, W. Peters, W. Ranson, S. McNeill, Determination of displacements using an improved digital correlation method, Image and vision computing, 1 (1983) 133–139.
[44] Z. Wang, S. Wang, Z. Wang, An analysis on computational load of DIC based on Newton–Raphson scheme, Optics and Lasers in Engineering, 52 (2014) 61–65.
[45] J.R. Rice, A path independent integral and the approximate analysis of strain concentration by notches and cracks, Journal of Applied Mechanics, 35 (1968) 379–386.
[46] Z.P. Baažant, Finite strain generalization of smallstrain constitutive relations for any finite strain tensor and additive volumetric-deviatoric split, International Journal of Solids and Structures, 33 (1996) 2887–2897.
[47] B. Pan, H. Xie, Z. Guo, T. Hua, Full-field strain measurement using a two-dimensional Savitzky-Golay digital differentiator in digital image correlation, Optical Engineering, 46 (2007) 033601-033601-033610.
[48] S. Yoneyama, Computing strain distributions from measured displacements on a three-dimensional surface, Journal of the Japanese Society for Experimental Mechanics, 10 (2010) s113–s118.
[49] B. Pan, A. Asundi, H. Xie, J. Gao, Digital image correlation using iterative least squares and pointwise least squares for displacement field and strain field measurements, Optics and Lasers in Engineering, 47 (2009) 865–874.
[50] B. Pan, K. Qian, H. Xie, A. Asundi, Two-dimensional digital image correlation for in-plane displacement and strain measurement: A review, Measurement Science and Technology, 20 (2009) 062001.
[51] J. Salençon, Handbook of continuum mechanics: General concepts thermoelasticity, Springer Science & Business Media, 2012.
[52] A.S. Khan, S. Huang, Continuum theory of plasticity, John Wiley & Sons, 1995.

[53] C. Chen, X. Qian, A non-contact approach to measure JR curve for side-grooved specimens using digital image correlation, Theoretical and Applied Fracture Mechanics, 121 (2022) 103549.

[54] T. Kawabata, T. Tagawa, Y. Kayamori, M. Ohata, Y. Yamashita, M. Kinefuchi, H. Yoshinari, S. Aihara, F. Minami, H. Mimura, Plastic deformation behavior in SEB specimens with various crack length to width ratios, Engineering Fracture Mechanics, 178 (2017) 301–317.

3 Ductile Tearing and Its Engineering Assessment

3.1 INTRODUCTION

The previous chapter presented various approaches for measuring fracture resistance in both material specimens, represented by standard fracture specimens, and welded structural components, exemplified by welded pipes. This chapter focuses on integrating the measured engineering fracture resistance into engineering assessments using established procedures, such as failure assessment diagrams (FAD) and their extensions, including the tearing assessment diagrams (TAD).

The sustainable operation of steel structures requires comprehensive integrity assessments of critical welded details [1], which are susceptible to fatigue-induced defects under cyclic loading. During an overloading event, such defects can lead to two competing failure mechanisms: fracture failure initiated from fatigue cracks and plastic collapse of the remaining intact material. The FAD approach, as outlined in design standards such as BS 7910 [2] and API 579–1 [3], employs failure assessment curves (FAC) derived from simplified cracked geometries, including Option 1 and Option 2 curves, to evaluate structural integrity. However, practical welded components, such as branch connections and various joints in piping and tubular systems, often have complex geometries that may limit the applicability of these generic curves. In such cases, geometry- and material-dependent FACs, such as the Option 3 curve in BS 7910 [2], are more suitable. These curves require both linear-elastic and elastic-plastic energy release rate solutions for fatigue cracks in realistic structural details, which are often analytically complex and challenging to determine.

The FAD, widely used in engineering critical assessments, is a versatile tool for evaluating the structural integrity of various engineering structures, including pressure vessels, pipelines, and tubular connections. Numerous international studies have investigated the applicability of generic FACs for cracked tubular connections [4, 5]. Early researchers integrated the FAD into the European FITNET procedure [6–8] for assessing realistic engineering structures. Additionally, studies on welded tubular joints have demonstrated the feasibility of the FAD framework [9–12].

Recent advancements continue to refine FAD applications. Arandjelovic *et al.* [13] incorporated FAD into the engineering critical analysis framework to evaluate multiple defects in welded connections, while Wang *et al.* [14] examined the influence of various parameters on the failure assessment of high-strength steel S690 welded connections. The latest developments in NDT [15, 16] enhance both the accuracy and efficiency of defect sizing, leading to improved failure assessment outcomes for welded joints. Looking ahead, the future application of FAD is expected to leverage artificial intelligence and digital twins for real-time assessment and predictive maintenance [17–21].

DOI: 10.1201/9781003509806-3

To streamline structural integrity assessments and integrate component-level evaluations into global structural analyses, recent studies have introduced the TAD approach [22, 23]. The TAD method provides an accurate estimation of load evolution during ductile crack extension in cracked structural components, which can be directly incorporated into global structural analyses. This estimation relies on material fracture resistance and linear-elastic FEA, eliminating the need for computationally intensive damage mechanics models. Unlike the FAD framework, which requires multiple failure curves to assess ductile tearing, the TAD approach simplifies the process by using a single tearing assessment curve (TAC) to characterize crack extension behavior.

3.2 BASIC FRAMEWORK OF FAILURE ASSESSMENT DIAGRAM

The FAD defines the failure conditions of a structural component by accounting for two competing failure modes: fracture failure and plastic collapse. In a typical FAD, as illustrated in Figure 3.1, the vertical axis represents the fracture ratio, which quantifies the component's susceptibility to fracture failure.

$$K_r = \frac{K_I}{K_{Ic}} \tag{3.1}$$

where K_{Ic} denotes the fracture toughness of the material, obtained either from standard laboratory testing or the methodologies outlined in Chapter 2.

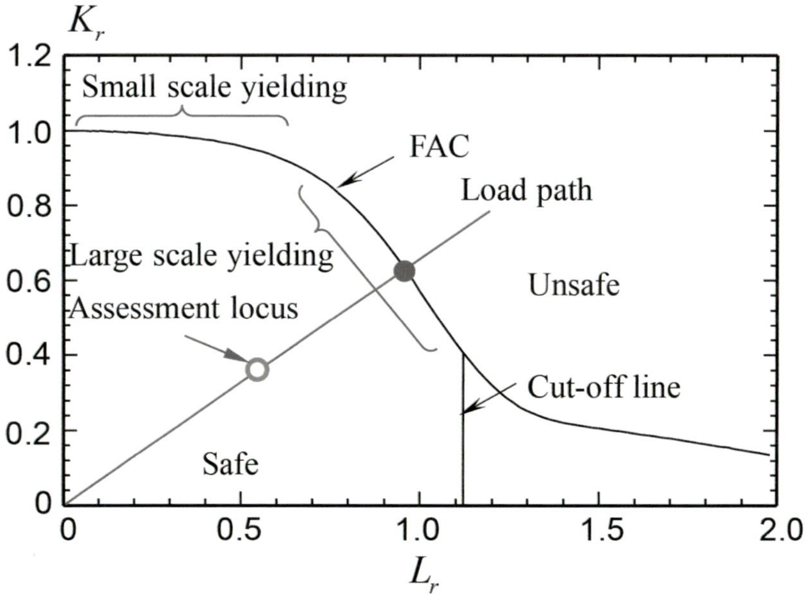

FIGURE 3.1 Definition of the failure assessment diagram (FAD).

The horizontal axis of the FAD represents the load ratio, which characterizes the component's proximity to plastic collapse.

$$L_r = \frac{P}{P_u} \tag{3.2}$$

The ultimate load P_u represents the maximum load resistance of a cracked structure before failure.

The FAD incorporates a FAC, $K_r = f(L_r)$, which defines the failure conditions for a stationary crack. The FAC, as described in BS 7910 [2], varies based on material properties and the required level of conservatism. It includes: (1) Option 1 FAC independent of structural geometry and with minimum material input; (2) the material-specific Option 2 FAC independent of structural geometry; and (3) the material and geometry-specific Option 3 FAC. The Option 1 FAC requires only the yield strength and ultimate strength of the material. For materials without a yield plateau, it follows the expression,

$$\begin{aligned} f(L_r) &= \left(1+\frac{1}{2}L_r^2\right)^{-0.5}\left[0.3+0.7\exp\left(-\mu L_r^6\right)\right] && for \quad L_r \le 1 \\ f(L_r) &= f(L_r = 1)\,L_r^{(N-1)/(2N)} && for \quad 1 < L_r < L_{r,\max} \end{aligned} \tag{3.3}$$

For material with a yield plateau, the Option 1 FAC reads,

$$\begin{aligned} f(L_r) &= \left(1+\frac{1}{2}L_r^2\right)^{-0.5} && for \quad L_r \le 1 \\ f(L_r) &= \left(1+\frac{1}{2\lambda}\right)^{-0.5} && for \quad L_r = 1 \\ f(L_r) &= f(L_r = 1)\,L_r^{(N-1)/(2N)} && for \quad 1 < L_r < L_{r,\max} \end{aligned} \tag{3.4}$$

In Eqs. (3.3) and (3.4), the material parameters μ and N are as follows,

$$\begin{aligned} \mu &= \min\left(0.001\frac{E}{\sigma_y}, 0.6\right) \\ N &= 0.3\left(1-\frac{\sigma_y}{\sigma_u}\right) \end{aligned} \tag{3.5}$$

The parameter λ in Eq. (3.4) depends on the extent of the yield plateau over the strain range $\Delta\varepsilon$,

$$\lambda = 1 + \frac{E\Delta\varepsilon}{R_{el}} \tag{3.6}$$

where R_{el} corresponds to the lower yield strength of the material.

The Option 2 FAC requires a detailed stress-strain curve, as described in Eq. (3.7),

$$f(L_r) = \left(\frac{E\varepsilon_{ref}}{L_r\sigma_y} + \frac{L_r^3\sigma_y}{2E\varepsilon_{ref}} \right)^{-0.5} \qquad for \qquad L_r \le L_{r,\max} \tag{3.7}$$

where ε_{ref} refers to the reference strain at the true stress of $\sigma_{ref} = L_r\sigma_y$.

Neither Option 1 nor Option 2 FACs consider the influence of structural geometry, which can significantly affect fracture behavior. To account for the constraint effect induced by geometry, Option 3 FAC is derived for a specific material, structural geometry, and loading condition. This curve is obtained through both linear-elastic and elastic-plastic numerical analysis of the cracked structural component, providing a more accurate assessment. The Option 3 FAC, as defined in BS 7910 [2], is expressed by the following equation for an increasing load ratio L_r,

$$f(L_r) = \frac{K_I}{K_J} = \sqrt{\frac{J_e}{J_{ep}}} \tag{3.8}$$

where J_e and J_{ep} represent the J-integral values obtained from linear-elastic and elastic-plastic analyses, respectively, at a load level corresponding to P. The stress intensity factor K_J is derived from J_{ep} using the following equation, assuming a plane-strain condition near the crack front,

$$K_J = \sqrt{\frac{J_{ep}E}{1-v^2}} \tag{3.9}$$

The FAD evaluates the integrity of a cracked structural component based on the assessment locus, as shown in Figure 3.1. The position of this locus, corresponding to a given applied load, is determined by Eqs. (3.1) and (3.2).

- If the assessment locus lies within the region bounded by the FAC, the structural component is considered safe.
- If the locus falls outside this boundary, the component is deemed unsafe.

Since the assessment locus is derived from a linear-elastic analysis, the straight line connecting the origin of the FAD ($L_r = 0$, $K_r = 0$) to the locus represents the load path of the structural component (see Figure 3.1). The point where the load path intersects the FAC indicates the load level at which failure is predicted to occur.

3.3 FAILURE ASSESSMENT CURVES FOR WELDED TUBULAR JOINTS

3.3.1 Definition of FACs for CHS Joints

As discussed earlier, geometry-independent FACs may face limitations when applied to cracked structural connections with complex geometries, such as welded tubular joints. This section presents geometry-specific failure assessment curves

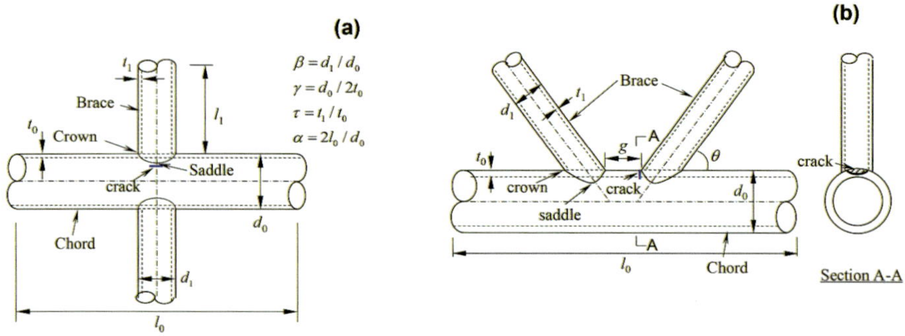

FIGURE 3.2 Geometric configuration of CHS: (a) X-joint; and (b) K-joint.

(Option 3 FAC) and demonstrates the non-conservatism of geometry-independent FACs for these connections under large-scale yielding.

This section considers two types of circular hollow section (CHS) joints: the X-joint and K-joint, as schematically illustrated in Figure 3.2. These figures depict the typical geometric configurations of CHS X-joints and K-joints.

Common non-dimensional geometric parameters used for CHS joints, as adopted in engineering design codes, include:

- Brace-to-chord diameter ratio ($\beta = d_1 / d_0$)
- Chord radius-to-wall thickness ratio ($\gamma = d_0 / 2t_0$)
- Brace-to-chord wall thickness ratio ($\tau = t_1 / t_0$)
- Chord length-to-outer radius ratio ($\alpha = 2l_0 / d_0$)

The failure assessment of tubular joints requires a clear strength definition for CHS joints, such as those provided in the International Institute of Welding (IIW) strength formulation [24] for intact CHS joints. This formulation is adjusted using an area reduction factor, as described below,

$$\frac{P_u \sin\theta}{\sigma_y t_0^2} = Q_u Q_f A_{RF} \tag{3.10}$$

where θ denotes the brace-to-chord intersection angle and Q_u considers the geometric effects on the joint strength and equals,

$$Q_u = \begin{cases} 3.16\left(\dfrac{1+\beta}{1-0.7\beta}\right)\gamma^{0.15} & \text{for X-joint} \\ \left(2+16\beta^{1.6}\right)\gamma^{0.3}\left[1+\dfrac{1}{1.2+\left(g/t_0\right)^{0.8}}\right] & \text{for K-joint} \end{cases} \tag{3.11}$$

The Q_f function in Eq. (3.10) quantifies the effect of chord stresses on the joint capacity,

$$Q_f = (1-\bar{n})^{0.2} \tag{3.12}$$

where $\bar{n}$ defines the ratio of the chord stress over the chord yield capacity, or, $\left|\sigma_{chord} / \sigma_y\right|$.

The area reduction factor A_{RF} in Eq. (3.10) provides a quantitative assessment of the crack area on the ultimate strength of the tubular joint,

$$A_{RF} = \left(1 - \frac{A_{crack}}{L_w t_0}\right)\left(\frac{1}{Q_\beta}\right)^m \tag{3.13}$$

A_{crack} defines the crack area. For semi-elliptical cracks, A_{crack} equals $0.5\pi ac$. L_w refers to the weld length along the brace-to-chord intersection with the following analytical expression [4],

$$L_w = 2\beta d_0 \int_0^{\pi/2} \left[\frac{\sin^2 \omega}{\sin^2 \theta} + \frac{\cos^2 \omega}{1-\beta^2 \sin^2 \omega}\right]^{0.5} d\omega \tag{3.14}$$

where ω refers to the position angle in the plane of the brace cross-section.

The Q_β factor in Eq. (3.13) follows [25],

$$Q_\beta = \begin{cases} 1 & \beta \le 0.6 \\ \dfrac{0.3}{\beta(1-0.833\beta)} & \beta > 0.6 \end{cases} \tag{3.15}$$

The exponent m in Eq. (3.13) can take a value of either 1.0 or 0, as reported by [25]. When m = 1.0, incorporating the area reduction factor (ARF) in Eq. (3.13) significantly reduces the predicted strength of cracked joints, particularly for joints with high β ratios, where the crack area approaches zero.

Previous studies [9, 26, 27] have shown that an ARF with m = 1.0 provides a conservative estimate of strength for joints containing either surface cracks or through-thickness cracks. However, Wang *et al.* [28] found that this approach results in overly conservative strength predictions for X-joints with semi-elliptical surface cracks near the weld toe at the saddle point. In Wang's study, ultimate strength was determined using Lu's deformation limit [29], a widely accepted criterion for predicting the ultimate strength of tubular joints.

The numerical investigation also examines the effect of strain hardening on the FAC. The material model used herein follows the power-law hardening rule implemented in WARP3D [30],

$$\frac{\varepsilon}{\varepsilon_{ys}} = \begin{cases} \dfrac{\sigma}{\sigma_{ys}} & \varepsilon \le \varepsilon_{ys} \\ \left(\dfrac{\sigma}{\sigma_{ys}}\right)^n & \varepsilon > \varepsilon_{ys} \end{cases} \tag{3.16}$$

where n refers to the hardening exponent, ε_{ys} and σ_{ys} define the reference yield strain and stress, respectively (based on the limit of proportionality). This study adopts a high-strength steel S690 as the base material, with the yield strength equal to 690 MPa, Young's modulus of 205 GPa and a Poisson's ratio of 0.3.

Figure 3.3a compares Wang's FE results with Eq. (3.13), demonstrating that the net area reduction factor (m = 0) closely aligns with numerical data for X-joints with surface cracks. Figure 3.3b further compares experimental results from previous studies [26, 31, 32] against the area reduction factor in Eq. (3.13). In this figure:

- The strength of cracked joints is normalized against test results of intact joints with the same geometry and material properties.
- Solid symbols represent specimens with through-thickness cracks, while hollow symbols indicate surface cracks.

The results confirm that an ARF with m = 1.0 provides a conservative strength estimate for joints with through-thickness cracks, while m = 0, that is, linear reduction in joint strength with respect to the crack area, aligns more closely with experimental data for surface-cracked joints. Burdekin [9] reviewed both numerical and experimental data on cracked joint strength, comparing them against the intact joint strength equations from the HSE guidance notes [33]. The findings reaffirm that m = 1.0 generally yields conservative strength estimates.

The next section will explore the effect of the area reduction factor on the FAC.

3.3.2 Proposed FACs for CHS X- and K-Joints

The FACs proposed in this chapter are derived from Option 3 FACs computed for CHS X-joints and K-joints under brace axial loading. This investigation examines the geometric dependence of FACs, including crack-front positions, local crack sizes and aspect ratios, and global joint dimensions. Additionally, the study explores the effects of material yield strength and strain-hardening properties on the FACs.

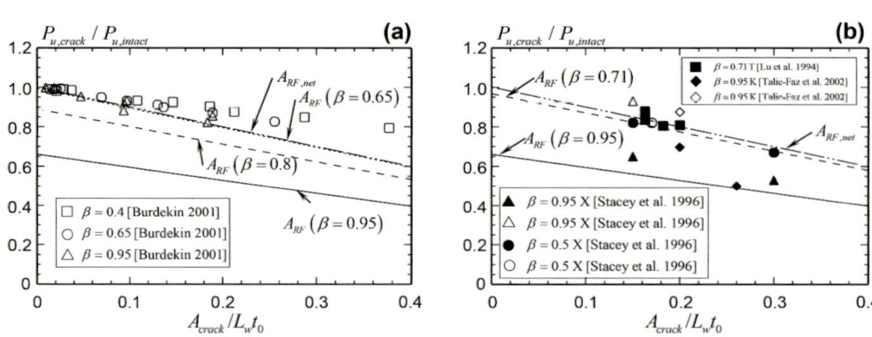

FIGURE 3.3 Comparison of the area reduction factor for the cracked CHS joints with the data in the literature.

The numerical study covers a practical range of joint geometric parameters, considering five β ratios (β=0.4, 0.6, 0.8, 0.9, and 0.95), three γ ratios (γ =5, 9, and 20), and a fixed τ ratio (1.0) for all joints. To eliminate chord end effects, the chord length factor is set to 16, following [34]. The brace length is fixed at four times the brace's outer diameter, while the chord outer diameter remains constant at 406 mm. For K-joints, the gap ratio is set at 2.0. The study assumes a nominal shallow crack, with a crack depth ratio $a/t_0 = 0.22$ and a crack aspect ratio $a/c = 0.25$. The surface crack is located near the weld toe at the saddle point in X-joints and at the crown position in K-joints, as illustrated in Figure 3.2.

Figure 3.4 presents typical FACs computed using Eq. (3.8) at the deepest point along the semi-elliptical crack front in near-toe cracks for X-joints and K-joints. The load ratio in Figure 3.4 incorporates the area reduction factor with a defined parameter $m = 1.0$. The Option 1 FAC from BS 7910 [2] and the R6 Option 1 curve [35] provide conservative estimates for thin-walled X-joints with ($\gamma = 20$) small β ratios, as shown in Figure 3.4b. For thick-walled X-joints with large β ratios, the FACs deviate significantly downward, falling below the BS7910 Option 1 [2] and R6 Option 1 [35] curves. The saddle region in an X-joint predominantly experiences membrane action in thin-walled joints. However, in thick-walled joints, increased brace and chord thickness introduces eccentricity between axial stresses in the brace wall thickness and the chord wall thickness (near the saddle point). This eccentricity-induced bending causes a greater deviation in the elastic-plastic energy release rate compared to the linear-elastic J-value, particularly in thick-walled joints.

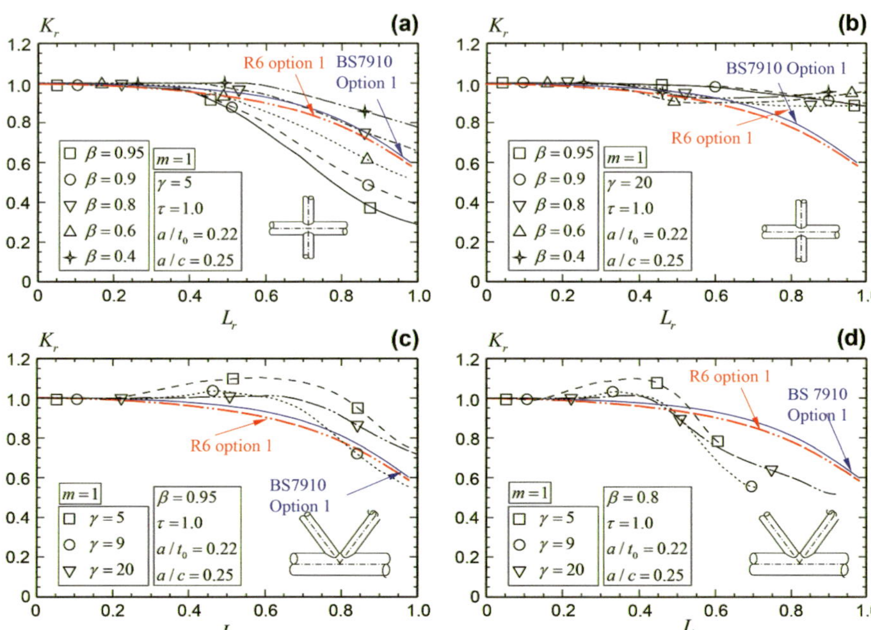

FIGURE 3.4 Comparison of failure assessment curves (FACs) computed from Eq. (3.8) for CHS joints with the generic FACs in engineering standards.

Figures 3.4c and d illustrate FACs for K-joints, showing a distinct behavior as the load ratio increases compared to X-joints. Equation (3.8) assumes plastic deformations are localized near the crack tip, which holds for laboratory-scale fracture specimens. However, in structural joints with complex geometries, geometric notch effects can introduce stress concentrations near weld-toe regions far from the crack location. The K_r ratio exceeds 1.0 when the load ratio exceeds 0.2 for K-joints, as shown in Figures 3.4c and d, indicating elastic-plastic energy dissipation in materials near weld-toe notches away from the crack. Thick-walled K-joints, subject to high transverse shear stresses, exhibit significant local plastic deformation over the gap region, influencing the elastic-plastic energy release rate. As the load increases, the elastic-plastic energy release rate surpasses the linear-elastic *J*-values, as seen in Figures 3.4c and d. The Option 1 FAC from BS 7910 and the R6 Option 1 curve remain conservative for K-joints with large β ratios.

Figure 3.5 shows FACs for the same X- and K-joints as shown in Figure 3.4, with the load ratio defined using the net area reduction factor ($m = 0$). The net area reduction factor reduces conservatism in strength estimation but results in a more conservative safe area below the FAC with $m = 1$, as seen in Figure 3.5 compared to Figure 3.4. Using the net area reduction factor decreases the β dependence of K-joint FACs, as shown in Figures 3.5c and d, compared to Figures 3.4c and d.

At high applied displacement levels, gross yielding occurs in brace members for thick-walled joints and joints with small β ratios. This leads to a slower increase in the elastic-plastic energy release rate with applied loading [36, 37]. Consequently,

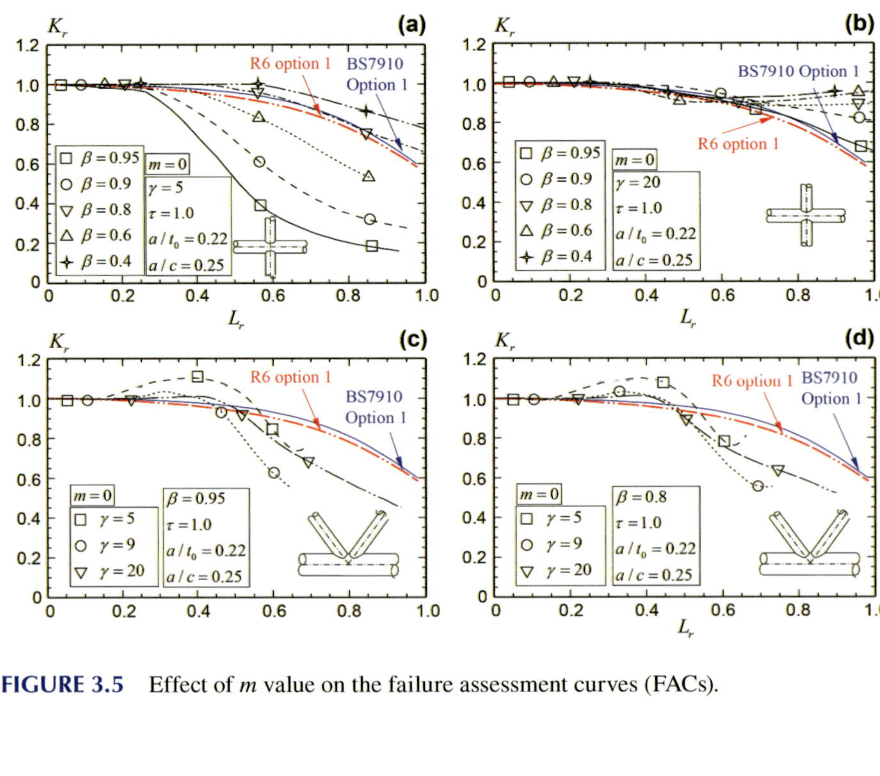

FIGURE 3.5 Effect of *m* value on the failure assessment curves (FACs).

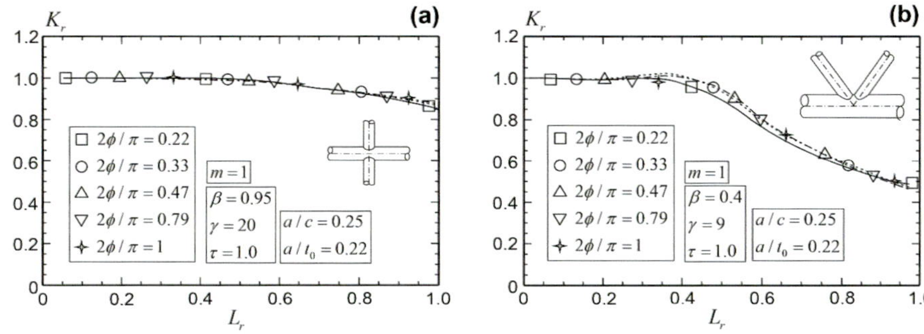

FIGURE 3.6 FACs derived at different crack-front locations in the CHS joints.

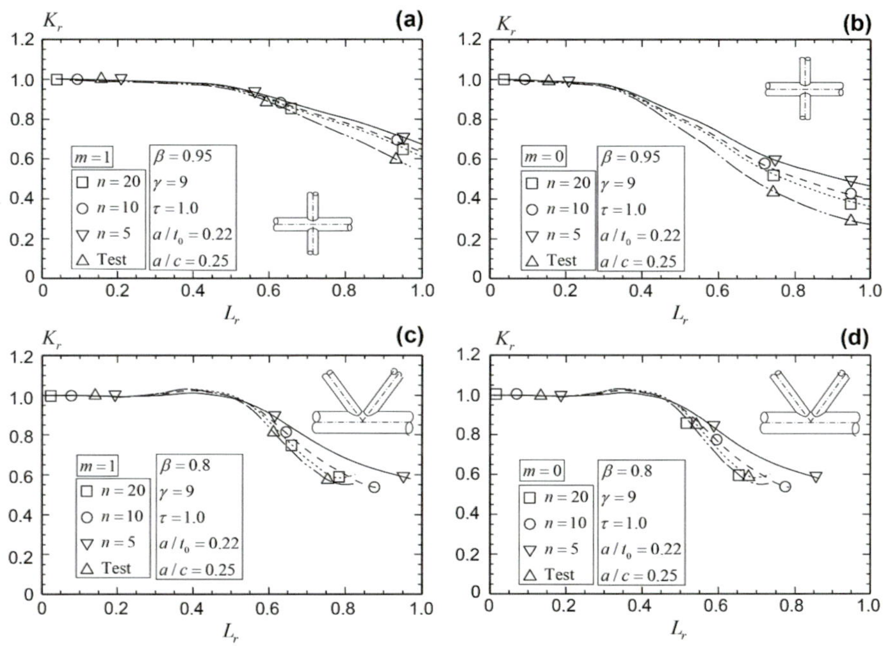

FIGURE 3.7 Effect of strain hardening on the FACs.

the FAC increases again after reaching a minimum value, as observed in K-joints with $\gamma = 5$ in Figure 3.5.

Figure 3.6 compares FACs along the semi-elliptical crack front for X-joints and K-joints. The results indicate that FACs do not significantly vary with crack-front location under brace axial loading, despite variations in linear-elastic and elastic-plastic crack driving forces along the crack front.

Figures 3.7–3.9 examine the effects of strain hardening, yield strength, and crack depth ratio on the FACs. In Figure 3.7, FACs for S690 steels with three different strain-hardening exponents show that strain hardening has minimal

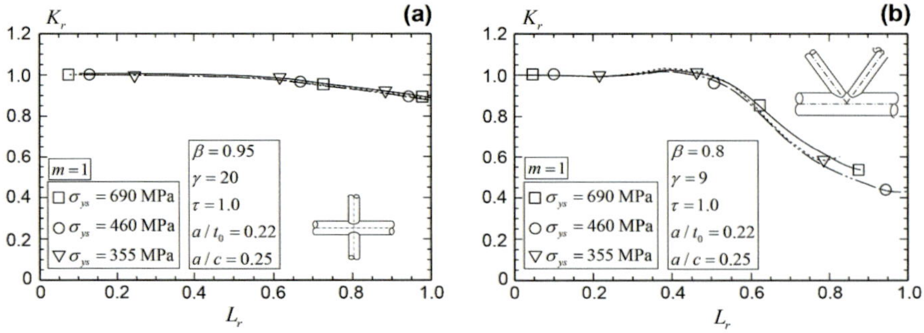

FIGURE 3.8 Effect of yield strength on the FACs.

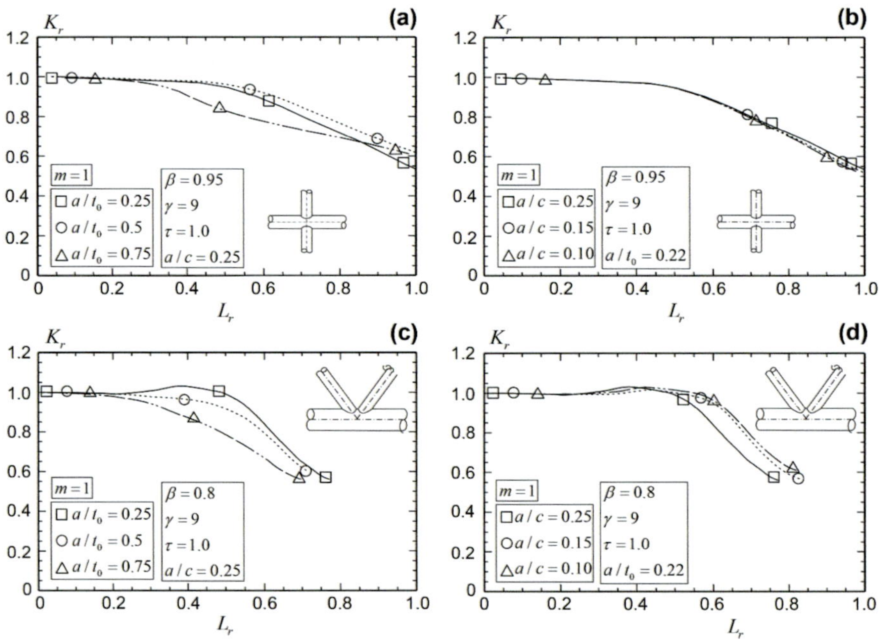

FIGURE 3.9 Effect of crack size on the FACs.

impact on X-joints but a stronger effect on K-joints due to shearing actions over the gap region. Figure 3.7b also demonstrates that when using the net area reduction factor (m = 0), the FACs for X-joints show a greater dependence on strain hardening than in Figure 3.7a. Figure 3.8 demonstrates that FACs for different yield strengths (690 MPa, 460 MPa, 355 MPa) indicate little dependence on yield strength for the geometric configurations considered. Figure 3.9 analyzes the effect of crack depth and crack aspect ratio on FACs. Figures 3.9a and c indicate that increasing the crack depth affects FACs in two ways: 1) larger crack areas reduce the ultimate strength, increasing the load ratio; and 2) greater crack depths elevate stress fields near the crack tip, leading to strong deviations in the elastic-plastic energy release rate from the linear-elastic J-value, particularly in deep cracks. Figures 3.9b and d show that the crack aspect ratio has minimal effect on FACs, as stress redistribution occurs over a longer crack front, resulting in slightly larger FAC areas.

In summary, the above results demonstrate that geometry-specific FACs (Option 3 FACs) provide more accurate failure assessments for CHS X- and K-joints under brace axial loading. The findings highlight the importance of incorporating geometry-dependent effects, particularly for thick-walled joints, high β ratios, and complex stress distributions in K-joints.

The proposed FACs for tubular joints derive from nonlinear regression analyses, following an approach similar to that used for estimating stress-intensity factors and T-stress solutions for embedded flaws in pipe structures [38, 39]. The evolution of the K_r ratio with increasing load ratio L_r is modeled using a log-logistic function:

$$K_r = \frac{1+C_1}{1-\left(L_r / C_2\right)^{C_3}} - C_1 \tag{3.17}$$

where C_1 to C_3 are regression coefficients determined through nonlinear regression analysis. Equation (3.17) is a widely used fitting function in engineering applications [40–42].

For both X- and K-joints, regression analyses are conducted using two different definitions of the load ratio L_r, based on the area reduction factor with $m = 1$ and based on the area reduction factor with $m = 0$.

The coefficients C_1 to C_3 exhibit either linear or quadratic dependence on the β or γ ratios. All regression analyses are performed using FAC data up to the point before the final upturn in the FAC (Figure 3.5), which occurs due to gross yielding in the brace cross-section under large applied deformations.

For CHS X-joints subjected to brace axial tension, the regression formula is given in Eq. (3.18)

$$K_r = \frac{1-\left(A_1 + A_2\beta + A_3\beta^2\right)}{1+\left(L_r / \left(A_4 + A_5\gamma\right)\right)^{A_6 + A_7\beta}} + \left(A_1 + A_2\beta + A_3\beta^2\right) \tag{3.18}$$

Table 3.1 provides the values of the coefficients A_1 to A_7 for X-joints, corresponding to both definitions of the area reduction factor $m = 1$ and $m = 0$. The low coefficient of variation (COV) values indicate a strong correlation between the FE data and the proposed FAC.

Figures 3.10a and b compare the FE data with the predictions from Eq. (3.18). The FACs based on the net area reduction factor ($m = 0$) exhibit slightly more scatter than those based on the area reduction factor ($m = 1$), suggesting a marginally higher variation in strength estimation when using the net area reduction approach.

TABLE 3.1
Values of the Coefficients for the Proposed Failure Assessment Curves.

Coefficients	A_1	A_2	A_3	A_4	A_5	A_6	A_7	A_8	FE/Prediction No. of data	Mean	COV
X (m = 1)	0.494	1.15	−2.04	0.670	0.0512	3.68	0.017	–	575	1.01	0.035
X (m = 0)	−0.280	4.59	−4.92	0.44	0.0575	12.5	−11.1	–	850	1.00	0.075
K (m = 1)	1.13	−0.120	0.00454	1.24	−1.95	1.55	3.82	6.05	2858	1.00	0.027
K (m = 0)	1.30	−0.180	0.0069	1.05	−0.837	0.409	1.35	6.67	4768	1.00	0.042

For CHS K-joints, the proposed FACs are expressed as

$$K_r = \frac{1-\left(A_1 + A_2\gamma + A_3\gamma^2\right)}{1+\left(L_r / \left(A_4 + A_5\beta + A_6\beta^2\right)\right)^{A_7+A_8\beta}} + \left(A_1 + A_2\gamma + A_3\gamma^2\right) \tag{3.19}$$

The regression analysis for K-joints using the area reduction factor ($m = 1$) excludes large β values ($\beta \geq 0.9$), since large β K-joints exhibit larger safe areas than both the Option 1 curve in BS 7910 [2] and the R6 Option 1 curve [35]. Conversely, the regression analysis using the net area reduction factor ($m = 0$) includes all FE data, as the FACs derived from the net area reduction factor result in smaller safe areas compared to those obtained using the A_{RF} with $m = 1$. The regression analysis for K-joints only incorporates FE data with $K_r \leq 1$. Table 3.1 lists the coefficients for Eq. (3.19). The low COV values indicate strong agreement between the FE data and the proposed failure assessment curves.

Figures 3.10c and d compare the FACs derived from FEA with the estimates from Eq. (3.19). Similar to X-joints, the FACs using the net area reduction factor ($m = 0$) exhibit slightly more scatter compared to those based on the area reduction factor ($m = 1$). This suggests a higher variability in strength estimation when employing the net area reduction approach.

Figure 3.11 compares the FAC estimates from Eq. (3.19) for K-joints with $\beta = 0.95$ and $\beta = 0.9$ values that were not included in the nonlinear regression analysis for A_{RF} with $m = 1$. The results indicate that Eq. (3.19) provides a conservative estimate for the FACs of K-joints with large β ratios. Additionally, both the R6 Option 1 curve and the Option 1 assessment curves in BS 7910 offer generally conservative predictions for K-joints with high β ratios.

3.4 DUCTILE TEARING ASSESSMENT VIA FAD

The FACs in BS 7910 [2] for all three options are based on a stationary, non-growing crack and therefore do not consider crack propagation. This section outlines the ductile tearing assessment procedure for high-strength welded CHS joints using the

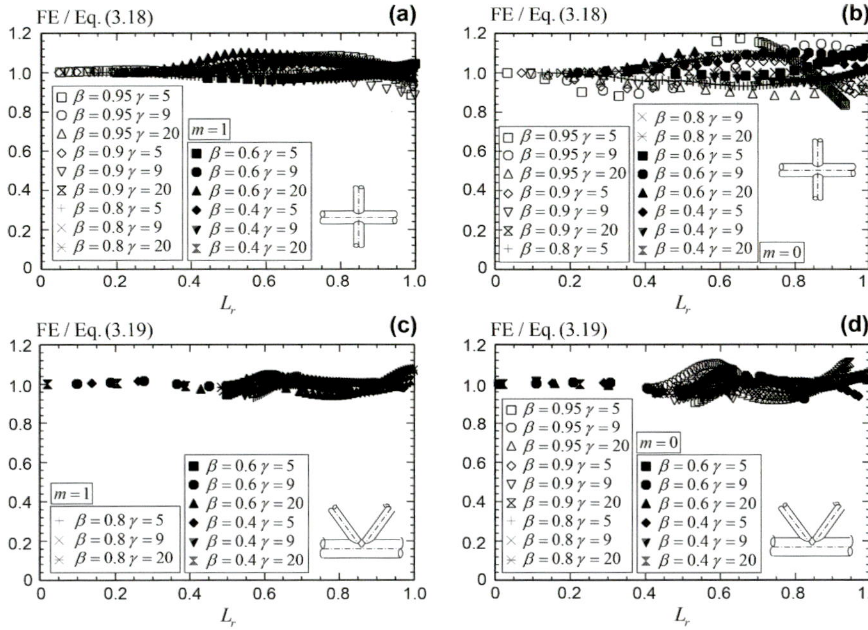

FIGURE 3.10 Comparison of the proposed FACs with FE data.

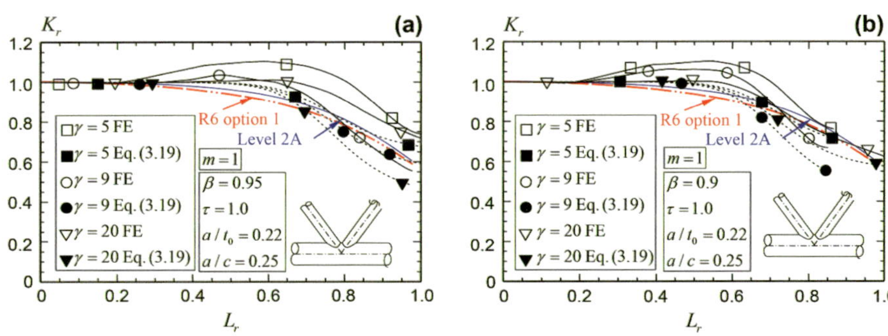

FIGURE 3.11 Comparison of proposed FAC with FE data not included in the regression database.

three FAC options in BS 7910, providing a more comprehensive evaluation of crack growth behavior.

The ductile tearing assessment procedure using the FAD requires experimental data on (1) the stress-strain curve of the material, obtained following engineering testing standards, such as ASTM E8/E8M [43]; and (2) the fracture resistance curve, measured in accordance with material testing standards, such as ASTM E1820 [44]. To validate the ductile tearing assessment for CHS X-joints, experimental data from large-scale X-joint tests with a predefined crack is also essential.

Figure 3.12 illustrates the geometric configuration of the CHS X-joint specimen, which is fabricated from high-strength steel S690. Figure 3.13a presents the true stress versus the true strain curve measured from coupon specimens for both the brace and the chord members, while Figure 3.13b shows the measured fracture resistance *J-R* curve measured from the standard C(T) specimens.

The thick-walled X-joint shown in Figure 3.12 features a prefabricated V-shaped notch near the weld toe at the crown position. The prefabricated notch has a depth of $d = 10$ mm, while its surface opening width h measures 11.5 mm. The notch root has a radius of $r = 0.2$ mm, and its total length on the outer surface of the chord is 120 mm. The two free faces of the V-notch form an angle of $\theta = 60$ degree, as shown in Figure 3.12. The notch root runs perpendicular to the longitudinal axis of the chord member.

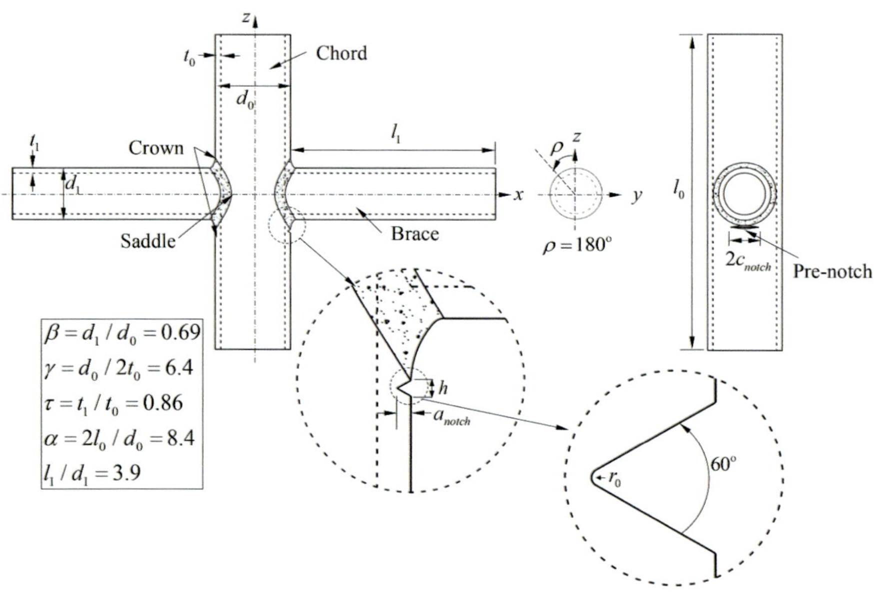

FIGURE 3.12 The geometric configuration of a notched CHS X-joint specimen.

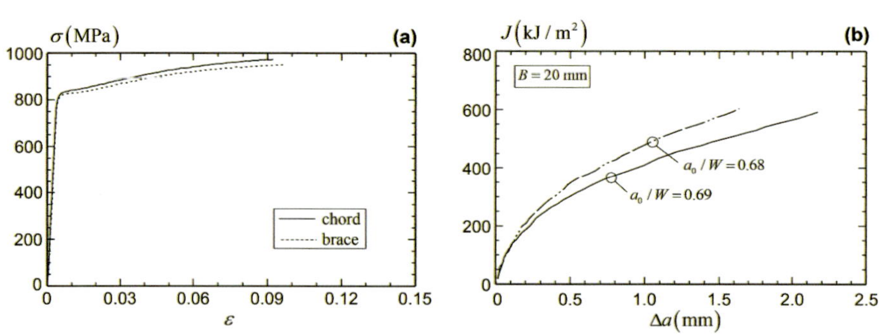

FIGURE 3.13 (a) Stress-strain curves; and (b) *J-R* curves for the S690 steels.

FIGURE 3.14 Test setup for the S690 CHS X-joint.

Figure 3.13a presents the stress–strain curves for the chord and brace materials. Figure 3.13b provides the *J-R* curves for S690 steels with two initial crack lengths.

The experimental setup applies vertical compression loads to the top end of the chord member, while the brace members are supported on two saddle supports – one resting on a roller support and the other on a pin support. This arrangement, as shown in Figure 3.14, induces an in-plane bending moment near the brace-to-chord intersection region. The chord member is fitted with a 40 mm-thick end plate welded to its top end, while each brace member contains a 25 mm-thick internal diaphragm located 30 mm from the brace end, positioned directly above the saddle support.

The testing procedure consists of two loading stages: (i) a cyclic pre-cracking phase followed by (ii) a monotonic loading phase with unloading-reloading events, similar to the loading scheme described in ASTM E-1820 for standard fracture specimens. The cyclic load is designed to generate a sharp fatigue crack along the root of the prefabricated notch. The cyclic loading magnitude is 180 kN, with a maximum load of 200 kN and a load ratio of 0.1. The maximum fatigue pre-cracking load generates a crack driving force significantly lower than the material's measured fracture toughness.

To monitor crack propagation, the experiment measures *CMOD* using three displacement transducers mounted across the prefabricated notch, as shown in Figure 3.14. A conventional crack-opening displacement (COD) gauge, commonly used in laboratory fracture specimens, is impractical due to challenges in implementing a knife-edge detail in the large-scale X-joint. The three transducers are spaced 30 mm apart (center-to-center) in a direction perpendicular to the longitudinal axis of the chord. One transducer is positioned at the crown point, while the other two are symmetrically placed around it. All three measure *CMOD* over a 26 mm gauge length.

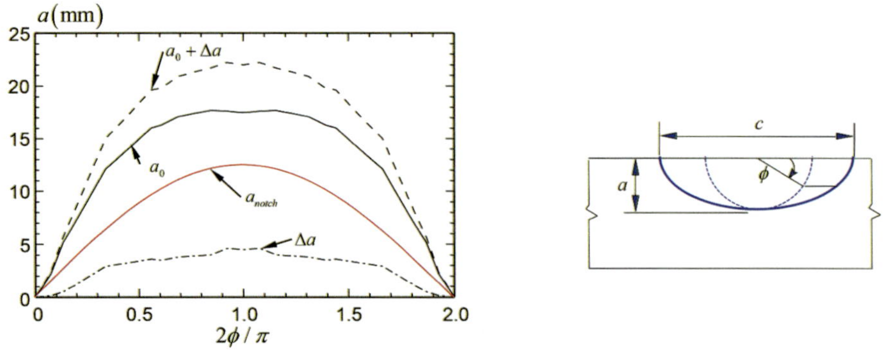

FIGURE 3.15 Crack-front profiles in the CHS X-joint.

The cyclic pre-cracking test concludes after approximately 100,000 cycles, when the measured increase in maximum *CMOD* exceeds 25% under constant-amplitude cyclic loading.

Figure 3.15 illustrates the profile of the prefabricated notch (a_{notch}), with its center located at the crown point of the chord. After 100,000 cycles of pre-cracking loads, a pre-fatigued crack (a_0) develops, as shown in Figure 3.15, and is measured through post-test sectioning of the cracked material. The maximum fatigue pre-crack size near the crack center is approximately 7.7 mm, corresponding to an initial crack depth ratio (a/t) of 0.64. The total crack area at the end of the fatigue phase, calculated based on Figure 3.15, accounts for 7.4% of the intact area along the brace-to-chord intersection region,

$$A_{crack}\% = \frac{A_{crack}}{L_w t_0} \tag{3.20}$$

where L_w denotes the length of the weld along the brace-to-chord intersection (see Eq. 3.14).

Figure 3.16a presents the load versus LLD curves recorded during the experiment. The X-joint reaches a maximum load of 1864 kN at a LLD of 49 mm, corresponding to a brace rotation of 0.053 radians, which remains well below Lu's deformation limit [29]. Beyond this peak load, brittle failure occurs, characterized by through-thickness crack propagation accompanied by a loud sound. The crack area expands from an initial 7.4% (fatigue pre-cracked) to 9.0% of the brace-to-chord intersection area before brittle failure, as measured from post-test sectioning.

The unloading-reloading cycles recorded in Figure 3.16 measure the compliance of the X-joint, expressed as $C = \Delta CMOD / \Delta P$. Since *CMOD* is directly related to the crack depth, the increase in measured compliance indicates crack extensions along the crack front. A comprehensive parametric investigation of the compliance, covering a full range of geometric parameters, could establish a compliance-versus-crack-length relationship. This relationship, when implemented during testing, enables real-time monitoring of crack growth. ASTM E-1820 [44] employs a similar approach to determine crack extension during fracture resistance testing.

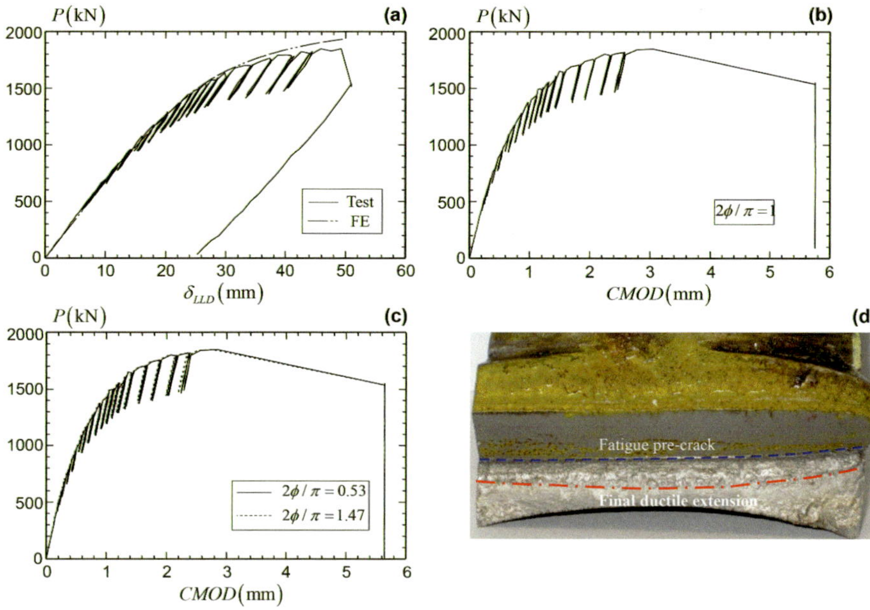

FIGURE 3.16 Load-displacement relationships measured in the experiment and the fracture surface from post-test sectioning.

The increase in compliance observed during testing suggests a finite amount of ductile crack extension before the final brittle failure, aligning with post-test sectioning observations. Figures 3.16b and c presents the recorded *CMOD*s at three different locations: $2\phi/\pi = 1.0$, $2\phi/\pi = 0.53$, and $2\phi/\pi = 1.47$. The *CMOD* values measured at $2\phi/\pi = 0.53$ and $2\phi/\pi = 1.47$ remain closely aligned, indicating an almost symmetric crack extension about the central crown point ($2\phi/\pi = 1.0$). This symmetry is further confirmed by the crack extension observed in the post-test sectioning, as shown in Figure 3.16d.

The post-test examination of the brace-to-chord intersection reveals the formation of three surface cracks in the chord material at the weld toe, in addition to the primary crack extension along the machined V-notch. These newly nucleated cracks are distributed on both sides of the intersection. One crack, designated as C1, is located on the left side (see Figure 3.14), while the other two, C2 and C3, appear on the right side (see Figure 3.14). A detailed assessment of these cracks, based on post-test sectioning, provides measurements of their depth in the through-thickness direction and their surface length along the chord. Among the three new cracks, C1 on the left side is the longest, with a total crack area accounting for 9.5% of the brace-to-chord intersection area. On the right side, the combined area of C2 and C3 reaches 16.8% of the intact brace-to-chord intersection area just before the final brittle failure. The presence of multiple cracks developed beyond the primary V-notch, further contributes to the structural failure of the X-joint.

The assessment of the ductile tearing requires numerical solutions for linear-elastic stress-intensity factors, obtained from multiple FE models with varying crack sizes [2]. This study focuses on the primary crack extending from the fatigue pre-cracked V-notch at the crown point. Figure 3.17a illustrates a typical, quarter-symmetric FE mesh used for analyzing the X-joint specimen. The numerical model employs a mesh-tying procedure, which couples a local-front model developed in FEA-Crack [45] with a global continuous model generated in Patran [46]. Previous studies [36, 37] have validated this mesh-tying approach for both linear-elastic and elastic-plastic analyses of tubular joints. The numerical simulations utilize 20-node brick elements with reduced integration, and all computations are performed using the open-source research code WARP3D [30].

Beyond computing linear-elastic stress-intensity factors along the crack front, this study also conducts a large-deformation, elastic-plastic analysis to validate the FE model against experimental load-deformation responses. Figure 3.17b demonstrates a close agreement between the numerical load versus LLD curve and the experimental data. The numerical model in Figure 3.17a represents a crack size matching the fatigue pre-cracked profile as shown in Figure 3.15a, but does not incorporate a material damage model to simulate crack extension. The FE model considers only the primary crack ahead of the fatigue pre-cracked V-notch, assuming that the remainder of the brace-to-chord intersection remains intact.

The experimentally measured variation in compliance, calculated using $C = \Delta CMOD / \Delta P$, is presented in Figure 3.17c. The numerical curve interpolates

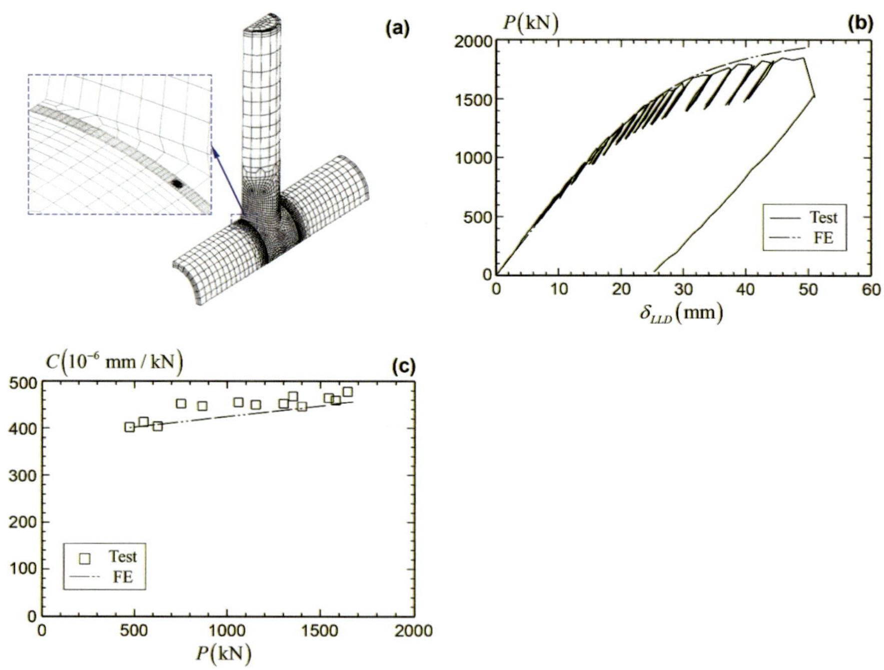

FIGURE 3.17 FE model for the cracked CHS X-joint and the compliance versus the crack size relationship.

between two linear-elastic compliance values derived from FE models – one corresponding to the initial fatigue pre-cracked profile and the other to the crack-front profile after ductile crack extension. The numerical compliance data suggests that integrating experimental compliance measurements with a numerically established compliance-crack size relationship offers a feasible approach for real-time estimation of crack extension during testing.

A comprehensive ductile tearing assessment of a structural component requires evaluating the crack driving force across multiple crack sizes, alongside a rising *J-R* curve that characterizes the increasing fracture resistance as the crack propagates. The validity of the *J-R* curve, derived from standard fracture specimens under SSY conditions, imposes strict constraints on its transferability to realistic flaws in structural components. For accurate assessment, the structural component must exhibit a well-contained crack front with high plasticity constraints. Additionally, the requirement for the *J*-integral to serve as a valid crack-tip parameter limits the assessment to a small crack extension before significant unloading occurs near the crack tip.

The ductile tearing assessment based on the BS7910 framework may adopt the Option 3 FAC in Eq. (3.8). The failure assessment computes the linear-elastic crack driving force for the primary crack ahead of a fatigue pre-cracked V-notch in an X-joint, considering six different crack-front profiles with varying crack depth ratios. Each crack-front profile is interpolated linearly between the fatigue pre-cracked profile (a_0) and the ductile crack-front profile at the end of crack extension ($a_0 + \Delta a$) in Figure 3.15a [4].

The fracture resistance for different crack extensions, that is, the denominator in Eq. (3.1), is determined using the lowest *J-R* curve obtained from C(T) specimens (Figure 3.13b). The corresponding strength of the cracked joint is calculated by multiplying the intact joint strength from Eq. (3.10) with a variable area reduction factor from Eq. (3.13) with $m = 1.0$, adjusted for different crack sizes. The calculation of the reduction factor considers both the crack area near the prefabricated notch and the newly nucleated weld-toe cracks along the brace-to-chord intersection.

Figure 3.18 presents the mixed-mode, linear-elastic stress-intensity factors computed using the interaction-integral approach in WARP3D [30]. The crack front in the X-joint is primarily subjected to a mode I driving force, with a minor mode II contribution and negligible mode III effects. The normalized mode I crack driving force, $K_I / \sigma\sqrt{\pi a}$, decreases as the crack size increases. However, the absolute magnitude of K_I increases with crack depth, though at a slower rate than the $\sqrt{a}$. The mode II normalized crack driving forces remain small and largely invariant with respect to crack depth, as shown in Figure 3.18b.

Figure 3.19 illustrates the ductile tearing assessment at four crack-front locations ($2\phi/\pi$=0.25, 0.45, 0.55, and 1.0) for five fixed load levels. Each connected assessment locus represents the ductile tearing evaluation at a fixed load for six different crack depth ratios (a/t_0=0.64, 0.67, 0.71, 0.74, 0.78, and 0.80). The FACs include the generic Option 1 curve, the material-dependent Option 2 curve, and the geometric and material-specific Option 3 curve derived from Eq. (3.16). The crack-front location at $2\phi/\pi = 0.25$ corresponds to the position near the free surface, while $2\phi/\pi = 1.0$ represents the deepest crack-front point. The position $2\phi/\pi = 0.55$ corresponds to the location with the maximum mode I crack driving force before ductile crack extension.

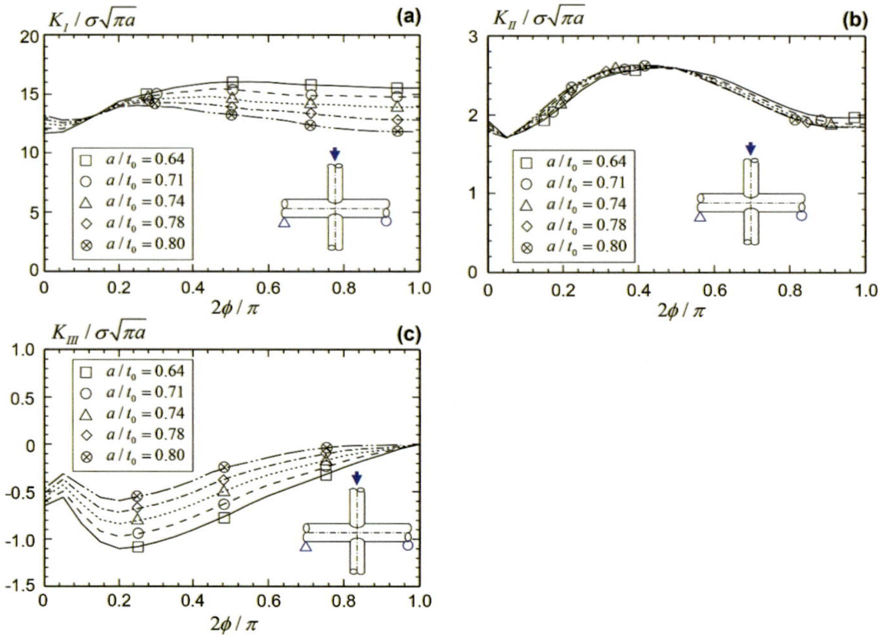

FIGURE 3.18 Mixed-mode stress intensity factors along the crack-front of the CHS X-joint.

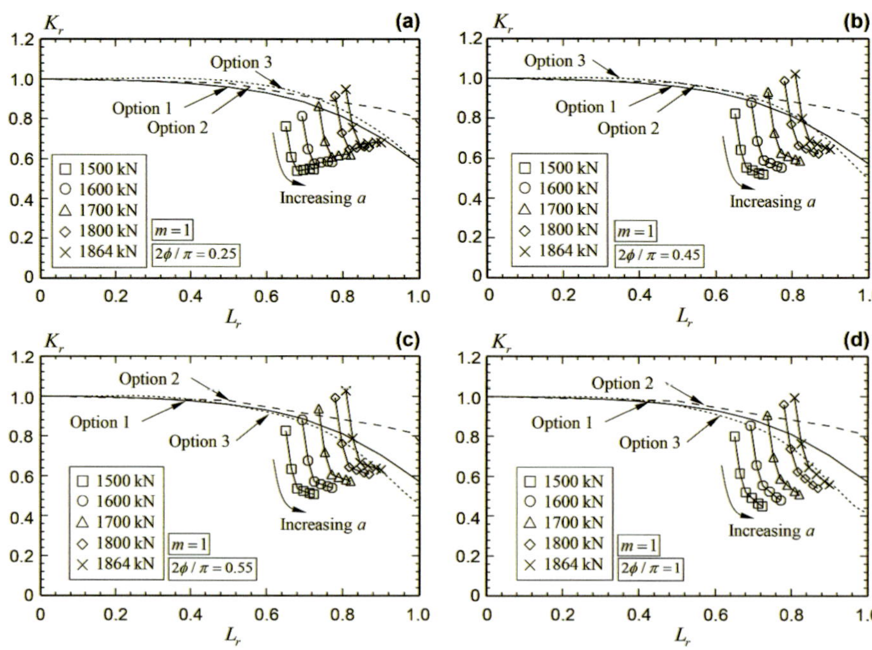

FIGURE 3.19 Ductile tearing assessment for the high-strength CHS X-joints.

As shown in Figure 3.19a, no crack extension occurs at $2\phi/\pi = 0.25$ until the applied load reaches 1700 kN, based on the Option 1 curve. Beyond this load level, the crack extension remains stable. Figures 3.19b and c indicate that crack extension initiates at $2\phi/\pi = 0.45$ and $2\phi/\pi = 0.55$ at approximately 1600 kN, while Figure 3.19d shows that crack extension at $2\phi/\pi = 1.0$ initiates at a slightly higher load, also around 1600 kN. The Option 1 curve predicts stable crack growth along the entire crack front at the maximum load level of 1864 kN.

However, the Option 3 FAC gradually descends below the Option 1 curve as the crack-front location transitions from near the free surface ($2\phi/\pi = 0.25$) to the deepest crack-front point ($2\phi/\pi = 1.0$). According to the Option 3 curve, crack extension initiates at approximately 1600 kN for both $2\phi/\pi = 0.55$ and $2\phi/\pi = 1.0$. At the maximum recorded load, the assessment loci indicate that fracture becomes unstable at $2\phi/\pi = 0.55$ and is approaching instability at $2\phi/\pi = 1$, at $a / t_0 = 0.80$, which is consistent with experimental observations. Meanwhile, crack extension at $2\phi/\pi = 0.25$ and $2\phi/\pi = 0.45$ remains stable at $a / t_0 = 0.80$ based on the Option 3 curve. Both Option 1 and Option 3 curves predict similar load levels for the onset of crack extension. The above example on the ductile tearing assessment using the BS7910 framework reveals that the Option 1 assessment curve does not provide a conservative estimate of the failure load associated with the brittle fracture observed in the experiment. In contrast, the Option 3 curve demonstrates strong agreement with experimental results.

3.5 TEARING ASSESSMENT DIAGRAM

Ductile tearing assessment evaluates the relationship between load resistance, crack extension, and fracture resistance. When using Option 3 FAC derived from a numerical model with a stationary crack, the FAC must be updated for each crack extension considered, necessitating a new FAC for each assessment with a different crack size. In contrast, ASTM E1820 [44] describes a single-specimen technique for measuring the fracture resistance curve using the load-deformation response of a specimen with an extending crack. The test data enables the establishment of a direct relationship between load resistance, crack extension, and fracture resistance, forming the basis for tearing assessment.

This study constructs a TAD using the load-deformation curve and *J-R* curve of fracture specimens to estimate the load-versus-crack-extension relationship during stable tearing. The TAD represents the fracture ratio (K_r) on the vertical axis and the load ratio (L_r) on the horizontal axis. The TAC incorporates crack extension into the assessment through the experimentally measured load-deformation curve. This section outlines the construction of the TAC, the assessment procedure using the load path, and the critical parameters required for the TAD. Additionally, it clarifies the scope of the TAD's application.

3.5.1 Construction of Tearing Assessment Curve

Figure 3.1 presents the typical FAC. To construct the TAC, Figure 3.20 presents the flowchart through a fracture test. The deformation parameter δ represents either the LLD or the *CMOD*. The partial unloading-reloading events at different displacement levels (Figure 3.21a) are used to measure compliance (C), which is defined as the

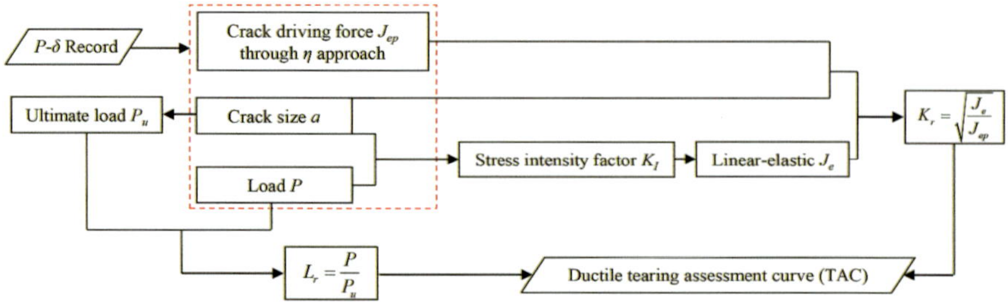

FIGURE 3.20 (a) A typical TAC; and (b) flowchart for the construction of TAC.

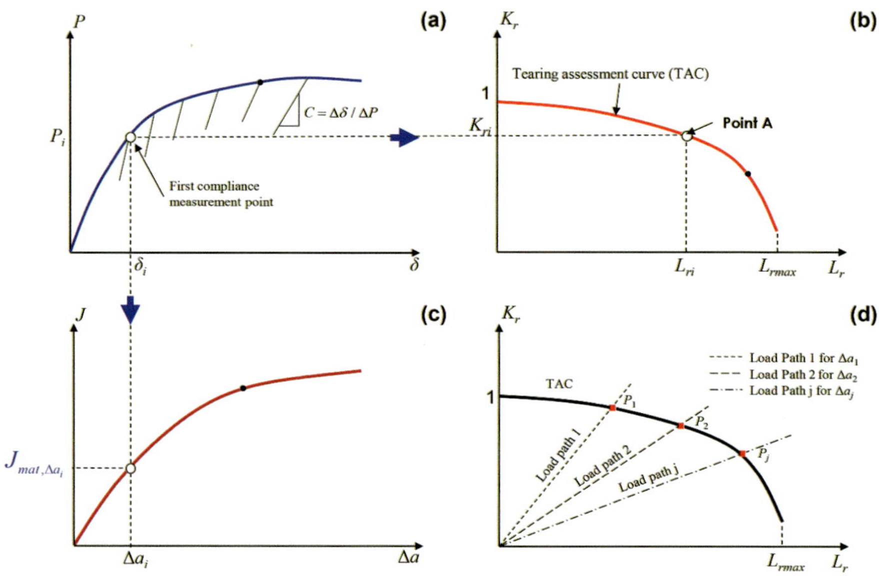

FIGURE 3.21 Derivation of TAC from the P-δ curve and J-R curve, as well as the load-path at different crack sizes for the tearing assessment.

ratio of deformation change to load change during the partial unloading-reloading stage, expressed as follows [44].

$$C = \frac{\Delta\delta}{\Delta P} \tag{3.21}$$

The compliance (C) enables the determination of crack size (a) at the unloading-reloading event through a unique relationship between crack size and compliance derived from linear-elastic analysis. Subsequently, the study calculates the J-value using the single-specimen technique (J-integral approach) [44] for standard fracture specimens and the modified J-integral approach [47–49] for other structural specimens. The J-integral approach separates the J-value into elastic and plastic components.

$$J_{ep} = J_e + J_p \tag{3.22}$$

where the J_ϵ derives from the linear-elastic analysis as shown in Eq. (3.9), while J_p follows the equation below based on the η-method,

$$J_p = \int_0^{\delta_p} \frac{\eta P}{A_{liagament}} d\delta_p \tag{3.23}$$

where η is a dimensionless parameter, $A_{liagment}$ denotes the net area of the remaining ligament, δ_p refers to the plastic deformation. The η approach builds a relation between the J-value and the area under the load-deformation curve of the fracture specimen.

Figure 3.21b presents the TAC [$f(L_r)$] derived from the load-deformation (P-δ) curve in Figure 3.21a. The TAC construction for a fracture specimen or a cracked structural component with an initial crack size (a_0) consists of two phases: before and after the onset of tearing.

- Before the first unloading-reloading event, the crack size is assumed to remain at its initial value, a_0. The load ratio L_r [Eq. (3.2)] is defined as the ratio of the applied load P (along the P-δ curve) to the load capacity P_u of the specimen with crack size a_0. The function $f(L_r)$ [Eq. (3.8)] is given by the square root of the ratio between the elastic crack driving force (J_e) and the elastic-plastic crack driving force (J_{ep}). Here, J_e is obtained from the existing equations [5], while J_{ep} is computed using the η-method, which integrates the area under the P-δ curve up to the applied load P. This establishes the FAC from $L_r = 0$ to L_{ri} (point A in Figure 3.21b).
- After the first unloading-reloading event, L_r and $f(L_r)$ are calculated at compliance measurement points. The load ratio L_r is now defined as the ratio of the applied load P to the load capacity P_u of the specimen at the current crack size a_i. The function $f(L_r)$ follows Eq. (3.8), where J_e remains the same as in the first phase, while J_{ep} is recalculated using the η-method for the updated crack size a_i. This extends the FAC beyond point A in Figure 3.21b.

Together, these two phases construct the TAC (as shown in Figure 3.21b) using the load-deformation curve of fracture specimens with an extending crack. The J-R curve (Figure 3.21c) is measured following the ASTM E1820 [44] procedure. The TAD is then formulated using the load path based on experimentally measured fracture resistance at different crack extensions.

3.5.2 TAD-Based Assessment through Load Path

Based on the TAC and J-R curve, the TAD assesses ductile tearing by constructing load paths for structural components with varying amounts of tearing, as illustrated in Figure 3.21d. The load path construction requires the linear-elastic solution for J in the cracked component and the fracture resistance curve obtained in Figure 3.21c.

Each load path corresponds to a fixed crack size a with a postulated crack extension Δa. The load path is represented by a straight line extending from the origin

($Lr = 0$, $Kr = 0$) to a point defined by (Lr, Kr) at the applied load. The corresponding fracture ratio [2] is given by:

$$K_r = \sqrt{J_e / J_{mat,\Delta a_j}} \tag{3.24}$$

where $J_{mat,\Delta a_j}$ is extracted from the J-R curve at the current crack extension Δa_j. The load level at the intersection of the load path and TAC indicates the onset of crack extension for the given crack size a_j. By incrementally increasing the crack extension, the assessment procedure compiles a series of intersection points (Δa_j, P_j), which together define the load-crack extension P–Δa relationship for the cracked component.

The TAD-based assessment begins with the construction of the TAC using the load-deformation curve of a fracture specimen with a growing crack. The calculation of the load ratio (L_r) [Eq. (3.2)] requires an expression for the ultimate load resistance (P_u) as a function of the current crack size (a). Meanwhile, the TAC function $f(L_r)$ [Eq. (3.8)] is derived from the elastic crack driving force (J_e) and the elastic-plastic crack driving force (J_{ep}), where J_{ep} is computed using the η method based on the load-deformation curve.

Following the TAC construction, the assessment proceeds to establish load paths corresponding to various postulated crack extensions. This stage requires the fracture resistance curve (J-R curve) and the elastic crack driving force (J_e) of the fracture specimen or structural component. Once these load paths are constructed, the TAD determines the load-crack extension relationship by identifying the load levels at which the TAC intersects the load paths associated with different crack extensions.

An essential component of the TAD-based assessment is the integration of the J-R curve while ensuring η-controlled crack growth, as specified by ASTM E1820 [44]. According to this standard, the maximum J-capacity for J-R curve measurement is given by:

$$J_{\max} = \min\{B\sigma_Y / 10, b_0\sigma_Y / 10\} \tag{3.25}$$

where B represents the specimen thickness, b_0 is the initial ligament, and σ_Y denotes the yield strength, respectively. Additionally, ASTM E1820 [44] defines the maximum crack extension capacity in the J-R curve by:

$$\Delta a_{\max} = 0.25 b_0 \tag{3.26}$$

The TAD-based assessment follows the requirements of J-controlled crack growth, as defined by Eqs. (3.25) and (3.26), to determine $L_{r,max}$. In this study, the TAD defines $L_{r,max}$ based on the maximum crack extension measured in the experiment.

The scope of TAD is limited to fracture specimens and cracked structural components undergoing a ductile tearing process. The load level predicted by TAD corresponds to a failure mode that combines fracture and plastic failure.

3.5.3 Validation of TAD-Based Assessments on Material Specimens

This section presents the validation of the TAD-based ductile tearing assessment using SE(B) specimens made of HY80 steel and SE(T) specimens made of aluminum alloy 6061-T6511.

Zhu and Joyce [50] conducted a series of fracture tests on SE(B) specimens made of HY80 steel. The thickness (*B*) of the specimen is equal to 25.4 mm, with a height (*W*) of 50.8 mm and a span of 101.6 mm. Figure 3.22a presents the true stress–true strain curve of HY80 steel, which has a Young's modulus of 207 GPa and a yield strength of 630 MPa. This study validates the TAD-based assessment using SE(B) specimens with a/W equal to 0.186 and 0.549. The specimen details are provided in Table 3.2.

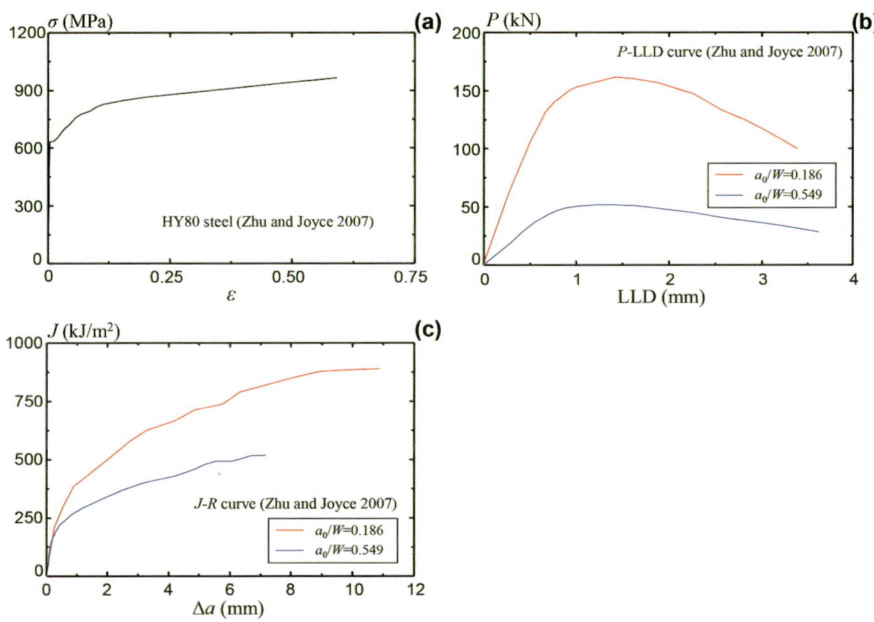

FIGURE 3.22 (a) Stress-strain curves for HY80 steel; (b) load-deformation curves for SE(B) specimen; and (c) *J-R* curve measured from SE(B) specimen.

TABLE 3.2
Information on the SE(B) Specimens.

Specimen No.	Initial crack ratio a_0/W	Width W (mm)	B (mm)	B_N (mm)
SEB-Shallow	0.186	50.8	25.4	20.32
SEB-Deep	0.549	50.8	25.4	20.32

To construct the TAC, this study expresses the ultimate load resistance P_u of the SE(B) specimen for calculating the load ratio L_r [Eq. (3.2)], as recommended in ASTM E1820 [44].

$$P_u = \frac{Bb^2\sigma_u}{S} \tag{3.27}$$

where B indicates the thickness, b refers to the remaining ligament, and σ_u denotes the ultimate strength (=735 MPa). The J_e in the TAC [$f(L_r)$ in Eq. (3.8)] derives from the linear-elastic J values,

$$J_e = \frac{(1-\nu^2)K_I^{\,2}}{E} \tag{3.28}$$

where the elastic crack driving force K_I follows ASTM-E1820 [44],

$$K_I = f(a/W)\frac{PS}{\left(BB_N\right)^{1/2} W^{3/2}} \tag{3.29}$$

where S refers to the span of the specimen, and $f(a/W)$ follows [5],

$$f(a/W) = \frac{3(a/W)^{1/2}\left[1.99-(a/W)(1-a/W)(2.15-3.93(a/W)+2.7(a/W)^2)\right]}{2(1+2a/W)(1-a/W)^{3/2}} \tag{3.30}$$

This study calculates J_{ep} in $f(L_r)$ using the η-approach and the P-LLD curves presented in Figure 3.22b. The J value follows the equation recommended by Zhu and Joyce [50]. Figure 3.22c displays the J-R curves for both specimens. Following the procedure in Section 3.5.2, this study integrates these solutions to construct the TACs through $K_r = \sqrt{J_e / J_{ep}}$.

Before the first compliance measurement point, the crack size is assumed to be equal to the initial crack size, and the (L_r, K_r) values are derived along the P-LLD curve. After the first compliance measurement point, the crack size is updated, and the (L_r, K_r) values are determined at compliance measurement points. Figures 3.23a and b present the TACs for the two specimens, with point A marking the first compliance measurement.

Subsequently, the assessment constructs the load path by integrating the J-R curves from Figure 3.22c, following Eq. (3.24). Here, J_e follows Eq. (3.28) and is extracted from the J-R curve. Figures 3.23a and b illustrate the TACs and load paths for the postulated crack extensions in the SEB-Shallow (a_0/W = 0.186) and SEB-Deep (a_0/W = 0.549) specimens, respectively. Figures 3.23c and d show the P-Δa curves derived from the intersection points between the load paths and TACs. The P-Δa relations determined from the TACs align well with the experimental data.

The peak load resistance predicted by the assessment closely matches the experimental record, with a difference of only 2%. Additionally, the evaluated load resistance for large crack extensions agrees well with experimental results, with a maximum

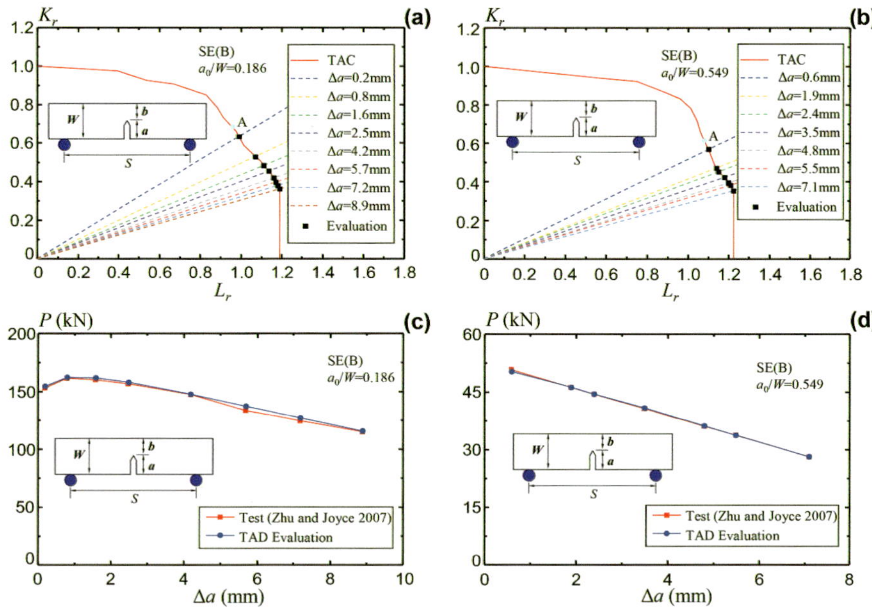

FIGURE 3.23 Tearing assessment for SE(B) specimens for (a) $a_0/W = 0.186$; (b) $a_0/W = 0.549$; and P-Δa relationships for SE(B) specimens with (c) $a_0/W = 0.186$; (d) $a_0/W = 0.549$.

difference of less than 2.5% for both specimens. This strong agreement in the P-Δa relationship validates the TAD-based assessment for the reported SE(B) specimens.

For the validation of SE(T) specimens, this study utilizes the fracture tests on a series of clamped SE(T) specimens reported by [51], as shown in Figure 3.24a. These SE(T) specimens feature a machined notch of either 10.5 mm or 15 mm, with a fatigue pre-cracking test extending the notch from 2 mm to 4 mm. The side groove on both sides of the SE(T) specimen leads to a net thickness (B_N) of 12 mm from a total thickness B of 15 mm. Figure 3.24b presents its true stress–true strain curve. The material properties include a Young's modulus of 68 GPa, a yield strength of 280 MPa and an ultimate strength of 340 MPa. During the fracture test, the experiment records the *P-CMOD* curve, as shown in Figure 3.24c, for two selected SE(T) specimens with an initial $a_0/W = 0.480$ and 0.576, respectively.

To obtain the load ratio L_r [Eq. (3.2)] required for the TAC, this study expresses the ultimate load resistance P_u of the clamped SE(T) specimen as:

$$P_u = B(W - a)\sigma_u \tag{3.31}$$

where B is the specimen thickness, W is the width, and σ_u is the ultimate strength (340 MPa). The determination of TAC [$f(L_r)$ in Eq. (3.8)] utilizes J based on the J-solution for clamped SE(T) specimens suggested by Liu *et al.* (2021),

$$K_I = f(a/W)\sigma\sqrt{\pi a} \tag{3.32}$$

where $f(a/W)$ is a dimensionless parameter dependent on the crack size:

$$f(a/W) = 3.66(a/W)^3 - 4.34(a/W)^2 + 3.57(a/W) + 0.49 \quad (3.33)$$

and σ represents the remote stress derived from the effective thickness:

$$\sigma = \frac{P}{W\sqrt{BB_N}} \quad (3.34)$$

where BB_N is the net thickness, excluding side grooves.

The study calculates J in Eq. (3.8) using the η-approach applied to the *P-CMOD* curves in Figure 3.24c and uses the *J-R* curves in Figure 3.24d. The η value follows the solutions provided by Liu *et al.* [51]:

$$\eta(a/W) = 1.68\left(a/W\right)^2 - 1.77\left(a/W\right) + 1.12 \quad (3.35)$$

Based on these solutions, this study constructs the TACs shown in Figures 3.25a and b, with point A representing the first compliance measurement.

Next, the load paths are obtained by coupling the linear-elastic J solution with the *J-R* curves in Figure 3.24d through Eq. (3.24). The load paths for specimen

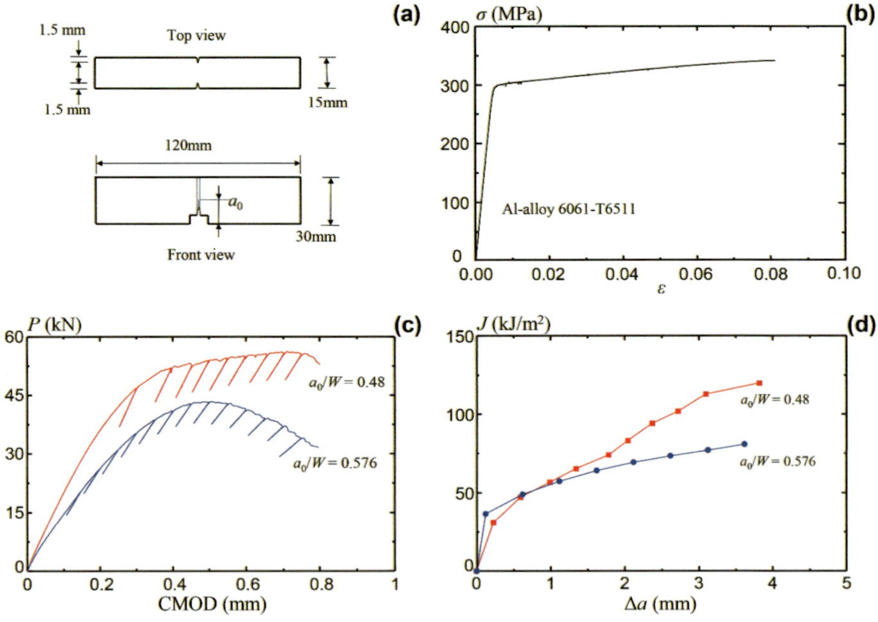

FIGURE 3.24 (a) Configuration of SE(T) specimens; (b) stress-strain curve for aluminium alloy 6061-T6511; (c) *P-CMOD* curves for two SE(T) specimens; and (d) *J-R* curves measured from the SE(T) specimens.

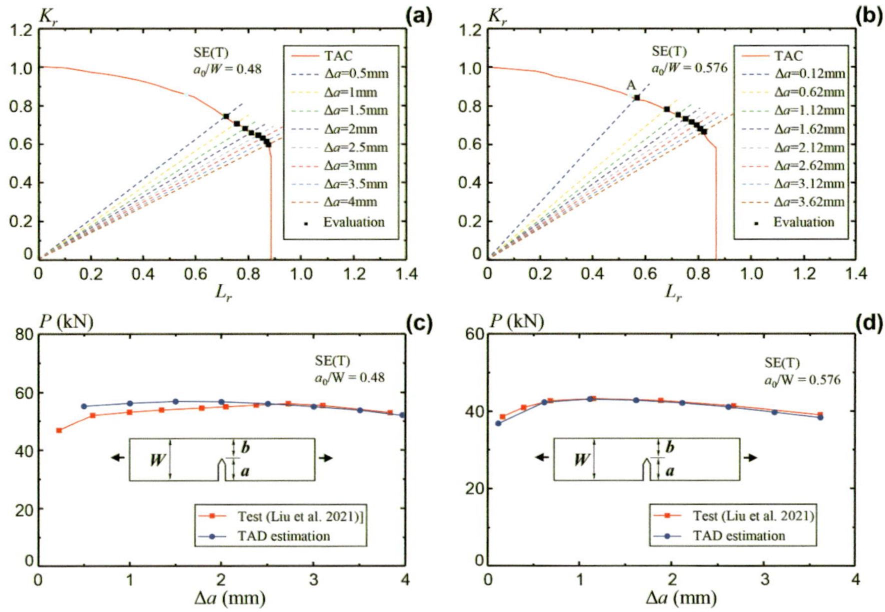

FIGURE 3.25 TAD assessments for the SE(T) specimens with: (a) $a_0/W = 0.48$; (b) $a_0/W = 0.576$; and P-Δa curves for the two SE(T) specimens with (c) $a_0/W = 0.48$; (d) $a_0/W = 0.576$.

SET10M-3 with $a_0/W = 0.480$ correspond to crack sizes ranging from 14.9 mm to 18.4 mm, with increments of 0.5 mm. Similarly, the load paths for specimen SET15M-1 $a_0/W = 0.576$ cover crack sizes from 17.4 mm to 20.9 mm, also in 0.5 mm increments.

Finally, Figures 3.25a and b present the ductile tearing assessments for specimens SET10M-3 ($a_0/W = 0.48$) and SET15M-1 ($a_0/W = 0.576$), respectively. Figure 3.25c compares the P-Δa curves from the TAD-based evaluation and experimental results for specimen SET10M-3. The maximum difference in load resistance between the evaluated and experimental curves is about 6% at peak load, while the discrepancy reduces to approximately 2% once the crack extends beyond 2 mm. For specimen SET15M-1 ($a_0/W = 0.576$) (Figure 3.25d), the TAD-based assessment closely predicts the P-Δa curve, with the difference in evaluated and experimental load resistances remaining within 2% for crack extensions exceeding 0.5 mm. The strong agreement in the P-Δa relationship validates the TAD-based assessment for SE(T) specimens.

This section validates the TAD-based assessment using SE(B) specimens made of HY80 steel and SE(T) specimens made of aluminum alloy 6061-T6511. The TAD assessments provide reliable predictions of the P-Δa relationship, closely matching experimental measurements. The TAC, derived from the experimental load-deformation curve, inherently accounts for crack extension effects.

Additionally, the TAC assessment is performed on the same fracture specimen from which the J-R curve is measured, ensuring consistency in constraint evolution under applied loading. As a result, the constraint conditions remain the same between the fracture specimen and the specimen used to derive the P-Δa relationship, enhancing the accuracy of the TAC-based assessment.

In the next section, the validation is extended to welded structural components, where the fracture specimen may experience a different constraint evolution than the welded structural components

3.5.4 Validation of TAD-Based Assessments on Welded Joints

This section validates the TAD-based assessment for two types of welded connections: (1) a surface-cracked pipe reported by Parool *et al.* [48] and (2) a welded tubular joint with a surface crack at the crown point, as reported by Qian *et al.* [4]. The TAC is derived from experimental records on fracture specimens, and the assessment is performed on the welded connections using the material-based TAC.

The TAC assessment requires the following key components:

1) The linear-elastic crack driving force J_e and the *J-R* curve of fracture specimens for TAC construction;
2) The *J*-value of the welded connections for calculating the fracture ratio K_r along the load path;
3) The load capacity P_u of the welded connections for determining the load ratio L_r in the load path.

Parool *et al.* [48] reported *J-R* curves for a series of pipes with surface cracks located at the toe of full-penetration circumferential welds. The fracture resistance in the *J-R* curve corresponds to the average *J*-value along the crack front. Figures 3.26a

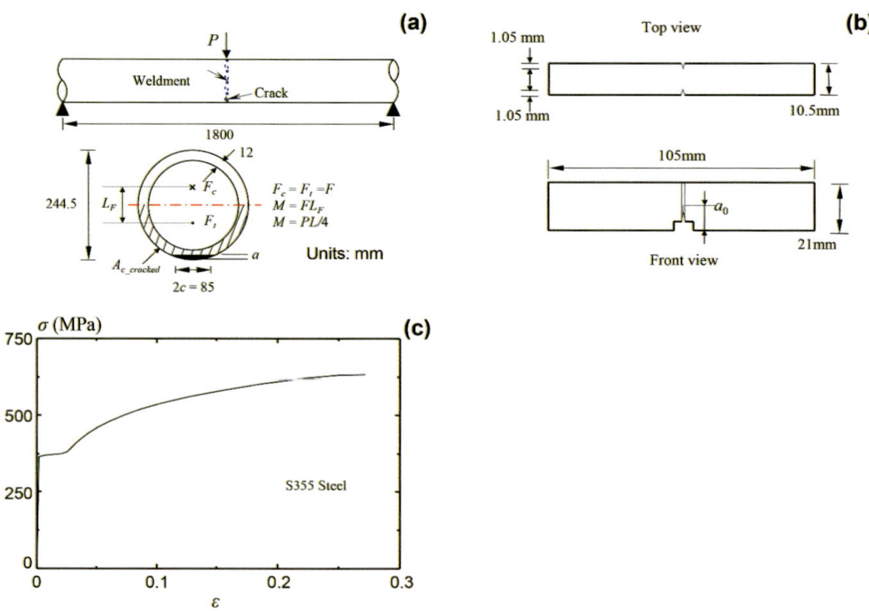

FIGURE 3.26 (a) Configuration of the surface-cracked pipe with a circumferential crack; (b) dimensions of the SE(T) specimen; and (c) true stress-strain curve for S355 steel.

TABLE 3.3
Information on the Surface-Cracked Pipes.

Case	Fabricated crack size a_N (mm)	Initial crack size a_0 (mm)	a_0/t	2c (mm)
PIPE01	7	9.14	0.762	85
PIPE02	7	9.43	0.786	85
PIPE03	7	8.98	0.748	85
PIPE04	7	9.27	0.773	85
PIPE05	7	9.30	0.775	85
PIPE06	7	9.42	0.785	85

and b illustrate the configurations of the surface-cracked pipe and the fracture specimen extracted from the pipe, respectively. The fracture specimens were taken from regions near the weld toe, where the surface crack is located. Figure 3.26c presents the true stress–true strain curve of the S355 steel material, which has a Young's modulus of 210 GPa, a yield strength of 365 MPa and an ultimate strength of 635 MPa. The surface-cracked pipe has a thickness of 12 mm with a machined notch of 7 mm depth at the toe of the circumferential weld. Fatigue pre-cracking extended the notch to an initial crack depth (including the machined notch) of approximately 9 mm. Table 3.3 provides details of the surface crack sizes in the pipes. After pre-cracking, the SE(B) and SE(T) specimens in this study had initial crack sizes of 13.13 mm and 13.07 mm, respectively.

This study predicts the P-Δa relationships for the surface-cracked pipes using TACs constructed from experimental data obtained from both pipes and fracture specimens. This allows a comparative assessment of the stable tearing process using material fracture properties versus fracture resistance directly measured from the pipes. The TAC derived from the pipe's load-deformation curve accounts for both material properties and geometric effects specific to each pipe. The specimen-based TAD assessment for the pipes validates the proposed approach, demonstrating that stable tearing can be evaluated using material fracture tests without requiring full-scale structural tests. Table 3.4 summarizes the details of assessments using three different sets of experimental records: pipes, SE(B) specimens, and SE(T) specimens.

The TAD-based assessment is first conducted directly using experimental data from surface-cracked pipes. To facilitate this, the study first derives the TAC from the load-deformation curve of the pipe. The TAC construction requires expressions for (1) the ultimate load resistance P_u in Eq. (3.2) and the linear-elastic crack driving force J_e in Eq. (3.10) for a surface-cracked pipe under three-point bending.

To derive the expression for the ultimate load resistance P_u, this study assumes a neutral axis at the horizontal centerline of the CHS, given that the surface crack area remains relatively small. The calculation assumes equal compressive and tensile resultant forces above and below the neutral axis:

$$F = \sigma A_{c\_cracked} \tag{3.36}$$

TABLE 3.4
Illustration on Different Approaches for Ductile Tearing Assessments of Pipes.

	Load ratio (L_r)		Fracture ratio [f (L_r)]	
Construction of TAC	**Applied load P of**	**Ultimate load resistance P_u of**	**J_e derived from the FE model of**	**J_{ep} derived from the P-δ curve of**
Pipe-based	Pipe	Pipe	Pipe	Pipe
SE(B)-based	SE(B) specimen	SE(B) specimen	SE(B) specimen	SE(B) specimen
SE(T)-based	SE(T) specimen	SE(T) specimen	SE(T) specimen	SE(T) specimen
	Load ratio (L_r)		**Fracture ratio (K_r)**	
Construction of load path	**Applied load P of**	**Ultimate load resistance P_u of**	**J_e derived from the FE model of**	**J_{mat} extracted from the J-R curve of**
Pipe-based	Pipe	Pipe	Pipe	Pipe
SE(B)-based	Pipe	Pipe	Pipe	SE(B) specimen
SE(T)-based	Pipe	Pipe	Pipe	SE(T) specimen

where $A_{c\_cracked}$ denotes the half-cracked cross-sectional area, represented by the dashed region in Figure 3.27a. Assuming a half-elliptical crack front, the crack area is estimated as:

$$A_{crack} = \frac{1}{2}\pi ac \tag{3.37}$$

where a is the crack depth and c is the crack length. The half-cracked cross-sectional area then becomes:

$$A_{c\_cracked} = \frac{1}{2}A_c - A_{crack} \tag{3.38}$$

where A_c is the total cross-sectional area of the pipe. The equivalent bending moment in the cross-section is determined using:

$$M = FL_F \tag{3.39}$$

where L_F represents the distance between the centroids of the upper and lower rings in Figure 3.26a. By equating the bending moment equations, the load equation for the pipe is obtained:

$$P = \frac{4L_F A_{c\_cracked}}{L}\sigma_{eff} \tag{3.40}$$

where σ_{eff} is an effective yield strength, taken as the average of the yield and ultimate strengths (500 MPa), which closely approximates the experimental load resistance.

To determine the linear-elastic crack driving force J_e, FE simulations were conducted using MSC PATRAN [52], as illustrated in Figure 3.27a. Eight quarter-symmetric FE

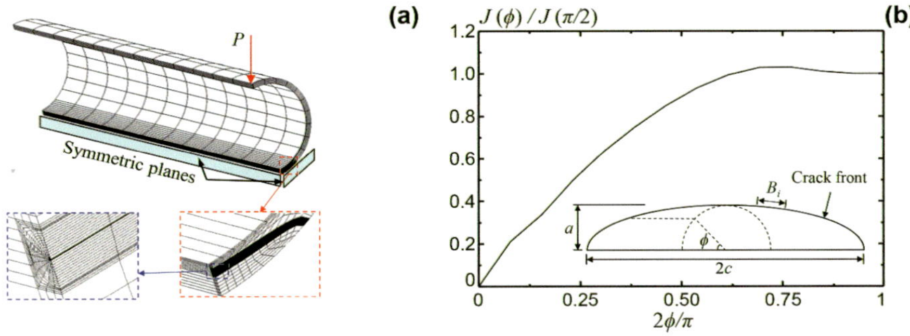

FIGURE 3.27 (a) A FE model for the surface cracked pipe; and (b) variations of J values along the crack front of a pipe.

models were developed, each containing a stationary crack with depths ranging from 9 mm to 10.4 mm in increments of 0.2 mm. A symmetric constraint was imposed along the pipe's axis, and the FE code WARP3D [30] was used to compute the elastic crack driving force J_e through linear-elastic analysis. Figure 3.27b shows the variation of J_e along the crack front for a 9 mm crack under a controlled displacement of 1 mm.

The average J-value along the crack front is computed as:

This study calculates $J_{el\text{-}ave}$ to represent the average elastic crack driving force along the crack front [22]. The relationship between $J_{el\text{-}ave}$ and the average stress intensity factor $K_{I\text{-}ave}$ follows:

$$K_{I-ave} = \sqrt{\frac{EJ_{el-ave}}{1-\nu^2}} = f(a/t)\sigma\sqrt{a} \tag{3.41}$$

where E is the effective modulus. The nondimensional function $f(a/t)$ to estimate the stress intensity factor is expressed as:

$$f(a/t) = 5.69(a/t)^3 - 14.39(a/t)^2 + 12.54(a/t) - 1.74 \tag{3.42}$$

The calculation of J_{ep} in the function $f(L_r)$ follows the η-approach for surface-cracked pipes proposed by Parool *et al.* [47]. The assessment constructs load paths corresponding to crack extensions based on the crack sizes in the numerical models. These load paths utilize the $J_{el\text{-}ave}$ solution from and extract $J_{mat,\Delta a_j}$ from the *J-R* curve of the pipes to conduct the tearing assessment.

In addition to the pipe-based assessment, this study also conducts two TAD assessments using TACs derived from the *P-CMOD* curves of SE(B) and SE(T) specimens.

- **SE(B) Specimen-Based Assessment:** The TAC is constructed using the SE(B) load-deformation curve, with the ultimate load resistance following Eq. (3.27). The elastic crack driving force J_e is obtained from Eqs. (3.28) and (3.29), while the J_{ep}-value is computed using the η-approach from Parool *et al.* [47]. The load path is then built using FE models, with crack extensions matching those in the numerical simulations of the pipes.

- **SE(T) Specimen-Based Assessment:** The TAC is constructed using the SE(T) load-deformation curve. The ultimate load resistance follows Eq. (3.31), and the elastic crack driving force is obtained from Eq. (3.32). The J_{ep}-value is computed using the η-approach from Parool *et al.* [47]. The load paths are constructed similarly to the SE(B) assessment, with the final tearing assessment performed using the TAD derived from the SE(T) specimen.

Figure 3.28 presents the experimental load-deformation curves and *J-R* curves for pipes PIPE01 and PIPE02 [48]. Figure 3.29 shows the TAD-based assessments

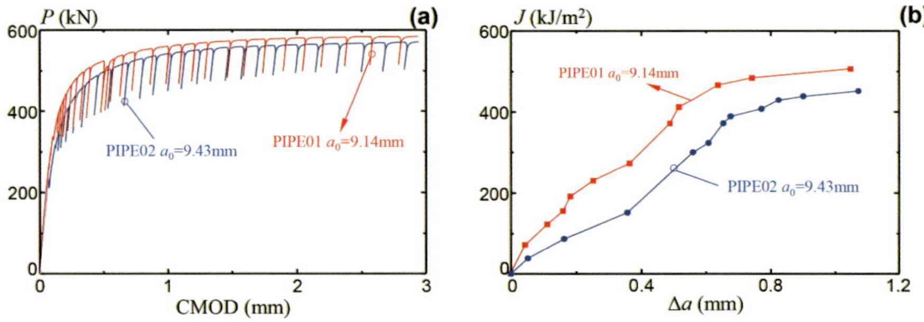

FIGURE 3.28 (a) *P-CMOD* curves for two cracked pipes; and (b) *J-R* curves measured from two cracked pipes.

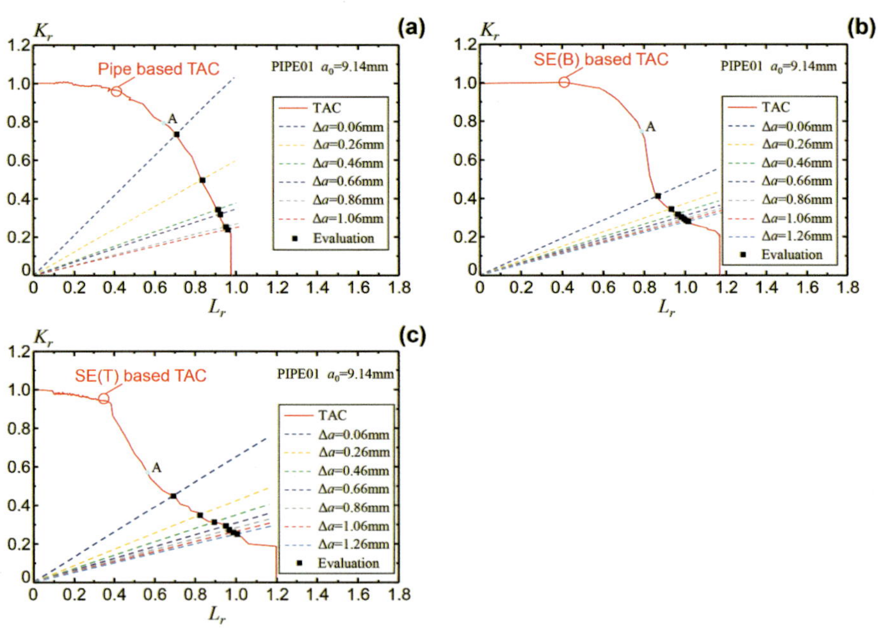

FIGURE 3.29 TAD assessments for the cracked welded pipe using: (a) TAC based on pipe data; (b) TAC based on SE(B) data; and (c) TAC based on SE(T) data.

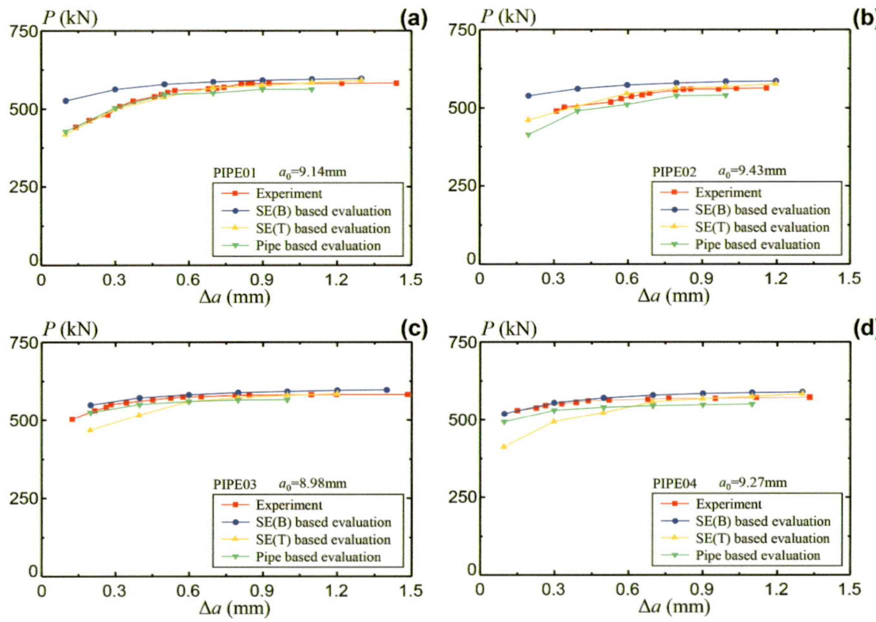

FIGURE 3.30 Comparison of the P-Δa curves for the cracked pipe specimens from the experimental measurement and those from TAD assessments for: (a) Pipe01; (b) Pipe02; (c) Pipe03; and (d) Pipe04.

for PIPE01 using TACs from pipes, SE(B), and SE(T) specimens, respectively. Point A marks the first compliance measurement.

Figure 3.30 compares experimental P-Δa curves with those predicted by TAD for four different pipe specimens. For the specimen Pipe01, the pipe-based assessment closely matches experimental results, with a maximum deviation of 3.2% in load resistance. The SE(T)-based assessment also aligns well with experimental data, with a maximum difference of 2.3%. However, the SE(B)-based assessment overestimates load resistance at crack initiation, though the discrepancy reduces to about 5% for crack extensions beyond 0.5 mm. Table 3.5 summarizes the discrepancies across different crack extensions for all pipes illustrated in Figure 3.30.

The findings indicate that pipe-based assessments provide a conservative estimate of P-Δa curves, underestimating load resistance by approximately 6%. SE(B) and SE(T)-based assessments perform well for crack extensions beyond 0.5 mm but show larger deviations at small crack extensions due to crack tip blunting effects.

Qian *et al.* [4] conducted a fracture test on an X-joint fabricated from high-strength S690 steel subjected to brace in-plane bending. Figure 3.31a illustrates the configuration of the welded tubular joint, consisting of a chord with an outer diameter of 356 mm along with other dimensionless parameters indicated. The chord contains a prefabricated notch with a depth of 10 mm, positioned near the weld toe at the crown point. The notch length ($2c$) measures 120 mm. Prior to the fracture test, a fatigue test was performed to extend the crack depth to 17.7 mm.

TABLE 3.5

Difference in the Loads from Experimental Record and Evaluations through Various TAD.

		$(P_{TAD} - P_{EXP})/P_{EXP}$		
Case	**Δ_a (mm)**	**Pipe-based evaluation**	**SE(B)-based evaluation**	**SE(T)-based evaluation**
PIPE01	0.26	1.40%	13.50%	−8.81%
$a_0 = 9.14$ mm	0.46	−0.66%	5.35%	−3.02%
	0.66	−2.76%	3.78%	0.73%
	0.86	−3.20%	1.79%	1.42%
	1.06	−3.30%	2.20%	0.57%
PIPE02	0.17	−6.36%	21.40%	4.02%
$a_0 = 9.43$ mm	0.37	−3.25%	10.77%	−0.29%
	0.57	−4.31%	7.51%	2.44%
	0.77	−3.33%	4.22%	1.16%
	0.97	−3.72%	4.13%	1.28%
PIPE03	0.22	−0.19%	4.37%	−11.03%
$a_0 = 8.98$ mm	0.42	−1.85%	1.89%	−8.08%
	0.62	−2.62%	1.21%	−2.62%
	0.82	−2.56%	1.58%	−1.35%
	1.02	−2.63%	2.01%	−0.22%
PIPE04	0.33	−3.28%	1.10%	−10.03%
$a_0 = 9.27$ mm	0.53	−3.95%	1.39%	−7.15%
	0.73	−3.74%	2.27%	−1.62%
	0.93	−3.60%	2.73%	−0.26%
	1.13	−3.56%	2.93%	0.65%

As depicted in Figure 3.31a, the experimental setup applies pin support and roller support at the two ends of the brace. A compressive load imposed at the top of the chord generates in-plane bending in the X-joint. The tubular joint is constructed from high-strength S690 steel with material properties detailed in Figure 3.31b, featuring a Young's modulus of 206 GPa, a yield strength of 827 MPa, and an ultimate strength of 899 MPa.

In addition to testing the welded tubular joint, Qian *et al.* [4] also performed fracture tests on side-grooved C(T) specimens (Figure 3.31c) extracted from the chord. Two types of C(T) specimens with thicknesses of $B = 10$ mm (CT10) and $B = 20$ mm (CT20) were prepared. The side grooves reduced the net thickness (BN) to 80% of the original specimen thickness. Detailed specifications of these two C(T) specimens are provided in Table 3.6.

Figure 3.32a depicts the material *J-R* curves measured from the C(T) specimens. The fracture assessment initially involves constructing the TACs based on the P-LLD relationships of the C(T) specimens. Subsequently, a load path is established

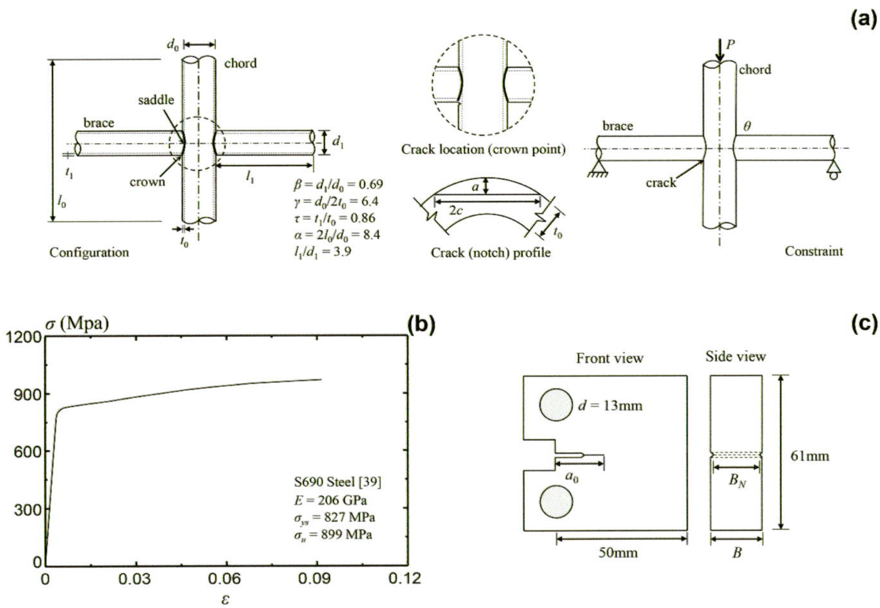

FIGURE 3.31 (a) Geometric configuration of CHS X-joint under in-plane bending; (b) true stress-true strain curve for S690 steel; and (c) configuration of the C(T) specimen.

TABLE 3.6

Information on the C(T) Specimens.

Specimen no	Fabricated crack size a_N (mm)	Initial crack size a_0 (mm)	B (mm)	B_N (mm)
CT10	30	33	10	8
CT20	30	34	20	16

by integrating the linear-elastic J_e solution of the cracked tubular joint with the material fracture resistance curves (*J-R* curves) from Figure 3.32a, thus predicting the load-crack extension (P-Δa) behavior of the X-joint.

The assessment constructs the TAC based on the load-deformation (P-LLD) curve of the C(T) specimen. The ultimate load resistance (P_u) for the C(T) specimen employs a formulation recommended by ASTM-E1820 [44], as shown in Eq. (3.43),

$$P_u = \frac{B(W-a)^2}{2W+a}\sigma_Y \tag{3.43}$$

where B is the specimen thickness, W is the specimen width, and σ_Y represents the effective yield strength (899 MPa) of the S690 steel. Subsequently, J_e and J_{ep} values in the $f(L_r)$ function are computed using solutions provided in ASTM-E1820 [44].

Following this, load paths for the joint are established. These load paths integrate J_e and J_{ep} solutions of the joint with $J_{mat,\Delta a_j}$-values derived from the material J–R curve. To determine the load ratio Lr [Eq. (3.2)] within the load path, the ultimate strength P_u is initially calculated using the bending capacity M_u according to guidelines by the IIW [24, 53] for intact X-joints subjected to brace-in-plane bending:

$$\frac{M_{ipb}\sin\theta}{\sigma_{ys}t_0^2 d_1} = 4.3\beta\gamma^{0.5}Q_f \tag{3.44}$$

Here, Q_f denotes the chord-stress function, calculated as:

$$Q_f = \left(1 - \left|\sigma_c / \sigma_y\right|\right)^{0.25-0.25\beta+0.01\gamma} \text{ for } \sigma_c < 0 \tag{3.45}$$

$$Q_f = \left(1 - \left|\sigma_c / \sigma_y\right|\right)^{0.2} \text{ for } \sigma_c > 0 \tag{3.46}$$

where σ_c represents axial stress in the chord, with negative values indicating compressive stress. The presence of a crack reduces the moment capacity to a modified value, $M_{ipb,cracked}$, by applying an area reduction factor, A_{RF}, expressed in Eq. (3.13).

Finally, $M_{ipb,cracked}$ is converted into the ultimate load capacity ($P_{u,crack}$) by

$$P_{u,crack} = \frac{2M_{ipb,crack}}{l_1} \tag{3.47}$$

where l_1 is the brace length.

To establish the fracture ratio K_r within the load path, the linear-elastic J_e solution for the joint is derived using the interaction integral approach [30] through six numerical models of the cracked tubular joint with stationary cracks (see Figure 3.33c). The crack depth-to-thickness ratios (a/t) are 0.64, 0.67, 0.7, 0.74, 0.77, and 0.79. The FE models employ a mesh-tie procedure for detailed crack modeling using FEA-Crack software [45], illustrated in Figure 3.32b. The FE software WARP3D [30] calculates

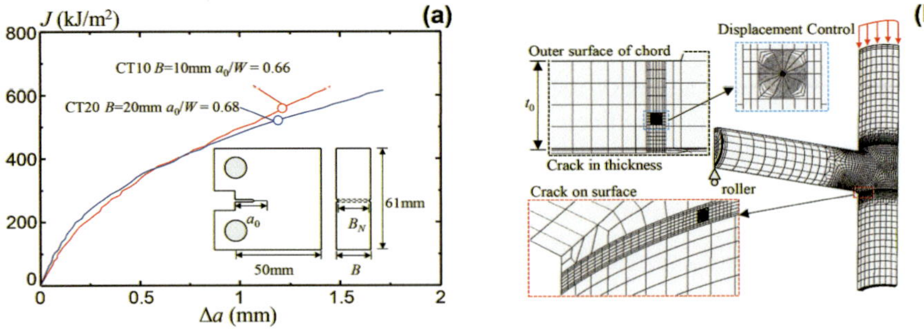

FIGURE 3.32 (a) J-R curves measured from the C(T) specimen; and (b) FE model for the cracked X-joint.

the elastic *J*-integral through linear-elastic analysis. Consistent with Qian *et al.* [4], the current study utilizes J_e at the deepest crack-front position, expressed as:

$$J_e = f(a/t)\frac{(P/A_{chord})^2 a}{E} \tag{3.48}$$

where A_{chord} denotes the chord cross-sectional area, t is the chord thickness, and $f(a/t)$ is a dimensionless parameter defined as,

$$f(a/t) = 36870.45(a/t)^3 - 78725.59(a/t)^2 + 53688.33(a/t) - 11070.59 \tag{3.49}$$

The fracture ratio K_r uses the derived J_e solution combined with the $J_{mat,\Delta a_j}$-value from the material *J-R* curve.

The study assesses the *P*-Δ*a* relationship of the X-joint based on experimental data from both C(T) specimens detailed in Table 3.10. Figure 3.33 compares the derived *P*-Δ*a* relationships using both specimens with results from Qian *et al.* [4]. In Qian *et al.*'s experiment, the load at crack initiation was approximately 1600 kN, and the ultimate load was 1864 kN. The discrepancy in crack initiation load between the TAD-based predictions and the experimental measurements is below 2%. Furthermore, the CT10- and CT20-based TAD evaluations slightly overestimate the ultimate load by about 3.7% and 4.8%, respectively. These close predictions affirm the validity of using material *J-R* curve-based TAD for structural integrity assessments.

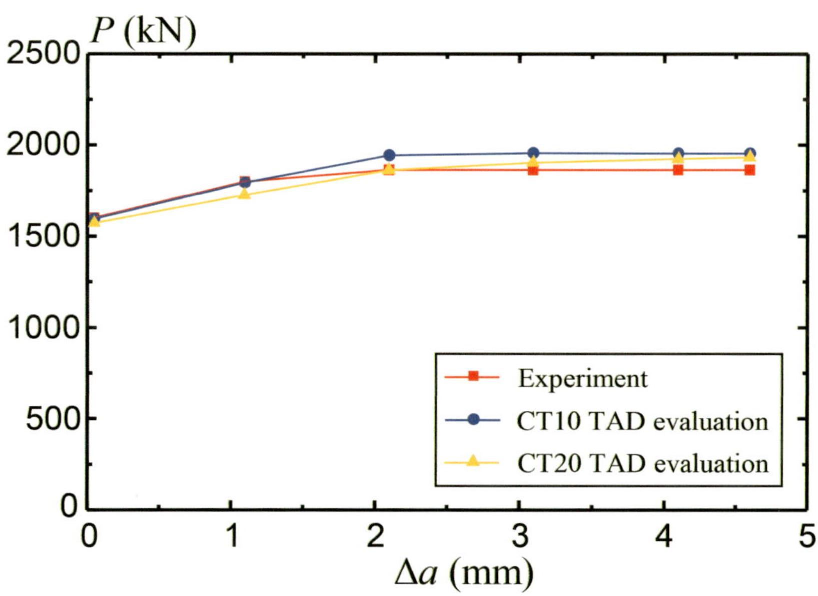

FIGURE 3.33 Comparison of the measured *P*-Δ*a* curves and the TAD estimations using TACs from C(T) specimens.

This section introduces a TAD approach for predicting the load versus crack extension behavior of cracked structural components. Utilizing material fracture tests and linear-elastic numerical analyses, the method avoids computationally intensive damage mechanics models. The results demonstrate that the TAD-based assessments closely match experimental data: load resistance predictions differ by a maximum of 6% for fracture specimens, remain within 10% for surface-cracked pipes after initial crack growth, and vary less than 5% for cracked X-joints. These findings validate the TAD method as an accurate and efficient tool for structural integrity evaluations based on material fracture resistance.

3.6 APPLICATION OF TAD FOR TUBULAR JOINTS UNDER REVERSED LOADINGS

This section extends the ductile TAD methodology to evaluate the load resistance and crack extension relationship in CHS joints subjected to reversed in-plane bending actions. By employing the envelope of load-deformation curves from fracture tests under reversed loading conditions, the TAD approach effectively connects fracture resistance, load capacity, and crack propagation. The validity of this method is demonstrated through experimental investigations on CHS X-joints fabricated from Q345 steel featuring surface cracks near the weld toe. These experiments highlight how reversed loading influences fracture resistance and confirm the accuracy of the TAD-based assessment for tubular joints. Additionally, cyclic fracture tests on single-edge-notched-tension [SE(T)] specimens are conducted to derive the TAD from their experimental data. The specimen-based assessment accurately predicts the load-crack extension behavior observed in CHS joints under reversed loading, establishing a reliable framework for integrating material fracture properties into structural response predictions.

3.6.1 Experimental Program

The experimental program consists of the following four key components. Tensile tests are conducted to determine the mechanical properties of Q345 steel under uniaxial tension. Fully reversed tensile-compression tests are performed to characterize the stress-strain behavior of Q345 steel under cyclic loading. Monotonic fracture tests are used to establish the fracture resistance curve of a cracked CHS X-joint under monotonic loading. Cyclic fracture tests are conducted on CHS X-joints subjected to reversed in-plane bending with different load ratios to evaluate cyclic fracture resistance.

Figure 3.34a illustrates the true stress-strain curve, which indicates that Q345 steel has a Young's modulus of 209 GPa, a Poisson's ratio of 0.3, a yield strength of 326 MPa, and an ultimate tensile strength of 632 MPa. Cyclic loading specimens are prepared according to ASTM-E606/E606M [54] and are tested under a displacement-controlled loading protocol at 0.5 mm/min. Figure 3.34b presents the true stress-strain behavior under cyclic loading.

To describe the cyclic hardening behavior, this study employs a combined isotropic and kinematic hardening model. The isotropic hardening (in Eq. (3.50)) governs

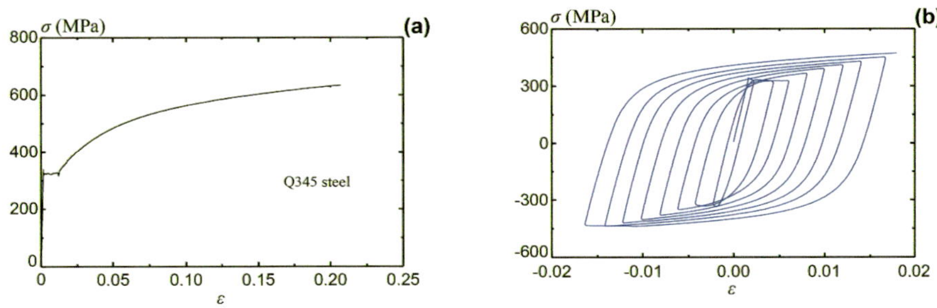

FIGURE 3.34 (a) Monotonic; and (b) cyclic; true stress-true strain curve for Q345 steel.

TABLE 3.7
Material Property of Q345 Steel.

Uniaxial tensile test			Reversed tension-compression test					
E (GPa)	σ_{ys} (MPa)	σ_u (MPa)	E (GPa)	σ_{ys} (MPa)	Q (MPa)	b	H_u (MPa)	γ_u
209	326	632	200	306	80	8	8000	70

the expansion of the yield surface, where the difference between yield and ultimate strength evolves with accumulated plastic strain.

$$k(\varepsilon^p) = k_0 + Q(1 - \exp[-b\varepsilon^p]) \tag{3.50}$$

where Q presents the difference between the yield strength and the ultimate strength, and b is the rate of the increase in the ultimate tensile stress with respect to the accumulated plastic strain, ε^p. The kinematic hardening [Eq. (3.51)] describes the translation of the yield surface, incorporating backstress evolution between loading cycles.

$$\alpha_k = \frac{H_u}{\gamma_u}(1 - e^{-\gamma_u \varepsilon^p}) + \alpha_{k,1} e^{-\gamma_u \varepsilon^p} \tag{3.51}$$

where H_u is a parameter in the unit of stress, γ_u is a dimensionless parameter, α_k and $\alpha_{k,1}$ indicate the current backstress and the backstress at the beginning of the current cycle, respectively. Table 3.7 summarizes the material parameters for cyclic behavior characterization.

Monotonic and cyclic fracture tests are performed on CHS X-joints to evaluate their ultimate capacity. Figure 3.35 presents the joint configuration, with an 8 mm prefabricated notch near the weld toe at the crown point. Prior to testing, a fatigue pre-cracking phase extends the notch by 3–5 mm over ~5000 cycles at a 150 kN load amplitude.

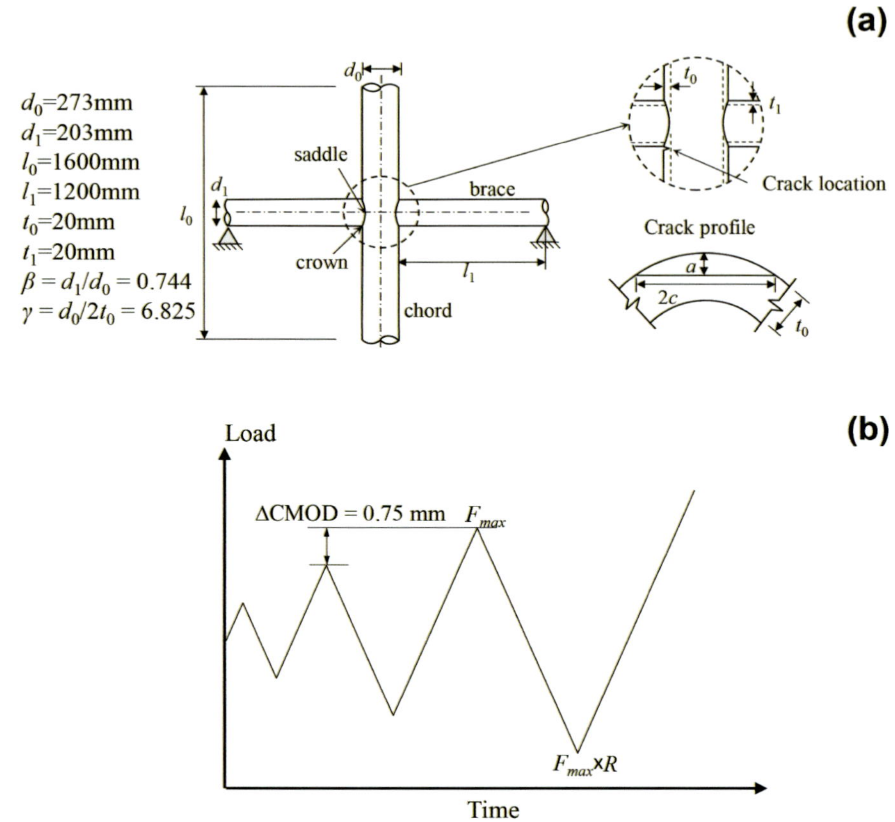

FIGURE 3.35 (a) Configuration of CHS X-joint under in-plane bending; and (b) reversed loading protocol.

For the final fracture tests, the setup applies vertical displacement-controlled loading at 5 mm/min, inducing in-plane bending in both monotonic and cyclic loading conditions (Figure 3.35b). One monotonic and two cyclic fracture tests are conducted with load ratios (R) of 0 and −1. A COD gauge measures *CMOD* during testing.

Post-fatigue, the crack front exhibits a semi-elliptical shape, which is assumed in subsequent tearing assessments. For deviations from this assumption, non-destructive evaluation techniques – such as ultrasonic phased array (UPA) [15] and strain-based methods [20] – can accurately determine crack-front geometry.

Figure 3.36 presents *P-CMOD* curves for cracked joints under reversed in-plane bending at varying load ratios. Monotonic and cyclic tests with $R = 0$ (Figures 3.36a and b) display comparable ultimate strength and ductility, while the $R = -1$ specimen (Figure 3.36c) exhibits reduced load capacity due to significant crack extension in the weldment after five cycles. These experimental findings provide critical insights into the fracture behavior of CHS X-joints under reversed loading, supporting further assessments using tearing analysis methodologies.

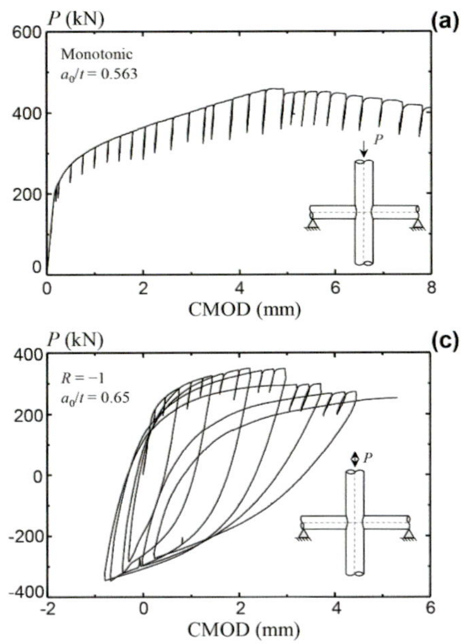

FIGURE 3.36 *P-CMOD* curves measured during the experiment for CHS X-joint under: (a) monotonic loading; (b) reversed loading with $R = 0$; and (c) reversed loading with $R = -1$.

3.6.2 Numerical Approach

This study extends the η-based approach to characterize the fracture resistance (i.e., the *J–R* curve) of a CHS X-joint under monotonic and reversed loadings. In this approach, the total J value is determined by combining the linear-elastic component (J_e) with the plastic component, J_{pl}, estimated using an η factor, the net area of the remaining ligament ($A_{liagment}$) and the plastic part of the potential energy (U_{pl}),

$$J = J_{el} + J_{pl} = \frac{(1-v^2)K_I{}^2}{E} + \eta \frac{U_{pl}}{A_{ligament}} \tag{3.52}$$

The investigation proceeds in two stages:

1. A linear-elastic FE analysis to obtain expressions for J_e and compliance.
2. A large-deformation, elastic-plastic FE analysis to calibrate and validate the η equation.

Detailed definitions of $A_{liagment}$ for surface-cracked pipes have been discussed in Parool [47]. Figure 3.37 presents the U_{pl} for specimens under monotonic loading and cyclic loadings.

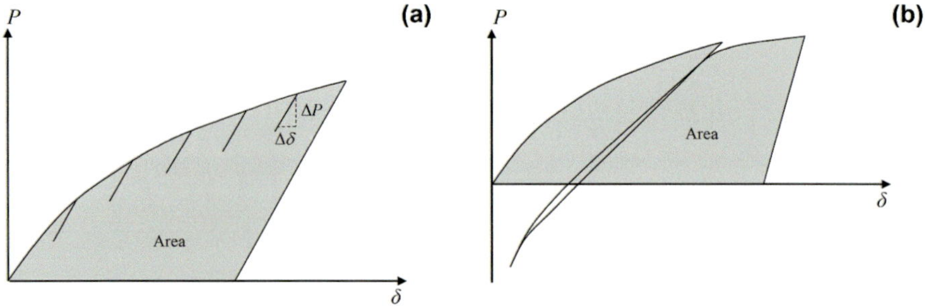

FIGURE 3.37 (a) Load-deformation curve with compliance measurement for single specimen technique; and (b) cyclic load-deformation curve from a fracture test

A series of three-dimensional FE models was developed in MSC Patran [52] using 20-node hexahedral elements. Each model represents a CHS X-joint containing a stationary crack of varying sizes (10–15 mm in 1 mm increments). Because the crack width (along $2c$) showed negligible growth during testing, the in-plane crack length was fixed at 92 mm. The numerical modeling follows a similar strategy as that shown in Figure 3.32b, with a small root radius of 0.1 mm at the crack tip and mesh tying to link the localized crack-front mesh to the global model. The domain integral approach in WARP3D [30] is employed for computing both the linear-elastic and elastic-plastic J values, with large-deformation theory adopted in the elastic-plastic simulations. During testing, the crack grows primarily in the depth direction (a), while the in-plane length ($2c$) experiences only minimal change. Hence, the J value at the deepest point is used as the representative driving force for crack growth in this geometry.

From the linear-elastic FEA, a relationship is established between J_e and the applied load P. This result is expressed in Eq. (3.48), where a dimensionless factor $f(a/t)$ characterizes the geometry, chord thickness (t), crack size (a), chord cross-sectional area (A), the applied load (P), and Young's modulus (E).

A polynomial curve fit of this dimensionless factor $f(a/t)$ yields Eq. (3.53).

$$f\left(a/t\right) = -1081.3\left(a/t\right)^2 - 3124.6(a/t) + 4202 \tag{3.53}$$

In addition, the *CMOD*-based compliance obtained numerically matches closely with the expression in Eq. (3.54).

$$C(a/t) = [-0.5679\left(a/t\right)^2 + 1.6345\left(a/t\right) - 0.0981] \times 10^{-3} \tag{3.54}$$

Next, large-deformation, elastic-plastic analyses are used to link the J-integral to the area under the load–deformation curve. Specifically, J is expressed via the plastic potential energy,

$$J_{pl} = \int_0^{\delta_{pl}} \frac{\eta P}{2c(t_0 - a)} d\delta_{pl} = \frac{\eta A_{pl}}{2c(t_0 - a)} \tag{3.55}$$

where $2c$ is the crack length, t_0 is the chord thickness, δ_{pl} is the plastic deformation, and A_{pl} is the plastic portion of the load–displacement area. LLD and *CMOD* both serve as deformation measures – LLD captures global deflection, while *CMOD* reflects local crack-tip opening. Because the experimental data records load versus *CMOD*, the elastic-plastic FEA focus on generating *P-CMOD* curves and computing the corresponding *J*-integral via domain integration. The plastic area beneath each *P–CMOD* curve is then used to calculate J_{pl}, and subsequently η from Eq. (3.55). A polynomial fit of these η values leads to,

$$\eta(a/t) = 15.9104\left(a/t\right)^3 - 28.2799\left(a/t\right)^2 + 15.0363\left(a/t\right) - 1.8477 \tag{3.56}$$

Because the plastic area is small at low loads, η can exhibit large fluctuations early in loading. However, η converges to a stable value as plasticity becomes more significant. Hence, an averaged η from the converged region is used to represent the behavior at each crack ratio. This η expression applies specifically to the tested CHS X-joint configuration.

3.6.3 Cyclic *J-R* Curves of Joints

The experimental program consisted of one monotonic fracture test and two cyclic fracture tests, conducted at load ratios $R = 0$ and $R = -1$. Figure 3.38a shows a representative crack surface through the thickness of the chord. Three boundaries on this surface divide the overall fracture process into four stages. The red boundary denotes the crack front of the machined notch before fatigue pre-cracking, while the blue boundary marks the crack front following the fatigue test. Initially, the material ahead of the fatigue crack front undergoes some ductile tearing. When the crack advances to the yellow boundary, new cracks form along the weldment at the chord–brace intersection. Beyond this boundary, the crack propagates in a different direction, placing the post-yellow-boundary crack surface in a different plane from the portion before it.

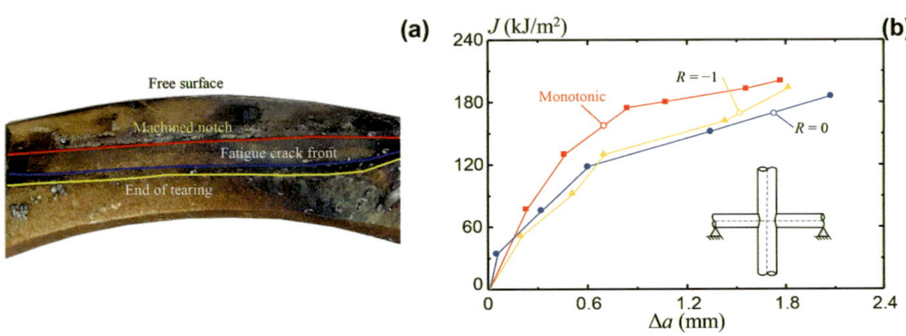

FIGURE 3.38 (a) Fracture surfaces of a CHS X-joint; and (b) *J-R* curves measured from the CHS X-joints.

The emergence of multiple cracks and a shift in crack orientation complicate the *J-R* curve measurements and demand considerable effort to capture their effects on fracture resistance. In fact, the resulting data cannot meet the requirements of the modified single-specimen approach. Therefore, in this study, the *J-R* curve is only calculated up to the point where the crack reaches the yellow boundary – prior to the change in crack orientation and the onset of multiple cracks.

Since the η value depends primarily on geometry and loading conditions, the same η value is used for both the cyclic and monotonic tests of the joint specimen. Figure 3.38b presents the *J-R* curve derived from the monotonic *P-CMOD* record and the cyclic *J-R* curves, which are constructed from the upper envelope of the cyclic *P-CMOD* data.

3.6.4 Tearing Assessment for Joints under Reversed Loadings

This study applies a TAD-based approach in Section 3.5 to evaluate the cracked CHS X-joint under both monotonic and reversed in-plane bending. The assessment proceeds in two ways: first, using the experimental TAC and *J-R* curves derived from tests on the CHS joints; and second, by coupling the numerical J_e solution for the joint with the TAC obtained from the experimentally measured *J-R* curve of fracture specimens. Table 3.8 lists the parameters used for each approach.

Following the TAD-based procedure, the assessment begins by constructing the TAC from the load–deformation curve of the joint. To determine the load ratio L_r for both the TAC and the load path, it is first necessary to calculate the ultimate load P_u of the cracked joint. According to the IIW [24], the bending capacity M_{ipb} for an intact X-joint subjected to brace in-plane bending is given by Eq. (3.44).

Once P_u is obtained, the TAD-based assessment proceeds. The load ratio L_r at each crack size is calculated by dividing the applied load by P_u. The fracture ratio K_r couples the joint's linear-elastic J_e with the driving force J_{ep} at the applied load. Six load paths are then constructed for crack extensions ranging from 0.2 mm to 1.7 mm in increments of 0.3 mm, using J_e from the joint and the fracture toughness $J_{mat,\Delta a_j}$ from the experimentally measured *J-R* curve. Figure 3.38a illustrates the intersection of the load paths with the TAC for the joint under reversed loading with $R = 0$.

Figures 3.39b and d compare the experimental load–crack extension (P–Δa) relationships with those predicted by the TAD approach for each load ratio R. Under

TABLE 3.8

Experimental Records and Numerical Solutions Used in TAD.

	Load path		TAC
Assessment	**J_e derived from the FE model of**	**J_{mat} extracted from the *J-R* curve of**	**Derived from the *P-δ* curve of**
Joint-based assessment	Joint	Joint	Joint
SE(T)-based assessment	Joint	SE(T) specimen	SE(T) specimen

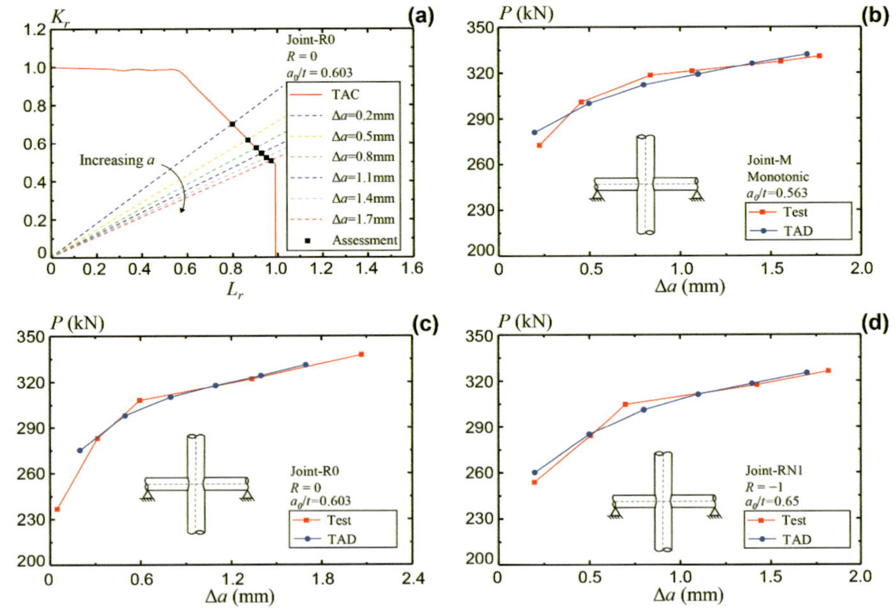

FIGURE 3.39 (a) TAD assessment for a CHS X-joint; and comparisons of P-Δa curves from test and TAD for CHS joints under: (b) monotonic load; (c) reversed load with $R = 0$; and (d) reversed load with $R = -1$.

monotonic loading, the evaluated curve shows good agreement with a maximum difference of about 5.5% at crack initiation. For Joint-R0, the estimated load at initiation is slightly higher than the experimental record; however, beyond 0.5 mm crack extension, the difference becomes negligible. Figure 3.39d also shows that Joint-RN1's TAD estimate matches the experimental response well, with a maximum discrepancy of only 4%. Overall, these comparisons demonstrate the validity of the proposed TAD-based method for the examined CHS X-joints under monotonic and reversed in-plane bending.

Since direct measurements of J-R curve on tubular joints are not always feasible, the below discussions present a TAD assessment approach using the J-R curve measured from fracture experiments. Here, the experimental study utilizes SE(T) specimens made of Q345 steel and demonstrates how their experimental records can be combined with the numerical J_e solution for CHS joints in a tearing assessment. Figure 3.40a shows the SE(T) specimen geometry, featuring a notch depth of 10 mm. Initially, cyclic loading at a frequency of 5 Hz is applied to induce a 2–4 mm fatigue crack ahead of the notch. After the pre-cracking, the fracture tests proceed under displacement control at a rate of 2 mm/min. Reflecting the loading conditions used for the joints, the SE(T) specimens are tested under monotonic loading and reversed loading with load ratios $R = 0$ and R=–1. As illustrated in Figure 3.40b, the displacement-controlled reversed loading uses three displacement increments, 0.2 mm, 0.4 mm, and 0.6 mm in *CMOD*, following Chowdhury *et al.* [55], who observed that smaller displacement increments increase the number of load cycles and can affect fracture resistance. Table 3.9 summarizes the test specimens under different loading conditions, while Figures 3.40c–d shows the resulting *P-CMOD* curves for monotonic and reversed loadings, respectively.

TABLE 3.9

Information of SE(T) Specimens.

Specimen no.	Notch size a_N (mm)	Initial crack size a_0 (mm)	Load ratio *R*	δ increment $\Delta CMOD$ (mm)	*J-R* curve $J = C_1(\Delta a)^{C2}$	
					C_1	C_2
SETM	10	12.75	Monotonic	\	697.64	0.89
SETR0	10	13.26	0	0.4	704.74	0.79
SETRN1–0.2	10	13.47	−1	0.2	255.55	0.69
SETRN1–0.4	10	13.05	−1	0.4	432.35	0.49
SETRN1–0.6	10	13.17	−1	0.6	408.11	0.57

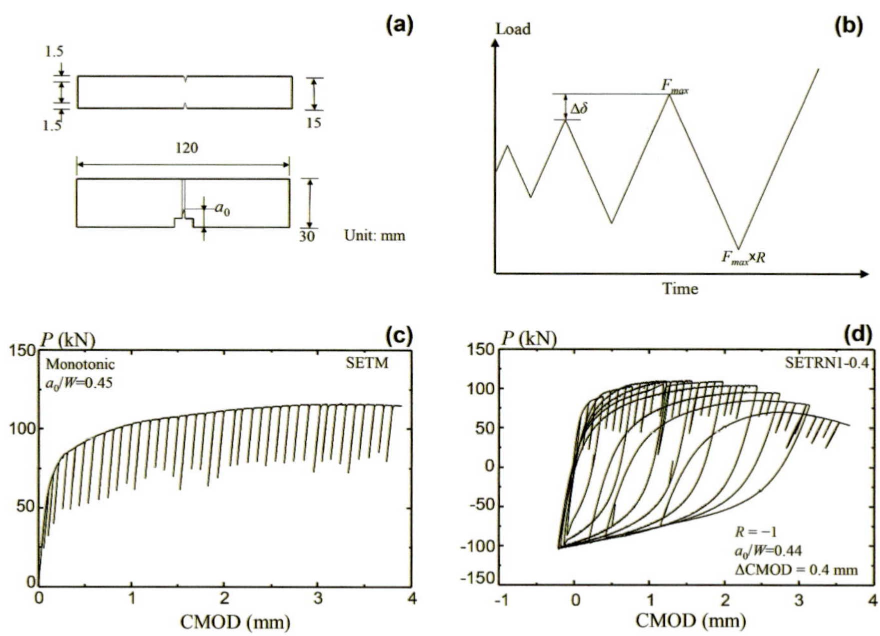

FIGURE 3.40 (a) Configuration of SE(T) specimens; (b) loading protocol for the SE(T); and *P-CMOD* curves for SE(T) under (c) monotonic loading; and (d) reversed loading with $R = -1$.

In the TAD-based assessment for the SE(T) specimens, the TAC is derived entirely from experimental data. To construct the TAC, the denominator in the ratio K_r is determined with the η-based method for SE(T) specimens:

$$J = J_e + J_{pl} = \frac{(1-v^2)K_I{}^2}{E} + \eta\frac{U_{pl}}{B_N b} \tag{3.57}$$

where B_N is the net specimen thickness excluding the side groove, and b is the ligament length. Liu *et al.* [51] calculate the stress intensity factor K_I using Eq. (3.31). For side-grooved SE(T) specimens, the stress σ follows Eq. (3.33),

Numerical analyses are then performed to obtain the linear-elastic K_I solution and the η value for the elastic–plastic J_{ep}. Figure 3.41a illustrates the quarter-symmetric FE model, which employs 20-node hexahedral elements in MSC Patran [52]. Each FE model considers a stationary crack with sizes ranging from 12.5 mm to 18.5 mm.

The computed dimensionless function *f*(*a*/*W*) and the compliance *C*(*a*/*W*) (mm/kN), based on the FE results of SE(T) specimen in Figure 3.40a, follow the equations below,

$$f(a/W) = 1.8232(a/W)^2 + 0.1339(a/W) + 1.1169 \tag{3.58}$$

$$C(a/W) = \left[1.0377(a/W)^2 - 0.3850(a/W) + 0.1009\right] \times 10^{-2} \tag{3.59}$$

Figure 3.41b shows that the derived η remains close to 0.6 for $0.4 \le a/W \le 0.6$. Consequently, $\eta \approx 0.6$ is adopted when computing the fracture resistance curves. As shown in Figure 3.41c, these *J-R* curves reveal that cyclic loading with $R = 0$ slightly

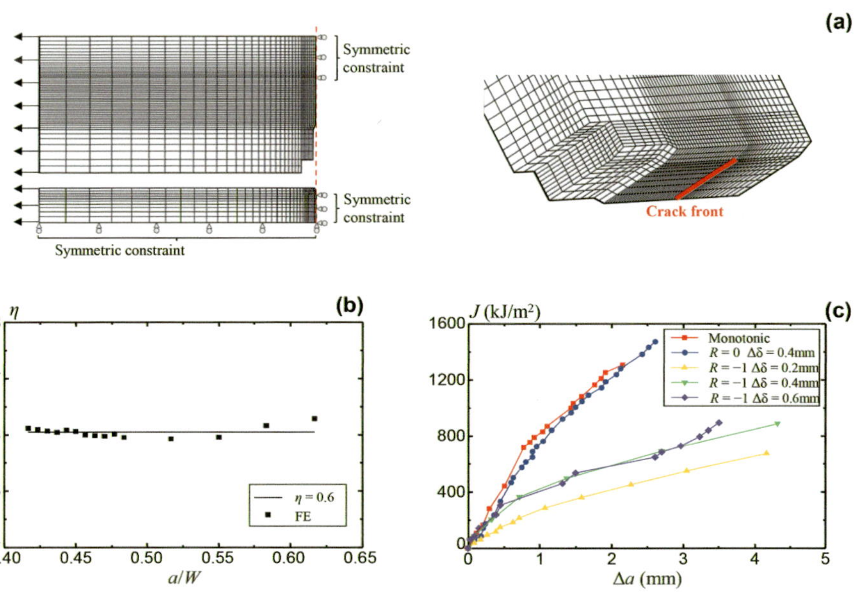

FIGURE 3.41 (a) FE mesh for SE(T); (b) η value for SE(T); and (c) *J-R* curves for the SE(T).

reduces the fracture resistance compared to monotonic loading, while fully reversed loading ($R = -1$) causes a significant reduction. Moreover, the chosen displacement increment strongly influences the curve: smaller increments introduce more cycles, promoting crack growth and altering the near-tip residual stresses, thereby reducing fracture resistance further than larger increments.

To extend the SE(T)-based TAD assessment to the CHS joints, the study constructs a TAC for the same load ratio (R) using only the SE(T) specimen data. The relevant parameters K_r and L_r for the TAC are calculated from the specimen's P-$CMOD$ response. According to Liu *et al.* [56],

$$P_u = \sigma_u \sqrt{BB_N}(W - a) \tag{3.60}$$

The fracture ratio couples the η-based linear-elastic solution K_I in (Eq. 3.31) with the elastic–plastic J. The resulting load path for the joint uses the structural ultimate load P_u [Eq. (3.47)], and the joint's fracture ratio incorporates the numerical J_e solution [Eq. (3.48)] and the SE(T)-derived J-R curve from Figure 3.41d. Table 3.8 summarizes all parameters and solution strategies used in the TAD-based assessments for both joint- and specimen-driven data. Because the specimen-derived TAD only requires the joint's linear-elastic solution – combined with SE(T) experimental results – it provides an efficient alternative to a fully numerical approach.

Figures 3.42a and b illustrate the specimen-derived TAD for monotonic loading and compare the experimental P-Δa of Joint-M with TAD predictions. The

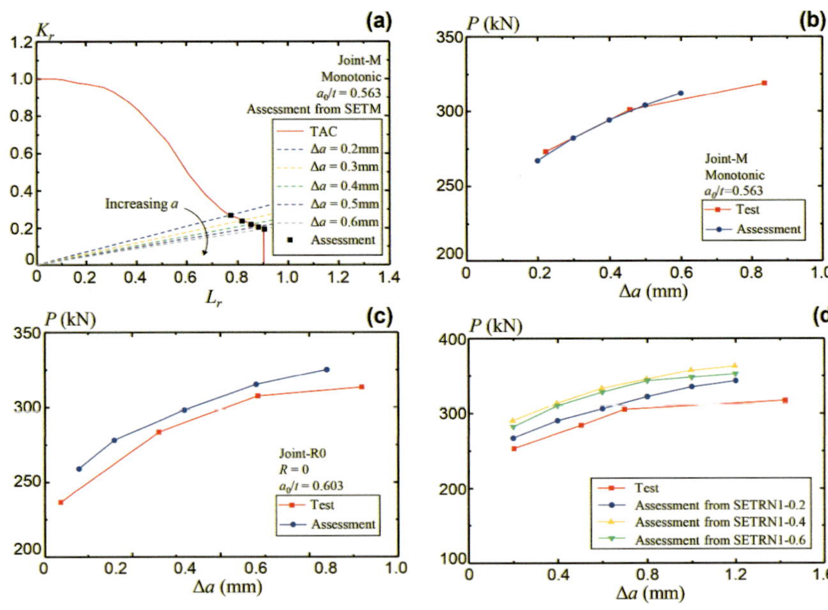

FIGURE 3.42 (a) TAD assessments on CHS X-joint under monotonic loading; and comparisons of the measured P-Δa curves and TAD estimations for CHS X-joints under: (b) monotonic loading; (c) reversed loading with $R = 0$; and (d) reversed loading with $R = -1$.

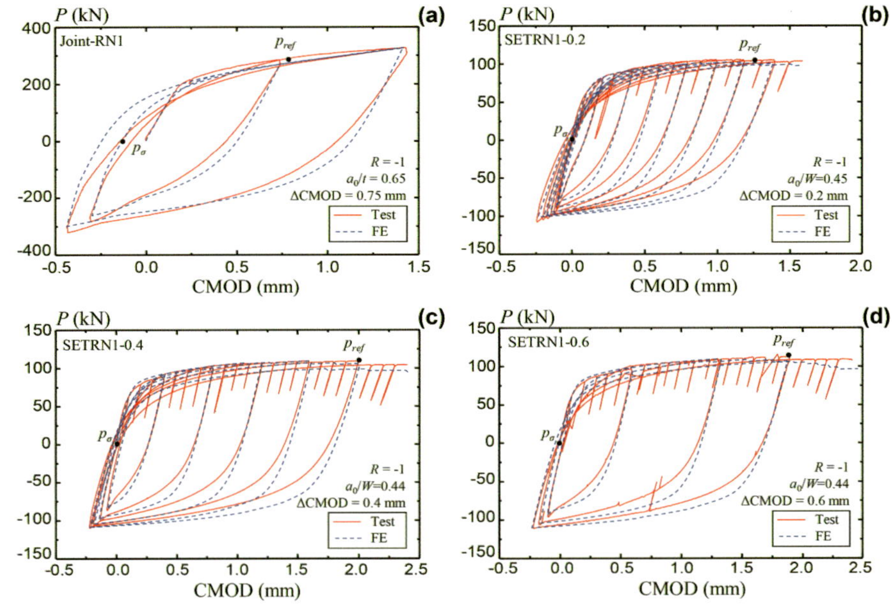

FIGURE 3.43 Comparison of the *P-CMOD* relationships from the test and FE under reversed loadings for: (a) X-joint; and SE(T) specimen with (b) $\Delta CMOD$ = 0.2 mm; (c) $\Delta CMOD$ = 0.4 mm; and (d) $\Delta CMOD$ = 0.6 mm.

excellent match validates the SE(T)-based approach. Under reversed loading with $R = 0$, Figure 3.42c shows that the TAD overestimates the joint's load resistance by approximately 3%. For fully reversed loading ($R = -1$), three TADs are constructed from SE(T) specimens tested at 0.2 mm, 0.4 mm, and 0.6 mm *CMOD* increments. As shown in Figure 3.42d, the SE(T)-based assessment using a 0.2 mm increment provides the closest prediction for Joint-RN1, overestimating the load resistance by about 7%, whereas using the larger increments yields overpredictions of 10%–15%.

Further numerical investigations clarify the causes of these predictive discrepancies for Joint-RN1 with $R = -1$ and different displacement increments. The FE models in Figures 3.32 and 3.41 were used for the joint and SE(T) specimens, respectively, simulating contact behavior to prevent crack-face interpenetration. Both the joint and specimens adopted the cyclic stress–strain curve in Figure 3.34b, and crack growth was introduced via the node-release method [51] in ABAQUS [57] (for the joint) and WARP3D [30] (for the SE(T) specimens). As shown in Figure 3.43, the numerical *P-CMOD* responses match the experimentally measured curves for Joint-RN1 and SE(T) specimens with different incremental *CMOD* values ($\Delta CMOD$) between adjacent reversed loading cycles.

When fully reversed fracture testing ($R = -1$) transitions from compressive to tensile loading, a high residual tensile stress remains at the crack tip even when the global load is zero [51]. This elevated tensile residual stress promotes crack growth early in the load cycle, reducing fracture resistance relative to monotonic loading. Hence, differences in the residual stress at the crack tip are a key factor in predicting fracture behavior under fully reversed cyclic loads.

REFERENCES

[1] L. Feng, T. Liu, X. Qian, C. Chen, A complete integrity assessment of welded connections under high and low cycle fatigue followed by fracture failure, Steel and Composite Structures, An International Journal, 43 (2022) 465–481.

[2] BS 7910: Guide to methods for assessing the acceptability of flaws in metallic structures: An engineering approach, in, British Standards Institution, 2019.

[3] API 579-1/ASME FFS-1, Fitness-For-Service, in, API Publishing, 2021.

[4] X. Qian, Y. Li, Z. Ou, Ductile tearing assessment of high-strength steel X-joints under in-plane bending, Engineering Failure Analysis, 28 (2013) 176–191.

[5] T.T. Seah, X. Qian, Y. Zhang, Constraint-based failure assessment for pipelines with coplanar circumferential cracks under biaxial loading, International Journal of Pressure Vessels and Piping, 190 (2021) 104320.

[6] S. Cicero, R. Lacalle, R. Cicero, D. Ferreno, Assessment of local thin areas in a marine pipeline by using the FITNET FFS corrosion module, International Journal of Pressure Vessels and Piping, 86 (2009) 329–334.

[7] B. Dogan, Fitness for service of welded components under creep and creep–fatigue loading, International Journal of Pressure Vessels and Piping, 87 (2010) 656–663.

[8] I. Dzioba, A. Neimitz, Application of the standard options of the FITNET procedure to the structural integrity assessment of welded specimens containing cracks, International Journal of Pressure Vessels and Piping, 84 (2007) 475–486.

[9] F. Burdekin, The static strength of cracked joints in tubular members. Offshore technology report OTO-2001/080, 2001.

[10] S. Lie, Z. Yang, W. Gho, Validation of BS7910: 2005 failure assessment diagrams for cracked square hollow section T-, Y-and K-joints, International Journal of Pressure Vessels and Piping, 86 (2009) 335–344.

[11] Z. Yang, S. Lie, W. Gho, Failure assessment of cracked square hollow section T-joints, International Journal of Pressure Vessels and Piping, 84 (2007) 244–255.

[12] U. Zerbst, J. Heerens, K.-H. Schwalbe, The fracture behaviour of a welded tubular joint – an ESIS TC1. 3 round robin on failure assessment methods: Part I: Experimental data base and brief summary of the results, Engineering Fracture Mechanics, 69 (2002) 1093–1110.

[13] M. Arandjelovic, B. Djordjevic, S. Sedmak, D. Radu, A. Petrovic, S. Dikic, A. Sedmak, Failure analysis of welded joint with multiple defects by extended finite element method and engineering critical analysis, Engineering Failure Analysis, 160 (2024) 108176.

[14] Y. Wang, A. MacDonald, L. Xu, M. Wright, R. Shenoi, Engineering critical assessment and variable sensitivity analysis for as-welded S690 steels, Engineering Failure Analysis, 109 (2020) 104282.

[15] L. Feng, X. Qian, Enhanced sizing for surface cracks in welded tubular joints using ultrasonic phased array and image processing, NDT & E International, 116 (2020) 102334.

[16] Y.-L. Zhou, X. Qian, A. Birnie, X.-L. Zhao, A reference free ultrasonic phased array to identify surface cracks in welded steel pipes based on transmissibility, International Journal of Pressure Vessels and Piping, 168 (2018) 66–78.

[17] E.W.W. Cheok, X. Qian, C. Chen, S.T. Quek, M.B.I. Si, A local digital twin approach for identifying, locating and sizing cracks in CHS X-joints subjected to brace axial loading, Engineering Structures, 299 (2024) 117085.
[18] E.W.W. Cheok, X. Qian, A. Kaps, S.T. Quek, M.B.I. Si, A strain-interfaced digital twin solution for corner fatigue crack growth using Bayesian inference, International Journal of Fatigue, 191 (2025) 108705.
[19] E.W.W. Cheok, X. Qian, S.T. Quek, M.B.I. Si, A digital twin approach toward unlocking elasto-plastic fatigue crack growth rates under negative load ratios, Engineering Fracture Mechanics, 301 (2024) 110034.
[20] L. Feng, X. Qian, Enhanced real-time crack monitoring and updating in welded structural components with limited measurement data, Mechanical Systems and Signal Processing, 198 (2023) 110403.
[21] L. Feng, X. Qian, An adaptive learning approach to determine and update crack sizes from strain relaxation data for welded plate joints, Engineering Fracture Mechanics, 259 (2022) 108165.
[22] T. Liu, X. Qian, W. Wang, Y. Chen, A ductile tearing assessment diagram to estimate load resistance versus crack extension for welded connections with surface cracks, Thin-Walled Structures, 169 (2021) 108435.
[23] T. Liu, X. Qian, W. Wang, Y. Chen, A ductile tearing assessment diagram for cracked circular-hollow-section joints under reversed loadings, Marine Structures, 89 (2023) 103394.
[24] I.I.o.W. (IIW), Static design procedure for welded hollow section joints – Recommendations, in, International Institute of Welding, 2009.
[25] M. Cheaitani, F. Burdekin, Ultimate strength of cracked tubular joints, in: Tubular structures, Routledge, 2021, pp. 607–616.
[26] I. Hadley, A. Dyer, G. Booth, M. Cheaitani, F. Burdekin, G. Yang, Static strength of cracked tubular joints: New data and models, in: OMAE 1998: 17 th international conference on offshore mechanics and arctic engineering, 1998, pp. 1998.
[27] A. Stacey, J. Sharp, N. Nichols, Static strength assessment of cracked tubular joints, in, American Society of Mechanical Engineers, New York, NY (United States), 1996.
[28] X. Wang, J.-F. Wu, X. Qian, Crack driving force and ultimate strength of thick-walled CHS X-joints with near-toe cracks, in: ISOPE international ocean and polar engineering conference, ISOPE, 2011, pp. ISOPE-I-11-203.
[29] L.d. Lu, G. De Winkel, Y. Yu, J. Wardenier, Deformation limit for the ultimate strength of hollow section joints, in: Tubular Structures, Routledge, 2021, pp. 341–347.
[30] A. Gullerud, K. Koppenhoefer, A. Roy, S. RoyChowdhury, M. Walters, B. Bichon, K. Cochran, R. Dodds, WARP3D: 3-D dynamic nonlinear fracture analysis of solids using parallel computers and workstations, Structural Research Series (SRS), 607 (2004).
[31] F. Burdekin, G. Yang, J. Cao, Assessment of fracture strength of cracked tubular joints, Report OTH 554, Health and Safety Executive, 1998.
[32] B. Talie-Faz, W. Dover, F. Brennan, Static strength of cracked high strength steel tubular joints, Report OTH 562, Health and Safety Executive, 2002.
[33] H.a.S.E. HSE, Health and safety executive offshore installations: Guidance on design construction and certification, in, HSE, 1995.
[34] G. Van der Vegte, Y. Makino, Further research on chord length and boundary conditions of CHS T-and X-joints, Advanced Steel Construction, 6 (2010) 879–890.
[35] I. Milne, R. Ainsworth, A. Dowling, A. Stewart, Assessment of the integrity of structures containing defects, International Journal of Pressure Vessels and Piping, 32 (1988) 3–104.

[36] X. Qian, R.H. Dodds Jr, Y. Choo, Mode mixity for circular hollow section X joints with weld toe cracks, Journal of Offshore Mechanics and Arctic Engineering, 127 (2005) 269–279.
[37] X. Qian, R.H. Dodds, Y. Choo, Elastic–plastic crack driving force for tubular K-joints with mismatched welds, Engineering Structures, 29 (2007) 865–879.
[38] X. Qian, KI-T estimations for embedded flaws in pipes–Part I: Axially oriented cracks, International Journal of Pressure Vessels and Piping, 87 (2010) 134–149.
[39] X. Qian, KI-T estimation for embedded flaws in pipes – Part II: Circumferentially oriented cracks, International Journal of Pressure Vessels and Piping, 87 (2010) 150–164.
[40] F. Ashkar, F. Aucoin, Discriminating between the lognormal and the log-logistic distributions for hydrological frequency analysis, Journal of Hydrologic Engineering, 17 (2012) 160–167.
[41] D.D. Joseph, B.H. Yang, Friction factor correlations for laminar, transition and turbulent flow in smooth pipes, Physica D: Nonlinear Phenomena, 239 (2010) 1318–1328.
[42] F.K. Wang, T.P. Chu, The mean time between failures for an LCD panel, Quality and Reliability Engineering International, 27 (2011) 203–208.
[43] ASTME8, ASTM E8–22, Standard test methods for tension testing of metallic materials, in, ASTM International, West Conshohocken, PA, 2022.
[44] ASTME1820, ASTM E1820–22, Standard test method for measurement of fracture toughness, in, ASTM International, West Conshohocken, PA, 2022.
[45] Quest Integrity Group, FEA CRACK version 3.2 users' manual, 2011.
[46] X. Qian, A. Romeijn, J. Wardenier, Y. Choo, An automatic FE mesh generator for CHS tubular joints, in: ISOPE international ocean and polar engineering conference, ISOPE, 2002, pp. ISOPE-I-02-392.
[47] N. Parool, X. Qian, C. Koh, An η-compliance method to estimate the J-Δa curve for pipes with a circumferential surface crack, Fatigue & Fracture of Engineering Materials & Structures, 40 (2017) 1624–1639.
[48] N. Parool, X. Qian, C. Koh, A modified hybrid method to estimate fracture resistance curve for pipes with a circumferential surface crack, Engineering Fracture Mechanics, 188 (2018) 1–19.
[49] Y. Zhang, X. Qian, An eta-approach to evaluate the elastic–plastic energy release rate for weld-toe cracks in tubular K-joints, Engineering Structures, 51 (2013) 88–98.
[50] X.-K. Zhu, J.A. Joyce, J–Resistance curve testing of HY80 steel using SE (B) specimens and normalization method, Engineering Fracture Mechanics, 74 (2007) 2263–2281.
[51] T. Liu, X. Qian, W. Wang, Y. Chen, Fracture resistance curve for single edge notched tension specimens under low cycle actions, Engineering Fracture Mechanics, 211 (2019) 47–60.
[52] MSC Software Corporation, MSC. Patran user's guides and reference manuals, 2001.
[53] G. Van der Vegte, J. Wardenier, X. Qian, Y. Choo, Re-evaluation of the moment capacity of CHS joints, Proceedings of the Institution of Civil Engineers-Structures and Buildings, 163 (2010) 439–448.
[54] ASTME606, ASTM E606–12, Standard test method for strain-controlled fatigue testing, in, ASTM International, West Conshohocken, PA, 2012.
[55] T. Chowdhury, S. Sivaprasad, H. Bar, S. Tarafder, N. Bandyopadhyay, Cyclic fracture behaviour of 20MnMoNi55 steel at room and elevated temperatures, Fatigue & Fracture of Engineering Materials & Structures, 38 (2015) 813–827.
[56] T. Liu, X. Qian, W. Wang, Y. Chen, A reversed η approach to estimate load-deformation curves for fracture specimens and surface-cracked pipes, Theoretical and Applied Fracture Mechanics, 106 (2020) 102485.
[57] A. Standard, User's manual, Hibbitt, Karlsson & Sorensen, Pawtucket, 1995.

4 Assessment of Brittle Fracture Failure

4.1 INTRODUCTION

Structural steels frequently experience cleavage fracture at low ambient temperatures with minimal or negligible prior plastic deformation. Such brittle fracture events often result in catastrophic consequences, causing significant economic, environmental, and human losses. Previous research has demonstrated that cleavage fracture occurrence is strongly influenced by the random distribution of inherent microscopic defects and the inhomogeneous microstructure ahead of the crack tip [1–4]. This microscopic variability contributes to significant scatter in measured macroscopic fracture toughness (K_{Jc} values), as extensively documented in experimental studies conducted at low temperatures.

In the ductile-to-brittle transition (DBT) region, accurately assessing fracture becomes increasingly complex due to the concurrent or competing mechanisms of cleavage and ductile fracture, particularly in specimens experiencing low plasticity constraints. Traditional statistical-based local methodologies for cleavage failure, though accurate, typically require complex procedures involving substantial experimental calibration to determine material-specific parameters. Therefore, structural integrity assessments necessitate simplified yet reliable approaches for determining these critical parameters in practical engineering applications.

Over the past decades, extensive research has developed statistical methodologies for assessing cleavage fracture through both macroscopic and microscopic (local) approaches. The master curve method has become a standardized procedure, widely used to characterize the temperature-dependent macroscopic fracture toughness across the DBT region. The ASTM E-1921 [5] standard provides guidelines to determine the reference temperature (T_0), crucial for quantifying median fracture toughness ($K_{Jc,mean}$). At a specific temperature within the DBT zone, fracture probability typically follows a three-parameter Weibull distribution model based on the macroscopic fracture toughness. Application of the master curve to specimens with reduced constraint conditions typically involves a constraint adjustment using the ratio of *T*-stress to the material yield strength.

At the microscopic level, the Weibull stress model, initially proposed by Beremin's group [6], is widely recognized as an effective approach to assess cleavage fracture probability. This model incorporates the weakest-link concept with the distribution of microscopic defects ahead of the crack tip, leading to the definition of a scalar microscopic fracture-driving parameter, the Weibull stress (σ_w). Subsequent modifications, notably the introduction of a threshold Weibull stress ($\sigma_{w\text{-}min}$) by Gao *et al.* [7], have enhanced the model's ability to account for observed fracture non-occurrences below specific macroscopic driving forces. Further developments

DOI: 10.1201/9781003509806-4

by Petti and Dodds [8] established correlations between microscopic Weibull stress (σ_w) and macroscopic fracture toughness (K_{JI}), incorporating a constraint function to address plasticity-induced constraint loss. Many studies have validated the three-parameter Weibull stress model through strong correlations with experimentally measured macroscopic fracture toughness.

However, calibration of the three-parameter Weibull stress model typically demands extensive experimental data and meticulous instrumentation. Minami *et al.* [9] simplified the process by using a single high-constraint dataset to determine the Weibull modulus (m) and the Weibull scale parameter (σ_u) via the maximum likelihood method. Nevertheless, Gao *et al.* [10] and Ruggieri *et al.* [11] have identified potential uniqueness issues associated with using a single dataset for determining these Weibull parameters. Subsequent improvements have proposed a rigorous calibration process involving experimental data from specimens with distinctly different crack-front constraints to robustly characterize cleavage fracture behavior.

4.2 ASTM AND WEIBULL STRESS APPROACH

4.2.1 Basic Framework of ASTM

ASTM E-1921 [5] recommends using a three-parameter Weibull distribution to characterize the observed scatter in experimental fracture toughness values:

$$P_f = 1 - \exp\left[-\left(\frac{K_{Jc} - K_{min}}{K_0 - K_{min}}\right)^4\right] \tag{4.1}$$

In this formulation, K_{min} represents the threshold fracture toughness below which cleavage fracture is not observed, typically set to 20 $\text{MPa}\sqrt{\text{m}}$ according to ASTM E-1921 [5]. This threshold value is generally considered invariant across different materials, specimen configurations, and temperatures.

The maximum likelihood method enables the estimation of the temperature-dependent scale parameter, K_0:

$$K_0 = \left[\sum_{i=1}^{n} \frac{(K_{Jc(i)} - K_{min})^4}{r}\right]^{1/4} + K_{min} \tag{4.2}$$

Here, n is the total number of specimens tested, and r represents the number of valid fracture toughness data points.

To ensure high-constraint conditions and SSY ahead of the crack front, ASTM E-1921 [5] sets a criterion to exclude specimens exhibiting significant plastic deformation and fracture toughness exceeding a limiting value, $K_{Jc(limit)}$:

$$K_{Jc(\text{limit})} = \sqrt{\frac{E b_0 \sigma_{ys}}{M_{\text{limit}}(1 - \nu^2)}} \tag{4.3}$$

where b_0, E, v, and σ_{ys} denote the initial remaining ligament, Young's modulus, Poisson's ratio, and material yield strength, respectively. Data points exceeding the limit defined in Eq. (4.3) are excluded from validity. ASTM E1921 [5] recommends a cut-off value of $M_{\text{limit}} = 30$ to ensure a uniform high-constraint, J-dominated crack-front stress field.

Fracture toughness measurements derived from specimens with thicknesses different from the standard 1T (1 inch or 25.4 mm) must undergo a thickness correction [12]:

$$K_{Jc(1T)} = K_{\min} + \left(K_{Jc(xT)} - K_{\min}\right)\left(\frac{B_{xT}}{B_{1T}}\right)^{1/4} \tag{4.4}$$

where B_{xT} indicates the specimen thickness in multiples of 1T.

Finally, ASTM E-1921 [5] defines the Master Curve – a transition temperature curve correlating median fracture toughness with the test temperature – as follows:

$$K_{Jc(\text{med})} = 30 + 70\exp\left[0.019\left(T - T_0\right)\right] \tag{4.5}$$

where T_0 is the reference temperature at which the median fracture toughness of a 1T-thick specimen equals 100 $\text{MPa}\sqrt{\text{m}}$.

4.2.2 Basic Framework of Weibull Stress Method

The two-parameter Weibull stress model proposed by Beremin's group [6] integrates the distribution of micro-crack sizes into the cumulative probability of cleavage fracture as follows:

$$P_f\left(\sigma_w\right) = 1 - \exp\left[-\left(\frac{\sigma_w}{\sigma_u}\right)^m\right] \tag{4.6}$$

In this equation, the Weibull modulus m describes the temperature-independent distribution of micro-crack sizes near the crack front. Typical reported values for ferritic steels range from 10 to 20 [9, 11, 13]. The Weibull stress scale parameter σ_u incorporates both micro-crack size distribution and micro-crack toughness. The scalar Weibull stress (σ_w) is computed as:

$$\sigma_w = \left[\frac{1}{V_0}\int_V \sigma_{eff}^m \, dV\right]^{1/m} \tag{4.7}$$

where V_0 represents a reference volume of 16,378 mm³ (1 inch³) and denotes the fracture process zone ahead of the crack front where the effective stress exceeds a specific threshold, or, $\sigma_{eff} > \lambda\sigma_{ys}$. Consistent with previous studies [14], this work adopts an λ value of 2.0 and defines the maximum principal stress as the effective stress, that is, $\sigma_{eff} = \sigma_1$.

Under high-constraint and SSY conditions prescribed by ASTM E-1921 [5], a uniform J- or K-dominant stress field emerges ahead of the crack front, establishing a direct correlation between the global crack driving force (K_J or J) and local Weibull stress (σ_w):

$$\sigma_w{}^m = CBK_J^4 \text{ or } \sigma_w{}^m = \bar{C}BJ^2 \tag{4.8}$$

Here, C and $\bar{C}$ are constants dependent on material properties and Weibull parameters. For low-constraint conditions, Gao *et al.* [7] introduced a non-dimensional constraint correction function. $g(M)$, to quantify geometry- and plasticity-induced constraint loss:

$$\sigma_w{}^m = CBK_J^4 g(M) \text{ or } \sigma_w{}^m = \bar{C}BJ_{avg}^2 g(M) \tag{4.9}$$

where J_{avg} represents the through-thickness averaged energy release rate, and $g(M)$ equals 1.0 under SSY and plane-strain conditions with zero T-stress.

Integrating Eq. (4.8) into the global Weibull model [Eq. (4.1)], Petti and Dodds [8] proposed a modified three-parameter Weibull stress model:

$$P_f(\sigma_w) = 1 - \exp\left[-\left(\frac{\sigma_w{}^{m/4} - \sigma_{w\text{-}min}{}^{m/4}}{\sigma_u{}^{m/4} - \sigma_{w\text{-}min}{}^{m/4}}\right)^4\right] \tag{4.10}$$

where $\sigma_{w\text{-}min}$ denotes the threshold Weibull stress below which cleavage fracture does not occur (i.e., $Pf(\sigma_w) = 0$ for $\sigma_w \leq \sigma_{w\text{-}min}$). Experimental Weibull stress data must be corrected to 1T thickness under SSY conditions for consistent comparisons across datasets:

$$K_{Jc}^{1\text{T}} = K_{min} + \left[K_{Jc}^{(x\text{T})} g\left(M^{(x\text{T})}\right)^{1/4} - K_{min} g\left(M_{min}^{(x\text{T})}\right)^{1/4}\right]\left(\frac{B_{x\text{T}}}{B_{1\text{T}}}\right)^{1/4} \tag{4.11}$$

Here, B_{xT} denotes the specimen thickness.

4.3 CALIBRATION OF WEIBULL STRESS PARAMETERS

4.3.1 Simplified Calibration Procedure

This section introduces a simplified calibration method that requires only two macroscopic fracture toughness values, each corresponding to a 63.2% cumulative probability of fracture, from two specimen sets exhibiting pronounced differences in crack-front constraint. The proposed approach is validated using C(T) specimens of varying thicknesses from the recent Euro fracture toughness dataset [15], as well as Charpy V-notch (CVN) specimens fabricated from the same Euro materials [16].

By significantly reducing the experimental demands typically required for Weibull stress parameter calibration, this method offers a practical and validated alternative for materials where fracture toughness data are limited or difficult to obtain.

This simplified calibration method significantly reduces the experimental burden by generating two fictitious fracture toughness datasets from two reported K_0 values. The method assumes that specimens governed by cleavage fracture (*i.e.*, without extensive prior plastic deformation) conform to the macroscopic Weibull model described in Eq. (4.1). To address the non-uniqueness issue in Weibull parameter calibration [7, 11], the proposed procedure requires two K_0 values corresponding to two specimen sets with clearly different crack-front constraints. The fictitious datasets are generated assuming a fixed cumulative probability value of 63.2%, as recommended in ASTM E1921 [5].

The key steps in the simplified calibration procedure of (m, K_{min}) are shown in Figure 4.1, as follows:

1. Experimental test data input: The method requires two fracture toughness values $(K_0^{\text{HC}}, K_0^{\text{LC}})$, each representing the 63.2% cumulative probability level, obtained from specimen sets with differing crack-front constraint levels (high constraint and low constraint). To ensure valid cleavage-dominated fracture behavior, an M-value threshold (M_{limit}) of 10 is applied, excluding data with excessive plastic deformation, following Wasiluk *et al.* [17]. If any specimens with $M < 10$ are included in determining K_0, the associated valid-data ratio (r/n, i.e., the number of valid data over the total number of data) must also be specified. For engineering applications, ASTM E1921 typically recommends using datasets with a valid-data ratio (r/n) of 1.0.
2. Generation of fictitious datasets: Using Eq. (4.1) and the two K_0 values, two synthetic fracture toughness datasets $(K_{Jc}^{\text{HC}fic(i)}, K_{Jc}^{\text{LC}fic(i)})$ are generated under the assumption value of K_{min} (20 $\text{MPa}\sqrt{\text{mm}}$). The corresponding rank probabilities $P_{i\text{-}rank}$ are computed as:

$$P_{i-rank} = \left(\frac{i-0.3}{n+0.4}\right)\left(\frac{r}{n}\right) \tag{4.12}$$

   where i is the rank number and n is the total number of fictitious data points.
3. Constraint function $g(M)$ determination by FEA: FEA are performed to determine the $\sigma_w - J_{avg}$ relationships for both specimen geometries [Eq. (4.9)] and the SSY model based on [Eq. (4.8)]. These are then used to derive the constraint correction functions $g(M)$ for each specimen set. The FEA uses stress-strain data measured at the same temperature as that used to determine the toughness values.
4. Scaling to SSY condition: The constraint correction functions $g(M)$ obtained in Step 3 are used to transform the two sets of fictitious toughness data to equivalent 1T SSY-based datasets, denoted as $K_{Jc}^{\text{HC}-SSY(1\text{T})}$ and $K_{Jc}^{\text{LC}-SSY(1\text{T})}$, for the high- and low-constraint datasets, respectively, using Eq. (4.11). Equation (4.1) and (4.12) generates a fictitious dataset of fracture

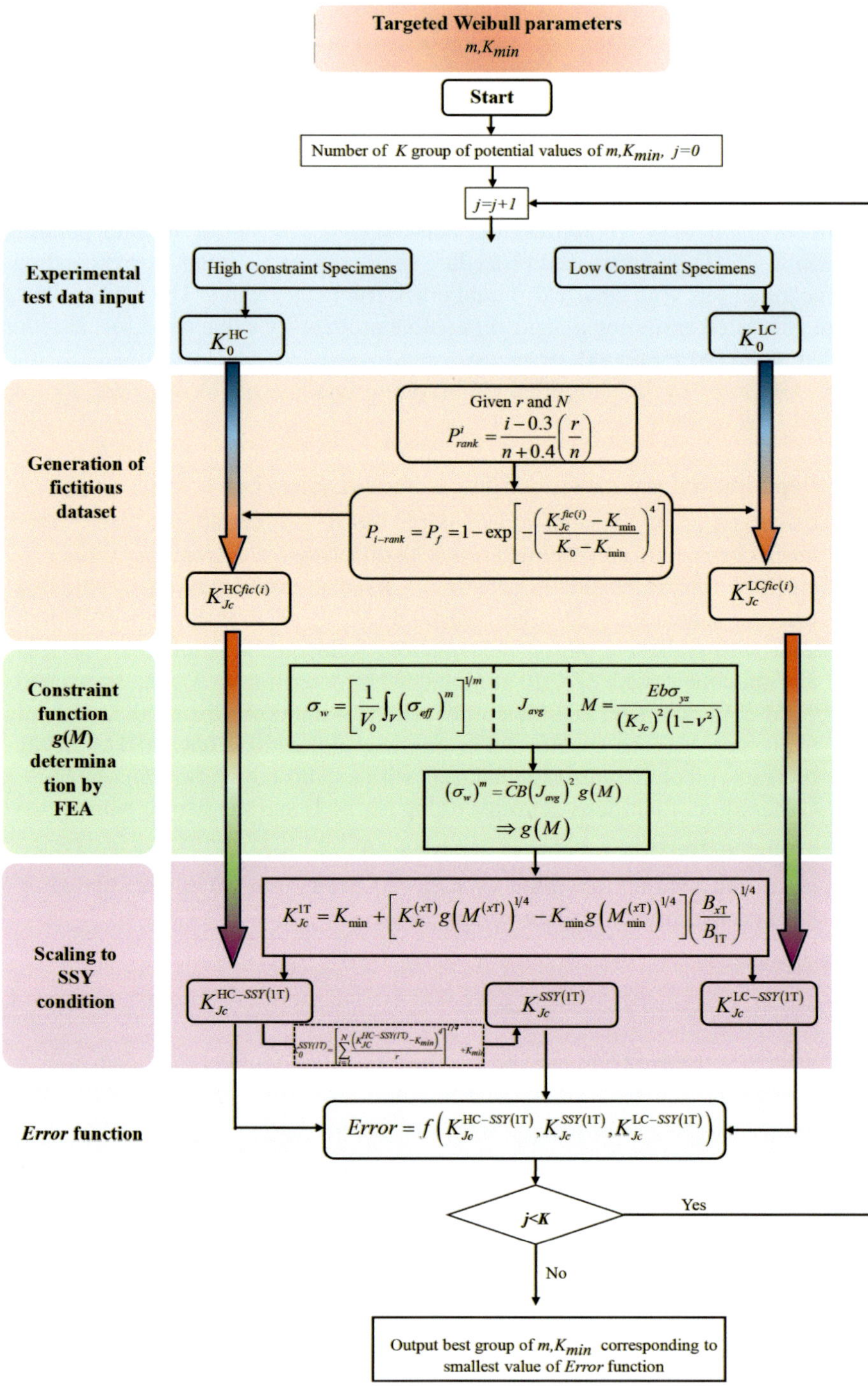

FIGURE 4.1 Simplified calibration procedure of Weibull stress approach.

toughness $K^{SSY(1T)}$ based on the scale parameter $K_0^{SSY(1T)}$ [Eq. (4.2)] computed from the corrected fracture toughness $K_{Jc}^{HC-SSY(1T)}$.

5. Parameter calibration via error minimization: For each trial pair of m and K_{min}, an error function is computed:

$$Error = \sum_{i=1}^{N_{HC}+N_{LC}} \left| K_{Jc(i)}^{xC-SSY(1T)} - K_{Jc(i)}^{SSY(1T)} \right| WF_{(i)} + \sum_{i=1}^{\min(N_{HC},N_{LC})} \left| K_{Jc(i)}^{HC-SSY(1T)} - K_{Jc(i)}^{LC-SSY(1T)} \right| WF_{(i)} \tag{4.13}$$

where $WF_{(i)}$ is a weighting function as defined in the earlier study by Wasiluk *et al.* [17]. Prior research [18] indicates that the choice of weighting function has minimal influence on the calibration outcome.

The main distinction between this approach and the original method by Wasiluk *et al.* [17] lies in the use of synthetic toughness datasets in place of raw experimental data. This enables calibration of Weibull stress parameters with minimal experimental effort, particularly valuable for materials with limited available fracture toughness data.

The calibrated values of m and K_{min} are obtained by minimizing the error function defined in Eq. (4.13). This minimization process effectively scales both the high-constraint and low-constraint toughness datasets to a common 1T SSY condition, ensuring that the resulting m and K_{min} provide consistent predictions of the cumulative probability of fracture across both specimen types. In the present study, the calibration process explores a range of m values from 1 to 30 and K_{min} values from 0 to 80 MPa.

The calibration procedure described above relies on detailed FE simulations to compute the Weibull stress. Figure 4.2a illustrates a half-symmetric modified boundary layer (MBL) model, representing the SSY condition – with a single layer of elements through the thickness. Figures 4.2b and c show quarter-symmetric finite element models for a C(T) specimen and a CVN specimen, respectively. The C(T) and CVN models are discretized using 14 and 16 layers of elements through the thickness, with progressively refined mesh density near the free surfaces to capture through-thickness constraint effects.

All models incorporate a finite initial crack-tip root radius to improve numerical stability under large deformation. To evaluate the constraint correction function $g(M)$ across a broad range of crack-front constraint levels, multiple FE models with varying R_0 values (ranging from 0.5 μm to 12.7 μm) are employed. The meshes use 20-node brick elements with reduced integration. Nodes on the symmetry planes are constrained in the displacement direction normal to the respective planes. Displacement-controlled loading is applied uniformly across all nodes on the loading surfaces.

To ensure high spatial resolution near the crack tip, the mesh includes approximately 16 elements surrounding each crack-tip node and around 100 elements radially distributed in the x–y plane. The smallest element near the crack tip is sized to

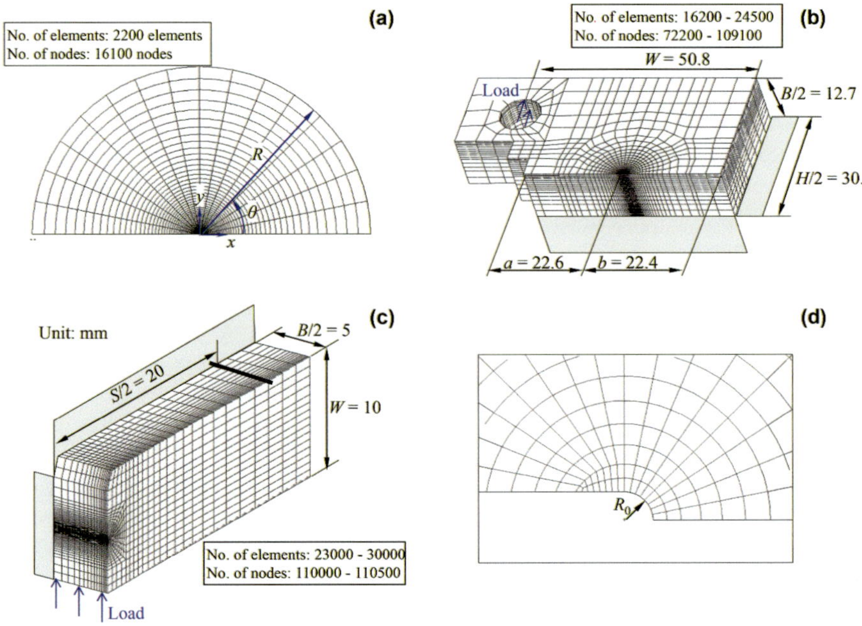

FIGURE 4.2 Typical finite element meshes for: (a) a half symmetric small-scale yielding (SSY) model; and (b) a quarter symmetric C(T) model; and (c) a quarter symmetric CVN model; and (d) a close-up view at the crack tip.

approximately 10% of the assigned initial root radius, ensuring accurate capture of stress and strain gradients critical for Weibull stress evaluation.

The numerical procedure imposes a remote mode-I displacement field on the circular boundary of the semi-circular MBL model. The applied displacement field is defined as:

$$u(R,\theta)=\frac{K_I(1+\nu)}{E}\sqrt{\frac{R}{2\pi}}\cos\left(\frac{\theta}{2}\right)(3-4\upsilon-\cos\theta) \quad (4.14)$$

$$v(R,\theta)=\frac{K_I(1+\nu)}{E}\sqrt{\frac{R}{2\pi}}\sin\left(\frac{\theta}{2}\right)(3-4\upsilon-\cos\theta) \quad (4.15)$$

where K_I is the applied mode-I stress intensity factor.

Figure 4.3a presents the true stress–strain curves of the Euro steels at four different temperatures. The curves at -20°C, -40°C, and -91°C follow material properties reported in prior studies [17, 19]. Since the stress–strain curve at 0°C is not available in the literature, it is generated using a parabolic extrapolation of the true stress values at each strain level based on the known data at the three lower temperatures. Figure 4.3b shows the variation of yield strength as a function of temperature. The Young's modulus is taken as 206 GPa and Poisson's ratio is assumed constant at 0.3 across all temperatures.

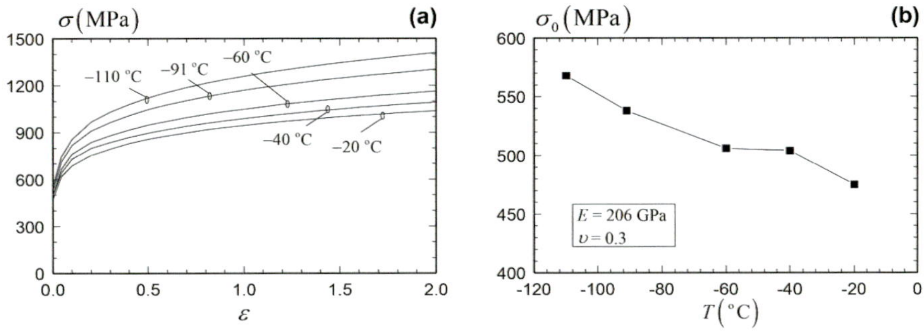

FIGURE 4.3 Material properties for Euro-dataset material 22Ni-MoCr37: (a) uniaxial true stress-true strain curves at different temperatures; and (b) variation of the material yield strength over the DBT region.

The numerical simulations are performed using the large-deformation algorithm implemented in the open-source finite element research code WARP3D [20]. The energy release rate is computed via the domain-integral method, which yields a through-thickness averaged *J*-integral for the C(T) and CVN specimens across different temperatures.

In calculating the Weibull stress, the analysis incorporates history effects, that is, damage accumulation is considered irreversible. As such, reductions in local stress due to unloading or redistribution near the crack tip do not reverse damage accrued in previous loading increments.

Figure 4.4a compares the evolution of the Weibull stress with increasing load for both the 1T SSY model and the 1T C(T) specimen at a representative temperature. As load increases, the constraint in the C(T) specimen progressively decreases, as evidenced by the growing divergence in Weibull stress between the C(T) and SSY models. This divergence indicates a softening of crack-tip constraint and implies a gradually decreasing constraint correction function $g(M)$ with increasing load for the C(T) specimen.

Figures 4.4b and c show the variation of the constraint function $g(M)$ as a function of the non-dimensional loading parameter M, computed using multiple FE models with different initial crack-tip root radii. At large M values (i.e., low applied loads, M from 400 to 600), the computed $g(M)$ values exhibit notable sensitivity to the initial root radius. However, at very small M (i.e., high load levels), FE models with small root radii fail to converge due to severe mesh distortion near the crack tip under large plastic deformation. To ensure numerical stability and a complete representation across the full M range, the final $g(M)$ function used in the calibration (shown in Figure 4.4d) is constructed by smoothly blending results from multiple FE simulations, each with a different root radius. This composite $g(M)$ function provides a continuous and physically consistent input for the Weibull stress calibration process.

This section compares the calibrated Weibull modulus m and threshold toughness K_{min} obtained using the proposed simplified procedure with those derived from the detailed calibration method of Wasiluk *et al.* [17], for Euro steel materials [15] at

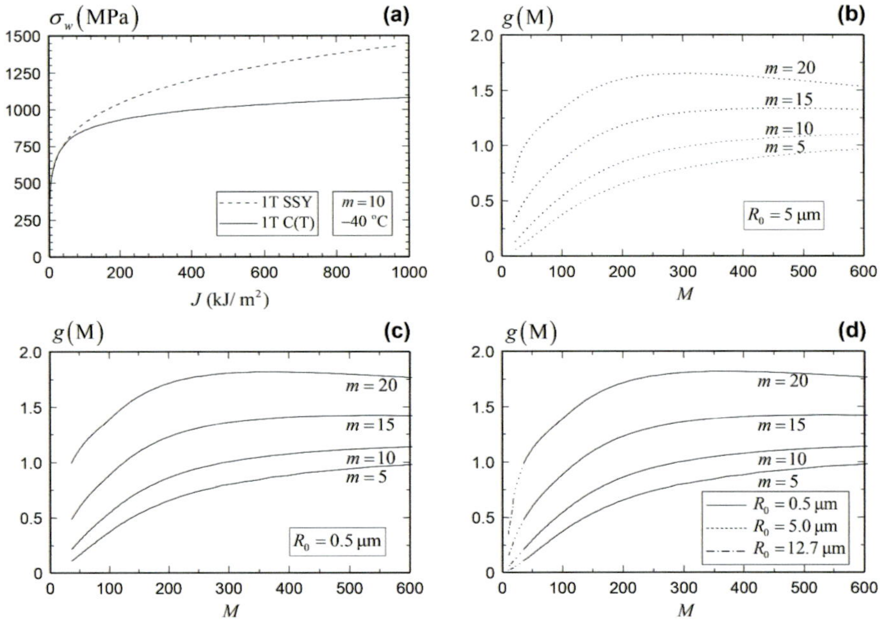

FIGURE 4.4 (a) Evolution of the Weibull stress with increasing loading at -40°C for a 1T C(T) model and a 1T SSY model; and constraint correction $g(M)$ functions at -40°C from: (b) a C(T) model with an initial root radius R_0 = 5 μm; (c) a C(T) model with an initial root radius R_0 = 0.5 μm; and (d) multiple C(T) models with different initial root radius.

TABLE 4.1
K_0 Values and r/n Employed in the Simplified Calibration Method.

Specimen	Thickness (mm)	Temperature(°C)	r/N ratio	K_0 (MPa$\sqrt{m}$)
1T C(T)	25.4	−20	27/30	335
2T C(T)	50.8	−20	30/30	257
4T C(T)	101.6	−20	15/15	203
0.5T C(T)	12.7	−40	18/30	251
1T C(T)	25.4	−40	30/30	230
2T C(T)	50.8	−40	30/30	181
0.5T C(T)	12.7	−110	55/55	88
0.4T CVN	10.2	−110	50/50	117

three temperatures: -20°C, -40°C, and -110°C. Table 4.1 summarizes the specimen types, corresponding K_0 values, and valid-data ratios (r/n) for each specimen set. All C(T) specimens listed in Table 4.1 are geometrically similar, implying a consistent $g(M)$ function for specimens tested at the same temperature.

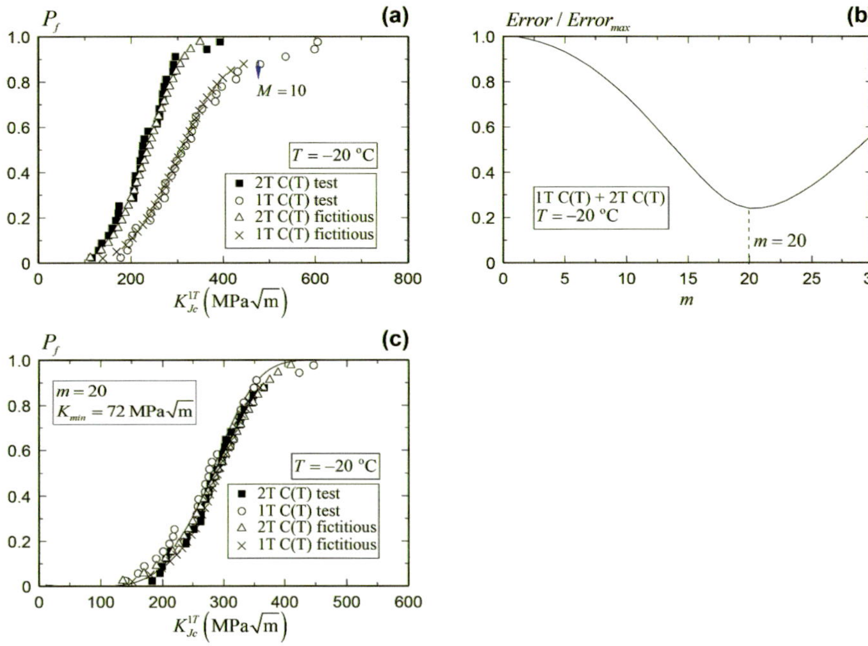

FIGURE 4.5 At -20°C (a) comparison of the experimental fracture data with the fictitious fracture data; (b) variation of the normalized error with respect to m; and (c) comparison of the estimated probability with experimental and fictitious data.

Figure 4.5a presents the rank probability plot for the original experimental fracture toughness data of 1T and 2T C(T) specimens tested at –20 °C, along with the fictitious datasets generated from the K_0 values in Table 4.1. The 1T C(T) dataset includes 27 valid data points, with three additional values falling below the $M = 10$ threshold. The generated fictitious dataset closely aligns with the experimental data and remains within the applied load envelope.

Applying the simplified calibration procedure described above yields a Weibull modulus $m = 16$ and a threshold toughness $K_{min} = 23$ MPa, as listed in Table 4.2. Figure 4.5b shows the normalized error function defined in Eq. (4.13), where the error is normalized by the maximum value across all trial m values. A clear minimum is observed, indicating the optimal m. Figure 4.5c compares the cumulative fracture probability, computed using Eq. (4.10) and the calibrated m and K_{min}, against both experimental and fictitious toughness data scaled to the 1T SSY condition via Eq. (4.11). The close agreement in Figure 4.5c confirms the accuracy of the calibrated parameters.

Table 4.2 compares the results of the simplified and detailed calibration approaches at -20°C. The Weibull modulus values are in excellent agreement, while some variation is observed in the calibrated K_{min}, which is further examined in subsequent discussions in the context of its influence on predicted fracture probability.

TABLE 4.2

Calibrated m and K_{min} Values at Three Different Temperatures for the Euro Steel Material.

Case	Temperature (°C)	Calibration following[34]		Simplified calibration method	
		m	$K_{min}(MPa\sqrt{m})$	m	$K_{min}(MPa\sqrt{m})$
1T C(T) + 2T C(T)	−20	20	48	20	72
1T C(T) + 4T C(T)	−20	20	80	20	72
2T C(T) + 4T C(T)	−20	20	80	20	56
0.5T C(T) + 1T C(T)	−40	21	43	21	56
0.5T C(T) + 2T C(T)	−40	19	52	22	64
1T C(T) + 2T C(T)	−40	18	55	19	48
0.5T C(T) + 0.4T CVN	−110	17	25	14	32

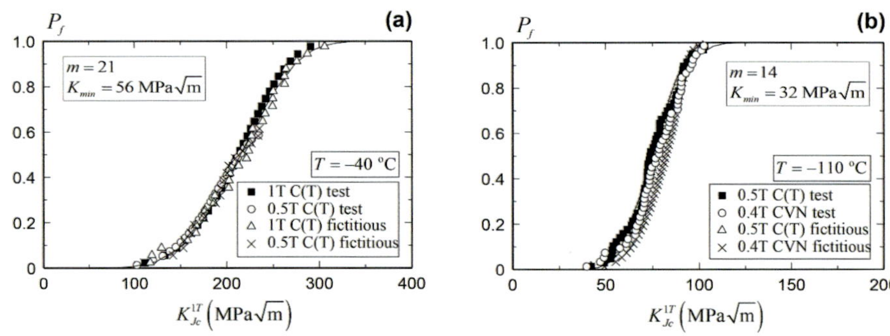

FIGURE 4.6 Comparison of the estimated cumulative probability with experimental and fictitious data at (a) -40°C and (b) -110°C.

The calibration at -40°C and -110°C follows the same methodology using the experimental data in Table 4.1. Figures 4.6a and b compare the fracture probability estimated using the calibrated m and K_{min}, as noted in Table 4.2 based on fictitious dataset generated from the K_0 values in Table 4.1.

The results in Table 4.2 reveal clear trends in both m and K_{min} with temperature. As temperature increases, the threshold toughness K_{min} rises notably, indicating that more energy is required to initiate cleavage fracture at higher temperatures. Conversely, the Weibull modulus m remains relatively insensitive to temperature changes, consistent with the theoretical basis of the Weibull model, where m reflects the distribution of microcrack sizes, a material property assumed to be independent of temperature.

The accuracy of the simplified calibration approach is highly dependent on the correct determination of the K_0 values used to generate the fictitious datasets. The

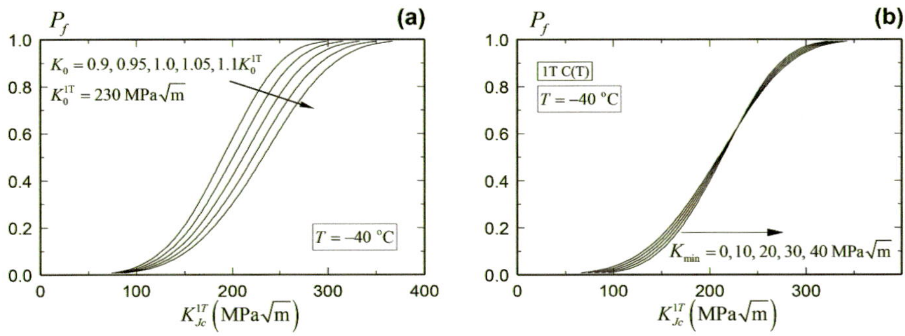

FIGURE 4.7 Effect of (a) K_0 and (b) $K_{\min}$ on the generated fictitious data for 1T C(T) specimens at −40°C.

TABLE 4.3
Effect of K_0 on the Calibration Results ($K_{min} = 20\text{MPa}\sqrt{\text{m}}$) for 1T C(T) and 0.5T C(T) Euro Steels at −40 °C.

0.5T C(T) + 1T C(T) at −40 °C	$0.9K_0^{1T}$ and $K_0^{0.5T}$	$0.95K_0^{1T}$ and $K_0^{0.5T}$	K_0^{1T} and $K_0^{0.5T}$	$1.05K_0^{1T}$ and $K_0^{0.5T}$	$1.1K_0^{1T}$ and $K_0^{0.5T}$
m	13	15	21	29	30
$K_{\min}$ (MPa$\sqrt{\text{m}}$)	80	80	56	40	21

TABLE 4.4
Effect of K_{min} on the Calibration Results (K_0 Remains Fixed for all Cases) for 1T C(T) and 0.5T C(T) Euro Steels at −40 °C.

0.5T C(T) + 1T C(T) at −40 °C	$K_{\min} = 0$	$K_{min} =$ 10 MPa$\sqrt{\text{m}}$	$K_{min} =$ 20 MPa$\sqrt{\text{m}}$	$K_{min} =$ 30 MPa$\sqrt{\text{m}}$	$K_{min} =$ 40 MPa$\sqrt{\text{m}}$
m	21	21	21	19	19
K_{min} (MPa$\sqrt{\text{m}}$)	42	49	56	80	80

subsequent text examines the sensitivity of the calibrated Weibull modulus mmm and threshold toughness K_{min} to variations in the assumed K_0 value.

Figure 4.7a illustrates this sensitivity using the 1T C(T) specimens at −40 °C as an example. The experimentally determined K_0 at this temperature is 230 MPa$\sqrt{\text{m}}$. Fictitious datasets are generated using five different K_0 values ranging from $0.9K_0$ to $1.1K_0$, while the corresponding K_{min} for the 0.5T C(T) specimens is fixed at 20 MPa$\sqrt{\text{m}}$, as listed in Table 4.3. Table 4.3 presents the resulting calibrated m and

K_{min} values. The variations in K_0 lead to noticeable differences in both parameters, but slightly affect the accuracy in the accumulative probability distribution [14].

Similarly, Figure 4.7b examines the effect of changing the assumed K_{min} value in generating the fictitious dataset for the 1T C(T) specimens. A range of K_{min} values from 0 to 40 MPa is used, while holding $K_0 = 230\ \text{MPa}\sqrt{\text{m}}$ for the 1T specimens and $K_0 = 251\ \text{MPa}\sqrt{\text{m}}$ for the 0.5T C(T) specimens. Table 4.4 summarizes the calibrated m and K_{min} values for each assumed K_{min}. The Weibull modulus m remains relatively stable across the range, while the calibrated threshold toughness K_{min} shows notable variation.

The above discussion presents a simplified calibration procedure for the three-parameter Weibull stress model to estimate the cumulative probability of cleavage fracture. Unlike traditional calibration approaches, which require two complete sets of experimental fracture toughness data, the proposed method uses only two discrete K_0 values corresponding to specimens with different crack-front constraint conditions.

4.3.2 Effect of Experimental Sample Size

Calibrating Weibull parameters from a single specimen set presents the issue of non-uniqueness [7, 11]. To address this, Wasiluk *et al.* [17] proposed employing two distinct sets of fracture toughness data characterized by significant differences in crack-front constraint conditions. However, their modified calibration method typically requires large datasets and can introduce biased estimates if insufficient specimen data are available. This section describes the procedure used to investigate the influence of sample size on the calibration of the Weibull modulus by utilizing randomly generated subsets derived from an extensive set of experimental fracture toughness data. Same to the previous section, the fracture toughness values analyzed in the session were obtained from experiments conducted at three distinct temperatures, as documented in the European Union project titled *"Fracture toughness of steel in the ductile to brittle transition regime"* [15, 16]. Figure 4.8 illustrates the cumulative probability of failure based on fracture toughness values measured experimentally using C(T) specimens and CVN specimens tested under three-point bending at these temperatures. The numerical procedure employed here mitigates this limitation by generating random subsets of varying sample sizes from both high-constraint (HC) and low-constraint (LC) specimen groups. Following Wasiluk *et al.* [17], the calibration approach adjusts for constraint loss in both datasets and converts the experimental fracture toughness data into an idealized plane-strain, SSY condition with zero T-stress. Subsequently, the Weibull modulus (m) and threshold toughness (K_0) are calibrated using these randomly selected HC and LC subsets.

The step-by-step methodology for examining how sample size affects calibrated Weibull parameters is summarized below:

1) From the experimental database, identify HC and LC specimen sets with corresponding fracture toughness values obtained at the same temperature within the ductile-to-brittle transition regime.

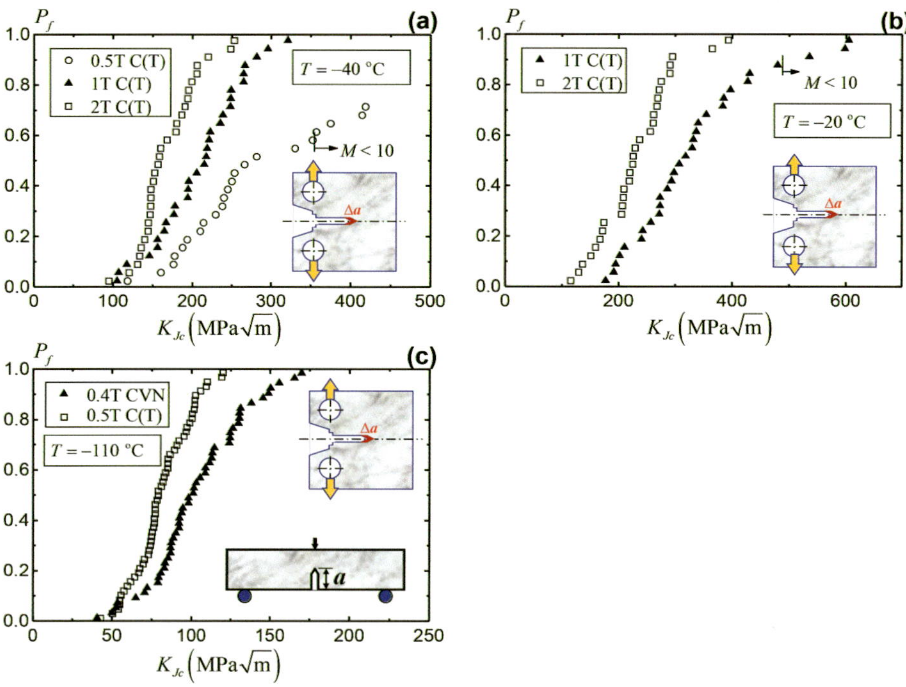

FIGURE 4.8 Cumulative probability of failure based on fracture toughness values measured experimentally using compact tension (C(T)) specimens and Charpy V-notched (CVN) specimens tested under three-point bending at different low temperatures.

2) Generate independent subsets by randomly selecting fracture toughness data from the HC and LC datasets identified in Step 1. The subset size n ranges from 6 up to N_{max}, where N_{max} denotes the smaller total size of the complete HC and LC datasets.
   - Assign random numbers (j) uniformly distributed between 0 and 1 to each specimen in both datasets from Step 1. Each random number corresponds uniquely to a specimen's fracture toughness value.
   - Sort the random numbers (j) in ascending order for each dataset.
   - Select fracture toughness values corresponding to the first n sorted numbers (j) to form new, independent subsets from both HC and LC datasets.
3) Rank the fracture toughness values within each subset, denoted as $K_{Jc}^{\mathrm{HC}(i)}, K_{Jc}^{\mathrm{LC}(i)}$, from smallest to largest. Assign the median rank failure probability according to the following formula:

$$P_{i\text{-}rank} = \frac{i - 0.3}{n_{x\mathrm{C}} + 0.4} \tag{4.16}$$

where i is the data point's rank number, and n is the subset size (equal for HC and LC subsets).

4) Using the calibration procedure "Construction function $g(M)$ determination by FEA," "Scaling to SSY condition," and " Error function" in Figure 4.1, determine the Weibull modulus (m) and threshold toughness (K_{min}).
5) Repeat Steps 2–4 to generate N pairs of independent subsets and calibrate the corresponding Weibull modulus (m). Ensure each generated HC subset contains a minimum of six valid toughness values, adhering to ASTM E-1921 requirements [5]. In this study, N = 5000 subset pairs were generated to ensure convergence of the calibrated Weibull parameters.
6) For each subset size n, calculate the mean (μ_m) and standard deviation (S_m) of the calibrated Weibull modulus (m). Evaluate the confidence levels (P_m and P_{K_0}) for the Weibull modulus (m) and high-constraint scale parameter (K_o) [defined in Eq. (4.2)], based on specified confidence intervals.

Previous studies by Wasiluk *et al.* [17] confirmed the temperature independence of the Weibull modulus (m) using "Euro" fracture toughness datasets, obtaining a calibrated modulus m = 20. Additional calibrations by Qian and Chen [14] and Qian *et al.* [18] across three temperatures produced a maximum deviation of ±3 from the reference modulus. Therefore, the current study adopts a slightly broader confidence interval for the Weibull modulus as $m_{true} - 4 \leq m \leq m_{true} + 4$. The m_{true} value refers to the modulus calibrated from the complete experimental dataset. The corresponding confidence level (P_m) is computed as:

$$P_m = \frac{N_{m,true}}{N} \tag{4.17}$$

where $N_{m,true}$ is the count of subset pairs yielding calibrated modulus values (m) within the defined interval $m \in [m_{true} - 4, m_{true} + 4]$.

Additionally, Qian and Chen [14] indicated that deviations exceeding 5% in the scale parameter (K_o) significantly influence calibrated modulus results. Therefore, this study adopts the following confidence interval for the high-constraint scale parameter, $0.95K_{0,true} \leq K_0 \leq 1.05K_{0,true}$. The corresponding confidence level (P_{K_0}) is calculated as:

$$P_{K_0} = \frac{N_{K_0,true}}{N} \tag{4.18}$$

where $N_{K_0,true}$ is the count of subsets with calibrated K_o values within the specified interval, $K_0 \in [0.95K_{0,true}, 1.05K_{0,true}]$.

The above assessment approach requires detailed FE modeling. In this study, large-deformation elastic-plastic FEA were performed using the open-source research code WARP3D [20], which calculates through-thickness energy release rates (J-integral values) via the domain-integral approach. The material properties and finite element models of C(T) and CVN specimens are the same as those in the previous section.

Wallin [21] previously applied the three-parameter Master Curve [Eq. (4.1)] to analyze the scale fracture toughness K_0 [Eq. (4.2)] for the "Euro" 22Ni-MoCr37

TABLE 4.5
Summary of the Constraint Differences for Different HC and LC Combinations.

Temperature (°C)	HC+LC	m_{true}	$K^{HC}_{0(1T)}$ (MPa$\sqrt{m}$)	Δ_{cstr} (%)
–40	1T + 0.5T C(T)	21	229	9.55
	2T + 0.5T C(T)	19	211	14.6
	2T + 1T C(T)	18	211	9.26
–20	2T + 1T C(T)	17	301	17.3
–110	0.5T C(T) + 0.4T CVN	17	77.2	17.2

pressure vessel steel across ductile-to-brittle transition (DBT) temperatures. Several researchers [13, 14, 17] calibrated the Weibull modulus (*m*) for this Euro-material at different temperatures using the modified three-parameter Weibull stress model [Eq. (4.10)]. Table 4.5 presents the calibrated reference *m*-values derived from complete datasets at three testing temperatures.

Since the calibration requires fracture datasets with significant differences in crack-front constraints, this constraint difference ($\Delta constraint$) between high-constraint (HC) and low-constraint (LC) specimens strongly affects calibration outcomes. This study quantifies the dimensionless constraint difference as:

$$\Delta_{constraint} = \sum_{i=1}^{r_{xC}} \left| K^{\mathrm{HC}}_{Jc(1\mathrm{T}),i} - K^{\mathrm{LC}}_{Jc(1\mathrm{T}),i} \right| \frac{1}{r_{x\mathrm{C}}} \frac{1}{K^{\mathrm{HC}}_{0(1\mathrm{T})}} \tag{4.19}$$

where $r_{xC} = \min(r_{\mathrm{LC}}, r_{\mathrm{HC}})$, $K^{\mathrm{HC}}_{Jc(1\mathrm{T}),i}$ and $K^{\mathrm{LC}}_{Jc(1\mathrm{T}),i}$ denote corrected 1T fracture toughness values at the same rank probability. Table 4.5 lists these computed dimensionless constraint differences for HC and LC specimen combinations.

Figure 4.9a shows the mean Weibull modulus (μ_m) calibrated from different specimen combinations as a function of subset size *n*. The mean μ_m remains stable across subset sizes for the 2T C(T)+1T C(T) combination, but decreases significantly for small *n* in the 1T C(T)+0.5T C(T) case. Figure 4.9b highlights that the standard deviation (S_m) of calibrated *m*-values strongly depends on the constraint difference between selected specimen sets, with larger constraint differences leading to smaller *Sm* and more stable calibration results.

Figure 4.9c illustrates confidence levels (P_m) for calibrated *m*-values within the interval $15 \leq m \leq 23$. Confidence levels increase with subset size *n* and the constraint difference between HC and LC sets. Notably, the 2T C(T)+0.5T C(T) combination achieves approximately 73% confidence with only 15 specimens per subset, despite the presence of invalid data points in the 0.5T dataset. Conversely, smaller constraint differences, coupled with invalid data points [1T C(T) + 0.5T C(T)], significantly reduce confidence and lead to underestimated mmm-values at smaller subset sizes (Figures 4.9a and b). Previous calibration approaches [7, 22, 23], based solely on statistical methods, neglected this constraint influence.

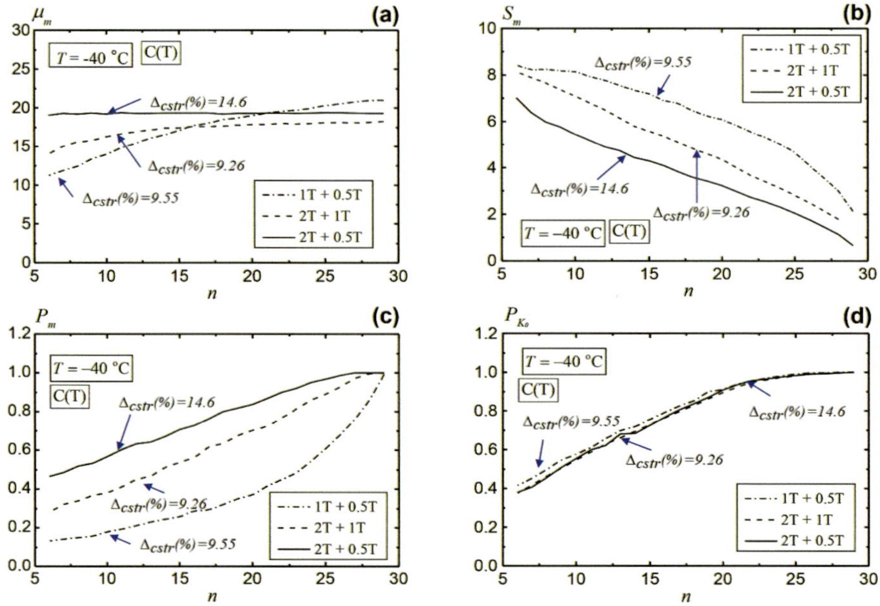

FIGURE 4.9 Effect of the sample size on the Weibull stress calibration at -40°C: (a) the mean Weibull modulus; (b) the standard deviation of the Weibull modulus; (c) the confidence level for the calibrated Weibull; and (d) the confidence level for the HC scale parameter K_0.

Figure 4.9d presents confidence levels P_{K_0} for the high-constraint scale parameter K_0, using an interval $0.95K_{0,true} \leq K_0 \leq 1.05K_{0,true}$. Minimal variation between specimen combinations confirms the consistently high constraint condition in both 1T and 2T specimens.

At –20°C, this study considers the 1T and 2T C(T) datasets (each with 30 specimens) from the Euro steel database. The 1T dataset includes three invalid specimens with excessive deformation ($M < 10$). Calibration from full HC and LC sets yields $m = 17$, consistent with previous research [18].

Figure 4.10a compares mean calibrated μ_m values at –20°C and –40°C for the 2T (HC)+1T (LC) specimen combinations. At both temperatures, μ_m remains relatively stable for subset sizes $n \geq 13$. Figure 4.10b shows standard deviations (Sm) decreasing with larger subset size and higher test temperature, caused by increased plastic constraint loss and thus greater constraint differences between datasets (Table 4.5). The 1T C(T) specimens exhibit minimum M-values of 24.3 (at –40°C) and 6.4 (at –20°C), highlighting significant constraint loss, whereas 2T specimens remain high-constraint ($M > 30$) at both temperatures.

Figure 4.10c demonstrates higher confidence levels (P_m) at –20°C compared to –40°C for the m confidence interval ($13 \leq m \leq 21$), primarily due to greater constraint differences. Approximately 21 specimens per subset are required at –20°C to achieve an 80% confidence level, compared to 23 specimens at –40°C. Figure 4.10d

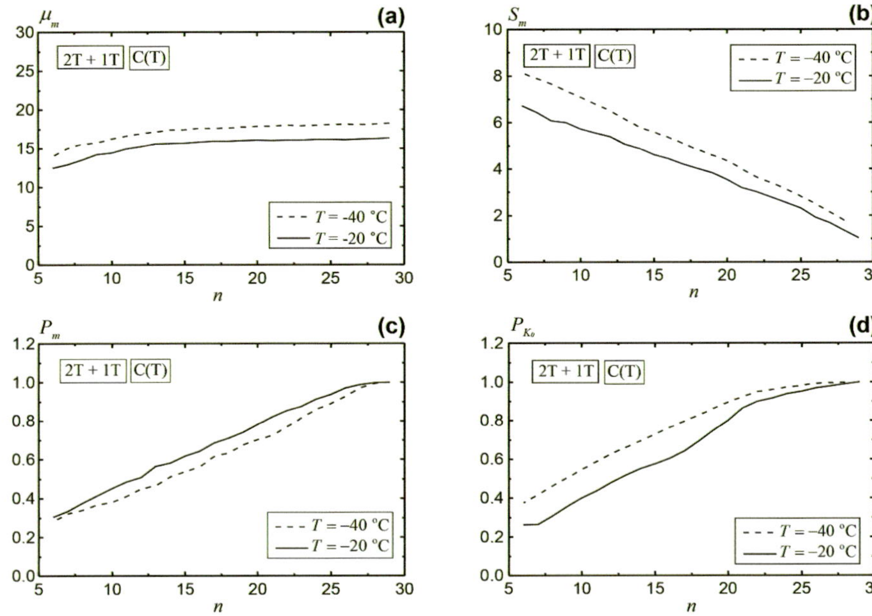

FIGURE 4.10 Effect of the sample size on the Weibull stress calibration at –20 °C and –40 °C: (a) the mean Weibull modulus; (b) the standard deviation of the Weibull modulus; (c) the confidence level for the calibrated Weibull modulus; and (d) the confidence level for the HC scale parameter K_0.

compares confidence levels (P_{K_0}) for the high-constraint scale parameter K_0 at both temperatures. Slightly lower confidence at –20°C arises from greater scatter in the 2T specimens, despite their high-constraint condition.

At –110°C, the analysis includes 0.5T C(T) and 0.4T CVN datasets with 55 and 50 specimens, respectively, all meeting $M > 10$ criteria (Table 4.1). Calibration using full datasets yields $m_{true} = 17$, matching previous studies [14, 18].

Figure 4.11a shows that mean calibrated μ_m-values decrease as subset size n decreases, reflecting greater sensitivity to smaller datasets. Figure 4.11b indicates increasing standard deviations (S_m) at smaller subset sizes. Figure 4.11c demonstrates a confidence level P_m = 50% when using subsets of 15 specimens each, highlighting the need for approximately 31 specimens per subset to reach an 80% confidence level. Figure 4.11d confirms confidence levels P_{K_0} > 80% for the HC scale parameter K_0 when subset sizes $n \geq 19$, underscoring the reliability of these calibrated values.

Figure 4.12a compares dimensionless bounding envelopes of m-values at 80% confidence level across different temperatures and specimen combinations, clearly indicating the strong dependence on constraint differences. Figure 4.12b highlights that the 2T+0.5T combination at –40°C exhibits the smallest bounding envelope, indicating the most stable calibration outcomes due to its large constraint difference.

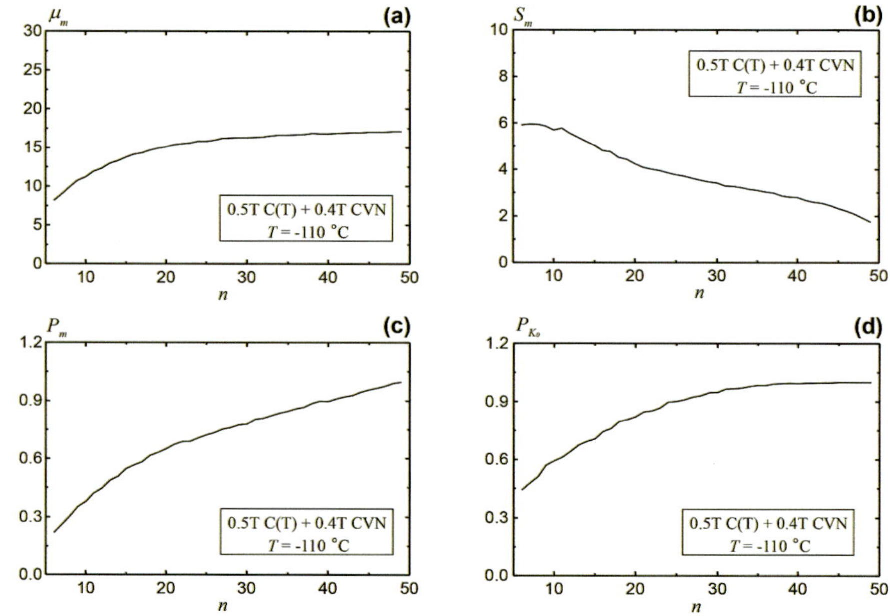

FIGURE 4.11 Effect of the sample size on the Weibull stress calibration at –110 °C: (a) the mean Weibull modulus; (b) the standard deviation of the Weibull modulus; (c) the confidence level for the calibrated Weibull modulus; and (d) the confidence level for the HC scale parameter K_0.

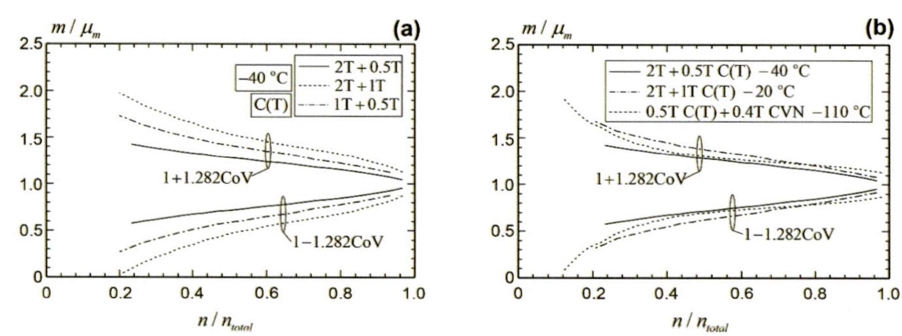

FIGURE 4.12 Dimensionless bounding envelopes for the Weibull modulus corresponding to the 80% confidence level for: (a) –40 °C; and (b) comparison among three different temperatures.

When the calibrated Weibull modulus falls near the upper bound of its confidence interval, the cleavage-fracture assessment can change appreciably. Because the constraint-correction function in Eq. (4.9) increases monotonically with the modulus m, a larger m yields a higher corrected fracture-toughness value in Eq. (4.11), which is a

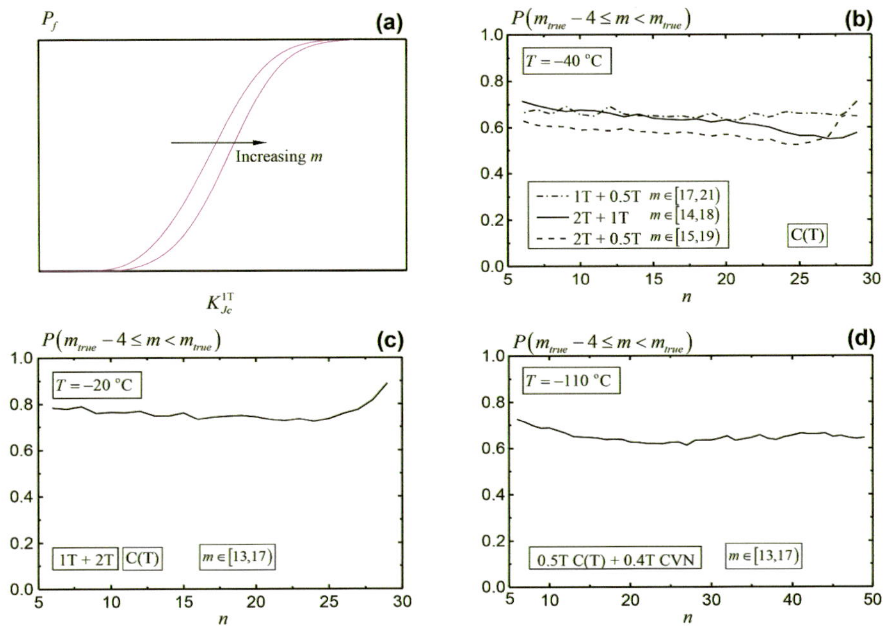

FIGURE 4.13 (a) Schematic variation of the estimated probability with an increasing m in the P_f-K_{Jc}^{1T} space; and the probability of underestimating the Weibull modulus within the confidence interval for: (b) –40 °C; (c) –20 °C; and (d) –110 °C.

trend confirmed in the literature [14, 17, 18]. Practically, an over-estimated m shifts the predicted *Kjc(1T)* curve toward higher toughness in the *K–T* space (Figure 4.13a). This upward shift overstates the fracture toughness associated with a given cleavage-failure probability, leading to an under-prediction of failure probability, particularly in the lower tail, and thus an unrealistically small safety margin.

Figures 4.13b–d show how the likelihood of under-estimating the Weibull modulus within its confidence band varies with specimen count n for the Euro-material at –110 °C, –40 °C, and –20 °C, respectively. That likelihood, denoted by P, is calculated by

$$P\left(m_{true} - 4 \leq m < m_{true}\right) = \frac{N\left(m_{true} - 4 \leq m < m_{true}\right)}{N_{m,true}} \tag{4.20}$$

where $N\left(m_{true} - 4 \leq m < m_{true}\right)$ is the number of subset pairs whose calibrated modulus m falls over the range $m_{true} - 4 \leq m < m_{true}$. Across all three temperatures, $P\left(m_{true} - 4 \leq m < m_{true}\right)$ stays above 0.5 for the sample-size range investigated, revealing a systematic bias toward smaller m values within the interval, and thus a conservative tilt in the resulting engineering assessment of cleavage-fracture risk.

This work examines how specimen count and crack-front constraint influence calibration of the Weibull modulus m for 22Ni-MoCr37 pressure-vessel steel, using the comprehensive "Euro" fracture-toughness database. A modified three-parameter Weibull-stress model [Eq. (4.10)] is fitted to thousands of randomly generated, independent HC/LC (high- and low-constraint) subset pairs drawn from the full dataset. The procedure allows the uncertainty in m and in the scale parameter σ_u to be expressed as functions of both sample size n and HC–LC constraint contrast.

4.4 BRITTLE FRACTURE ASSESSMENT FOR SURFACE-CRACKED PLATES

This section extends the previously described Weibull stress approach to evaluate cleavage fracture behavior in both through-thickness fracture specimens and surface-cracked plates made from S690 high-strength steel.

4.4.1 Experimental Details

The experimental program involves three types of specimens fabricated from the same high-strength S690 steel: (i) coupon specimens (Figure 4.14a) for uniaxial tensile tests; (ii) through-thickness fracture specimens (Figures 4.15a and b); and (iii) surface-cracked plate specimens (Figure 4.15c).

All specimens are extracted from a single S690 steel plate, maintaining a consistent material orientation. Table 4.6 lists the mechanical properties measured from uniaxial tensile tests at three different temperatures: 28 °C, –30 °C, and –60 °C. The true stress–logarithmic strain curves at these temperatures are shown in Figure 4.14b. Table 4.7 summarizes the complete experimental matrix. The through-thickness fracture specimens include three configurations: 11 SE(B) specimens of 0.5T thickness (12.5 mm), 17 C(T) specimens of 0.4T thickness (10 mm), and 18 C(T) specimens of 0.8T thickness (20 mm).

Figures 4.15b and c show the dimensions and geometry of the SE(B) and C(T) specimens, respectively. Additionally, the program includes 12 surface-cracked bend [SC(B)] specimens, each 17 mm thick (0.7T). These SC(B) specimens feature a wire-cut, semi-elliptical surface notch centrally located on the plate surface (Figure 4.15c). Each notch has a depth of 10 mm and a surface length of 40 mm.

All fracture specimens undergo two sequential loading stages: (i) cyclic fatigue pre-cracking, followed by (ii) monotonic fracture testing, both conducted using a 100-ton universal testing machine. For the through-thickness fracture specimens,

TABLE 4.6
Mechanical Properties of S690 Steel.

Steel	S690	S690	S690
Temperature (°C)	28	–30	–60
Young's modulus (GPa)	206	206	206
Yield strength (MPa)	810	853	952
Ultimate strength (MPa)	866	869	983

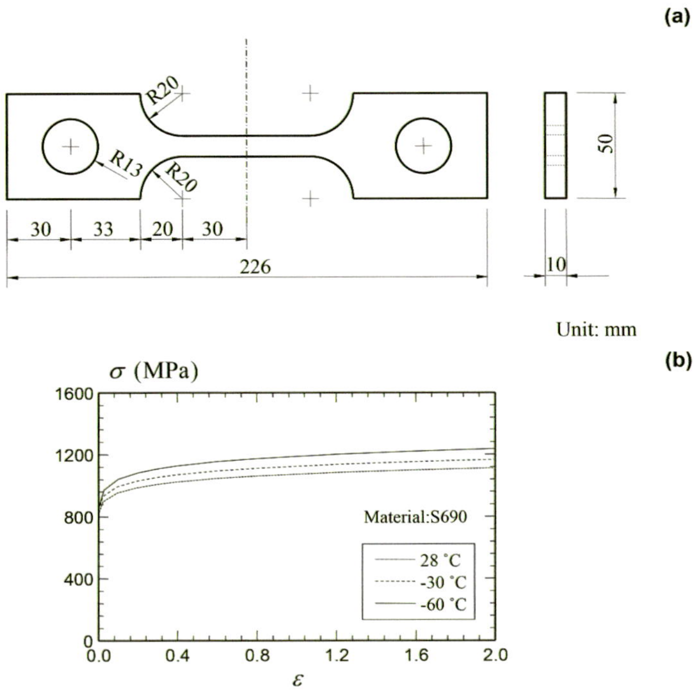

FIGURE 4.14 Dimensions of the uni-axial tension specimen (in mm); and (b) true stress-logarithmic strain curves measured at different temperatures of S690 steel.

TABLE 4.7
Experimental Program for S690 Steel.

Specimen type	material	No. of specimens	Loading condition	Test type	Test temperature (°C)
Coupon	S690	3	Uniaxial tension	Tension test	28
		3	Uniaxial tension	Tension test	–30
		3	Uniaxial tension	Tension test	–60
0.5T SE(B)	S690	11	Three-point bend	Fatigue pre-cracking ($K_I \in [12,63]\,\mathrm{MPa}\sqrt{\mathrm{m}}$)	28
			Three-point bend	Fracture test	–60
0.4T C(T)	S690	17	Tension	Fatigue pre-cracking ($K_I \in [8,65]\,\mathrm{MPa}\sqrt{\mathrm{m}}$)	28
			Tension	Fracture test	–60
0.8T C(T)	S690	18	Tension	Fatigue pre-cracking ($K_I \in [4,61]\,\mathrm{MPa}\sqrt{\mathrm{m}}$)	28
			Tension	Fracture test	–60
0.7T SC(B)	S690	12	Three-point bend	Fatigue pre-cracking ($K_I \in [8,55]\,\mathrm{MPa}\sqrt{\mathrm{m}}$)	28
			Four-point bend	Fracture test	–60

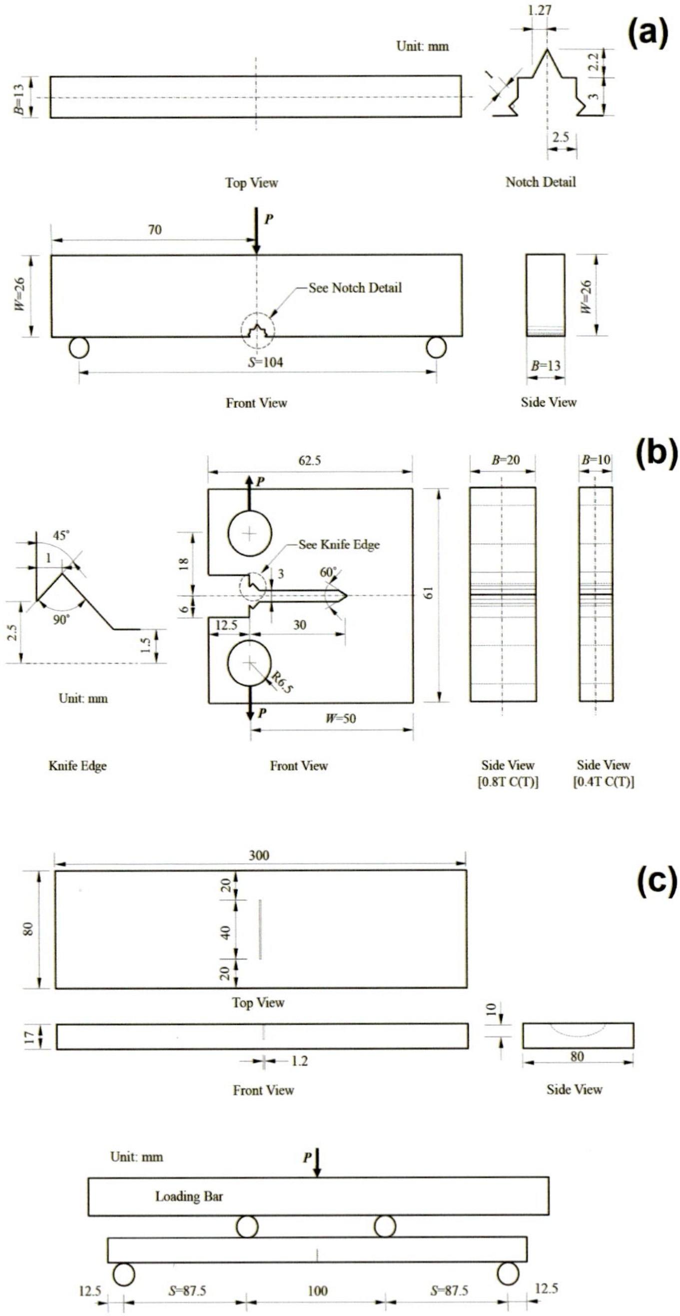

FIGURE 4.15 (a) Dimensions of the 0.5T SE(B) specimen (in mm); (b) dimensions of the 0.4T and 0.8T C(T) specimens (in mm); and (c) dimensions and test set-up of the 0.7T surface-cracked specimens of S690 steel (in mm).

the fatigue pre-cracking and fracture tests follow identical loading setups. For the surface-cracked specimens, fatigue pre-cracking is performed under a three-point bending configuration, whereas fracture testing is conducted under a four-point bending configuration.

The fatigue pre-cracking stage is conducted at room temperature (28 °C) using cyclic loading with low stress intensity factor ranges (ΔK_I), as detailed in Table 4.7. The loading conditions conform strictly to ASTM E1820 [24]. The minimum load during fatigue cycling is set at 3 kN or 10% of the maximum load (whichever is greater) to ensure actuator stability during high-frequency operation.

The fatigue pre-cracking stage generates a crack of 3–4 mm in the through-thickness specimens and approximately 0.5–1.5 mm in the surface-cracked specimens. Before fracture testing, the specimens are cooled to the desired test temperature (listed in Table 4.7) using an environmental chamber with liquid nitrogen as the cooling medium. Each specimen is held inside the chamber for 15 to 60 minutes to achieve a uniform temperature distribution. Three thermocouples, one placed at the crack tip and two at other locations, verify thermal uniformity prior to testing.

During the low-temperature fracture test, a calibrated COD gauge measures the *CMOD* as the load increases. Displacement-controlled monotonic loading begins at a rate of 0.1 mm/min and is incrementally increased to 0.5 mm/min until fracture occurs.

4.4.2 Experimental Results

Fracture toughness for the through-thickness SE(B) and C(T) specimens was evaluated following the procedure outlined in ASTM E1820 [24], using the experimentally measured load versus *CMOD* responses. The total energy release rate J is decomposed into elastic and plastic components as:

$$J = \frac{K_I^2\left(1-\nu^2\right)}{E} + J_{pl} \tag{4.21}$$

The elastic component is determined from the stress intensity factor K_I at fracture load based on the specimen configuration, as prescribed in ASTM E1820 [24].

The plastic component, J_{pl}, is computed using the energy-based η-approach originally proposed by Rice *et al.* [25]:

$$J_{pl} = \frac{\eta A_{pl}}{Bb_0} \tag{4.22}$$

where A_{pl} is the area under the load–plastic *CMOD* curve, B is the specimen thickness, and b_0 is the initial remaining ligament. The non-dimensional parameter η is a geometry-dependent function of the crack depth ratio a/W, as defined in ASTM E1820.

For surface-cracked [SC(B)] specimens, the η-based procedure is extended to account for semi-elliptical surface crack fronts, as developed in a prior study [26]. The plastic component is calculated as:

$$J_{avg} = \frac{\eta U}{A_{net}} \tag{4.23}$$

where A_{net} is the area of the remaining ligament in the cracked section, and U is the area under the bending moment versus crack-plane rotation curve, defined by:

$$U = \int_0^{\theta} PSd\theta \tag{4.24}$$

Here, P is the applied load, S is the moment arm (distance from loading point to roller support), and θ is the rotation of the crack plane, inferred from $CMOD$ measurements as detailed in Ref. [26].

The value of $\eta = 0.9$ for the SC(B) specimens is derived from large-deformation elastic–plastic analysis by matching Eq. (4.23) to the J-integral computed from the domain-integral approach [27], averaged along the entire crack front.

Figure 4.16 compares the calculated fracture toughness values across all four specimen types. Table 4.8 lists the corresponding values obtained from the experiments. The rank probability used in Figure 4.16 is calculated as:

$$P_{i-rank} = \frac{i-0.3}{n+0.4} \tag{4.25}$$

where i is the rank number and n is the total number of specimens in the test set.

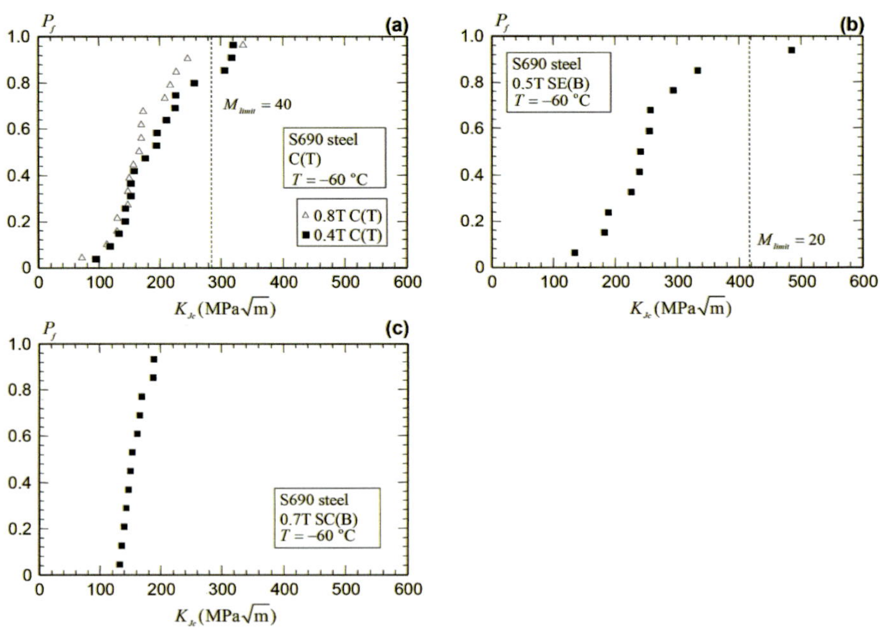

FIGURE 4.16 Rank probability of the fracture toughness data for: (a) 0.5T SE(B); (b) 0.4T and 0.8T C(T); and (c) 0.7T SC(B) of S690.

FIGURE 4.17 The fracture surface of a selected 0.7T SC(B) specimen under the high-magnification digital camera.

Figure 4.17 shows a representative fracture surface of a 0.7T SC(B) specimen captured with a high-resolution camera. Two prominent regions of river pattern convergence are visible near the free surfaces, suggesting cleavage initiation occurred near these locations.

To quantify the location of fracture initiation, a position parameter ϕ is introduced:

$$\phi = \frac{l}{B_{total}} \cdot 180^{\circ} \tag{4.26}$$

where l is the arc length measured from the free surface to the initiation point along the crack front, and L is the total length of the semi-elliptical crack front (see Figure 4.17). The data indicate that fracture initiation occurred near the free surfaces at $\phi \approx 0.08$ and $\phi \approx 0.92$. The measured crack-front angles at these locations range from 13.5° to 29° or from 151° to 166.5° across all 12 SC(B) specimens tested.

4.4.3 Weibull Stress Assessment

Fracture-probability estimation via the local approach requires evaluating the Weibull stress from detailed FE simulations. The Weibull material parameters [17] are calibrated under plane-strain, SSY conditions using a MBL model (Figure 4.2a). The half-symmetric, semi-circular MBL mesh comprises a single layer of 20-node quadratic brick elements through the thickness; out-of-plane translations are constrained to enforce plane strain.

Figures 4.18a and b show quarter-symmetry meshes for a single-edge-bend specimen SE(B) ($a/W = 0.32$) and a C(T) specimen ($a/W = 0.67$). Each model contains 12 element layers over the half-thickness, with element height gradually reduced towards the free surface; the smallest thickness-direction element measures 1.5 μm. Figure 4.18c presents the quarter-symmetry model of a surface-cracked plate with crack depth a (11 mm) and surface half-length c ($a/c = 0.428$). This geometry represents the average semi-elliptical crack produced by fatigue pre-cracking. The quarter-elliptical front is discretized with 100 uniformly sized elements (0.3 mm) along the crack front. An independent mesh-sensitivity study confirmed that this resolution yields converged Weibull-stress results.

All meshes employ 20-node quadratic solid elements with reduced integration. To avoid numerical divergence at large plastic strains, a finite root (key-hole) radius is

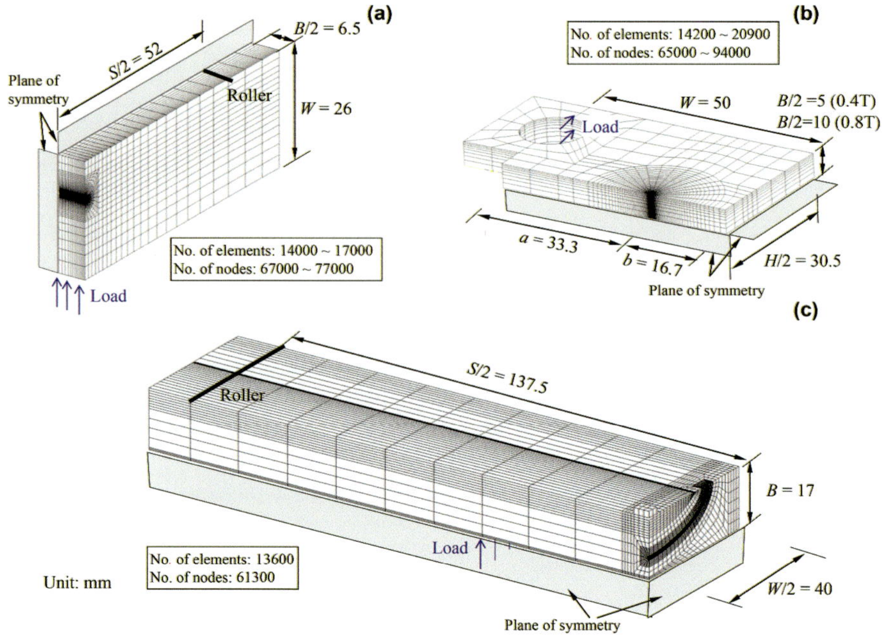

FIGURE 4.18 Typical FE meshes for: (a) a quarter-symmetric SE(B) model; (b) a quarter-symmetric C(T) model; and (c) a quarter-symmetric 0.7T SC(B) model.

imposed at the crack tip (Figure 4.2d). Accurate evaluation of the $g(M)$ function in Eq. (4.9) across the full loading range, therefore, relies on a set of FE models, each with a different crack-tip radius (R_0 = [0.5 μm, 6 μm]), ensuring stable near-tip fields throughout the analysis window.

The FEA apply the displacement loads to all models. The MBL model experiences an applied displacement along the circumference of the semi-circle given in Eqs. (4.14) and (4.15). For the SE(B), C(T), and surface-cracked plates, the entire thickness (or width) at the loading positions experiences a uniform displacement load. All simulations use the true stress-strain curve of Figure 4.14 with $E = 206$ GPa and $v = 0.3$ for the S690 steel. The energy-release rate J is obtained with the domain-integral formulation in WARP3D [20]. Figure 4.19a tracks J versus $CMOD$ for three through-thickness specimens [0.5 T SE(B), 0.4 T C(T), and 0.8 T C(T)].

History-dependent Weibull stresses σ_w were evaluated following Refs. [14, 17]. Figure 4.19b shows σ_w/σ_0 along the same crack front for three different load levels. Each σ_w represents the value integrated with Eq. (4.7) over a crack-front segment equal to the length of one finite element. Hence, σ_w is sensitive to the chosen element (process-zone) volume.). Figure 4.19c gives the normalized distribution (normalized by the maximum J along the semi-elliptical front) at three load levels, the lowest failure load at $K_{Jc,min} = 142\text{MPa}\sqrt{\text{m}}$ the average failure load at $K_{Jc,mean} = 177\text{MPa}\sqrt{\text{m}}$ and the largest failure load at $K_{Jc,max} = 225\text{MPa}\sqrt{\text{m}}$ from the twelve tests. In every case J peaks close to the free surface.

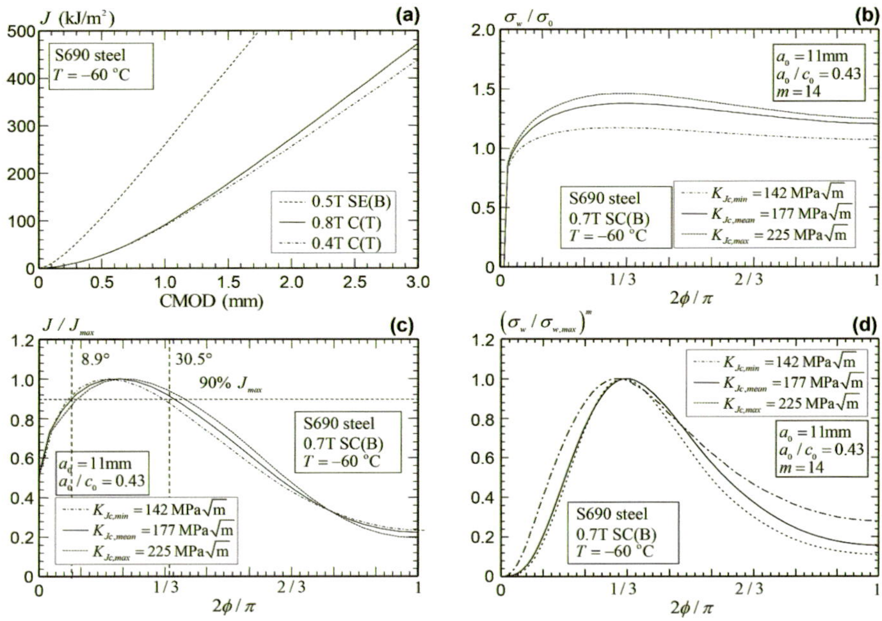

FIGURE 4.19 FE analysis results for: (a) J versus $CMOD$ for through-thickness specimens; (b) normalized Weibull stress along the crack front in S690 surface-cracked specimens; (c) normalized J values along the crack front in S690 surface-cracked specimens; and (d) normalized σ_w^m along the crack front in S690 surface-cracked specimens.

Post-test inspections confirmed that fracture nucleates where the macroscopic driving force is highest. Accordingly, the fracture-initiation zone is defined as all crack-front segments for which,

$$J \geq 0.9 J_{\text{max}} \tag{4.27}$$

This definition corresponds to a crack-front segment $\phi = 8.9^\circ$ to $\phi = 30.5^\circ$ at the average failure loads as shown in Figure 4.19c. This band coincides with the observed initiation sites over the range $\phi = 13.5^\circ$ to $\phi = 29^\circ$.

Gao *et al.* [28] proposed an alternative "risky-spot" indicator based on $\sigma_w^m / \sigma_{w-\max}^m$, as illustrated in Figure 4.19d, which leads essentially to the same zone as Eq. (4.27). Subsequent discussions show that predicted fracture probabilities are insensitive to the exact width of this critical segment.

Although J and σ_w peak at almost identical positions, the gradient of σ_w along the front is much shallower than that of J (compare Figures 4.19b and c). Moreover, Eq. (4.9) couples J to the Weibull modulus m; any σ_w-based fracture rule, therefore, requires an explicit, sometimes uncertain, choice of m. The simpler $0.9J_{max}$ rule avoids that complication and is adopted for the engineering assessment of surface-crack cleavage in this study.

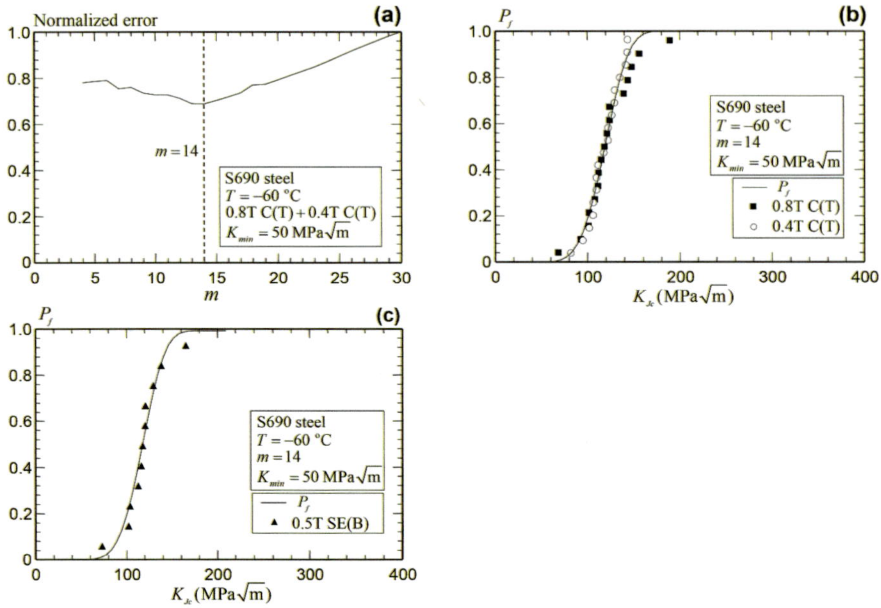

FIGURE 4.20 (a) Variation of the normalized error with respect to m; (b) estimation of P_f by the Weibull stress model using the calibrated m and K_{min} values for 0.8T and 0.4T C(T) specimens; and (c) estimation of P_f by the Weibull stress model using the calibrated m and K_{min} values for 0.5T SE(B).

Based on the flowchart in Figure 4.1, the Weibull stress-related parameters (m and K_{min} values) can be obtained by using the 0.8T C(T) specimens as the high-constraint specimen set and the 0.4T C(T) specimens as the low-constraint set. The third set of specimens, that is, the 0.5T SE(B) specimens, provides an independent set of test data to validate the calibrated Weibull material parameters. The calibration of the S690 steels at $-60\ ^{o}\text{C}$ covers a range of m values from 4 to 30, and a K_{min} range from 1 $\text{MPa}\sqrt{\text{m}}$ to 80 $\text{MPa}\sqrt{\text{m}}$. Figure 4.20a minimizes the calibration error function at $m = 14$ and $K_{min} = 50\ \text{MPa}\sqrt{\text{m}}$. Figure 4.20b confirms the accuracy in the probability of fracture for the 0.8T and 0.4T C(T) specimens. Furthermore, the independent set of fracture tests on 0.5T SE(B) specimens at $-60\ ^{o}\text{C}$ validates the calibrated parameters through the probability of fracture in Figure 4.20c.

4.4.4 Probability of Brittle Fracture

Surface-cracked plates do not behave like conventional through-thickness specimens: the crack-driving force J and the level of plastic constraint vary continuously along the curved crack front. Post-test fractography shows that cleavage always starts from a small, critical sector of the front rather than from its entire length. Consequently, a straightforward Weibull-stress analysis that integrates σ_w over the whole crack front tends to mis-predict the fracture probability. This limitation is evident in Figure 4.21a, where the probability obtained from Eq. (4.10), by treating the entire front as an "equivalent through-thickness" crack and using the calibrated Weibull parameters m and K_{min}, diverges markedly from the test data.

To overcome this issue, we localize the assessment. The initiation zone is defined as the portion of the crack front for which the local J exceeds 90% of the maximum J along the front. This study then treats that zone as an equivalent through-thickness specimen whose "thickness" B_{FZ} equals the arc length of the selected segment (64 mm in the present study). The Weibull stress is computed only over the fracture-process zone (FPZ) contained within this segment, yielding the representative value $\sigma_{w,FZ}$ $[\sigma_{w,FZ}^{m} = CB_{FZ}K_{FZ}^{4}g(M)]$ used for the probability calculation. This focused approach preserves the physics of cleavage initiation and produces a much closer match to the experimental fracture statistics.

The above definition is not ideal for routine engineering use because the key inputs, that is, the length of the fracture-initiation zone B_{FZ} and its zone-averaged toughness K_{FZ} cannot be obtained directly from standard test data. To keep the physics intact while simplifying the calculations, we introduce a modified definition of the Weibull stress, $\sigma_{w,FZ}$. This stress is still evaluated over the initiation zone, but it is expressed entirely in terms of test-accessible, whole-front quantities, the total crack-front length B_{total} and the toughness averaged along that front $K_{Jc,avg}$.

$$\sigma_{w,FZ}^{m} = CB_{total}K_{J,avg}^{4}g_{sc}(M) \tag{4.28}$$

Equation (4.28), therefore, embeds a surface-crack-specific constraint-correction function, $g_{sc}(M)$, which tailors the Weibull-stress formulation to the non-uniform triaxiality that characterizes curved crack fronts.

$$g_{sc}(M) = \frac{B_{FZ}K_{J,FZ}^{4}}{B_{total}K_{J,avg}^{4}}\,g(M) \tag{4.29}$$

For the S690 surface-cracked plates examined here, the curved crack front measures B_{total} = 64 mm, whereas the cleavage-critical initiation zone occupies only $B_{FZ} \approx$ 8 mm, roughly 12% of the front. By computing the Weibull stress (i) for a reference SSY model and (ii) for the initiation zone using Eq. (4.28), we obtain the surface-crack constraint-correction function $g_{sc}(M)$ directly from Eq. (4.29).

Applying the Weibull stress in conjunction with the new constraint-correction function defined in Eq. (4.28) allows the average fracture toughness measured along the crack front of a surface-cracked specimen to be translated to an equivalent 1T SSY condition, using the Weibull stress evaluated over the fracture-initiation zone as the scaling reference.

$$K_{Jc,avg}^{1T\ SSY} = K_{min} + \left[g_{sc}(M)^{1/4}K_{Jc,avg} - g_{sc}(M_{min})^{1/4}K_{min}\right]\left(\frac{B_{total}}{B_{1T}}\right)^{1/4} \tag{4.30}$$

The fracture-toughness of the plate specimen is first "scaled" with Eq. (4.30), employing the Weibull parameters $m = 14$ and $K_{min} = 50\ \text{MPa}\sqrt{\text{m}}$ that were calibrated with through-thickness (1T) specimens. Figure 4.21b then juxtaposes the cumulative probability of cleavage failure predicted by Eq. (4.10) with the fracture-toughness data for the 0.7T surface-cracked S690 plate after transformation to the equivalent 1T SSY reference state.

To gauge how sensitive the prediction is to uncertainty in crack size, three representative surface-crack sizes measured at the end of fatigue pre-cracking in the 12 surface-cracked plates were adopted in the Weibull-stress calculation: 1) the minimum crack size ($a_0 = 10.3$ mm, $a_0/c_0 = 0.43$); 2) the average crack size ($a_0 = 11$ mm, $a_0/c_0 = 0.43$); and 3) the maximum crack size ($a_0 = 11.2$ mm, $a_0/c_0 = 0.42$).

For each crack size, a FE model provided the local Weibull stress σ_w and the constraint-correction function $g(M)$ over the critical segment of the crack front. As Figure 4.21b shows, all three cases collapse onto virtually the same fracture-probability curve, demonstrating that the method is largely insensitive to moderate variations in crack size.

This agreement confirms the reliability of the Weibull-stress framework for cleavage-fracture assessment. In practical engineering, however, fracture-toughness data from surface-cracked specimens are seldom available. To overcome that limitation, we now propose a reverse-scaling procedure that predicts the probability of

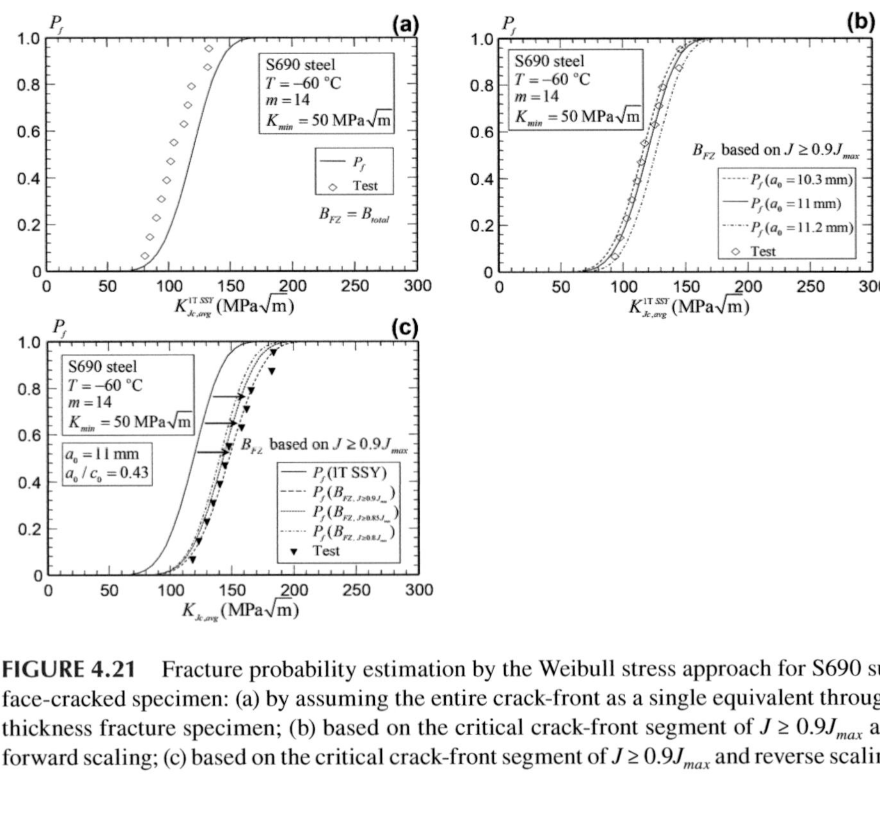

FIGURE 4.21 Fracture probability estimation by the Weibull stress approach for S690 surface-cracked specimen: (a) by assuming the entire crack-front as a single equivalent through-thickness fracture specimen; (b) based on the critical crack-front segment of $J \geq 0.9J_{max}$ and forward scaling; (c) based on the critical crack-front segment of $J \geq 0.9J_{max}$ and reverse scaling.

cleavage in surface-cracked plates using only numerically determined crack-driving forces (e.g., $K_{J,avg}$) and the associated Weibull stress field, eliminating the need for additional surface-crack toughness tests.

The probability of cleavage fracture in a surface-cracked specimen can be estimated with the calibrated Weibull shape parameter m and scale parameter K_{min} by following the workflow below.

1) FEA conduct large-deformation, elastic–plastic FE simulations for: a) the actual surface-cracked geometry of interest, and b) a reference 1T single-edge-crack model loaded under plane-strain, SSY conditions.
2) Identify the critical crack-front segment and compute Weibull stress: a) locate the critical portion of the crack front in the surface-cracked model with $J \geq 0.9J_{max}$; b) evaluate the Weibull stress over that segment, $\sigma_{w,FZ}$, as a function of the crack-driving force J; c) in parallel, calculate the corresponding Weibull stress σ_w for the 1T SSY model.
3) Derive the constraint-correction function $g(M)$: a) construct the $\sigma_{w,FZ}$ $K_{J,avg}$ curves for the surface-cracked specimens and 1T SSY condition; b) obtain the constraint-correction function $g_{sc}(M)$ by comparing the $\sigma_{w,FZ}$-$K_{J,avg}$ relationship for the surface-cracked plates and σ_w-$K_{J,avg}$ for the 1T SSY condition, which captures the difference in crack-tip constraint between the curved surface crack and the idealized SSY reference.
4) Reverse-scale to the surface-cracked toughness: a) choose a target cumulative probability of cleavage fracture P; b) using Eq. (4.10) and (4.30), together with the calibrated m and K_{min} values, find the crack-driving force $K_{Jc}^{1T\ SSY}$ that gives the probability P in the 1T SSY model; c) estimate the average fracture-toughness value $K_{Jc,avg}$ along the actual crack front via,

$$K_{Jc,avg} = \frac{1}{g_{sc}\left(M\right)^{1/4}}\left[\left(K_{Jc}^{1T\ SSY} - K_{min}\right)\left(\frac{B_{1T}}{B_{total}}\right)^{1/4} + g_{sc}\left(M_{min}\right)^{1/4} K_{min}\right] \qquad (4.31)$$

This reverse-scaling procedure provides a probabilistic fracture-toughness estimate for surface-cracked components without requiring dedicated surface-crack toughness tests, relying only on FE-derived crack-driving forces and the previously calibrated Weibull parameters.

Figure 4.21c contrasts the model predictions with the test data for the 0.7T surface-cracked S690 plates. The solid curve represents the probability of cleavage fracture for an idealized 1T, plane-strain SSY specimen, obtained directly from Eq. (4.10). The dashed curves indicate the probabilities recalculated for the surface-cracked geometry after reverse-scaling its J values to the 1T SSY reference with Eq. (4.31). The dashed curves fall almost exactly on the rank-ordered experimental fracture-toughness points, confirming that the reverse-scaling procedure captures the behavior of the real plates.

The analysis defines the fracture-initiation zone B_{FZ} by imposing a cut-off on the global driving force, $J \geq cJ_{max}$. Sensitivity was checked with $c = 0.80$, 0.85, and 0.90.

Lower cut-offs include a longer segment of the crack front, increasing the local driving force and raising the calculated failure probability slightly. As Figure 4.21c shows, the three curves differ by only a few percentage points, indicating that the predicted risk of fracture is essentially insensitive to the exact choice of cut-off level for the initiation zone.

4.5 ASSESSMENT OF CLEAVAGE ANGLES UNDER MIXED-MODE LOADING

4.5.1 Cleavage Angles

This section outlines a statistical framework for predicting both (i) the likely range of crack-initiation angles and (ii) the cumulative probability that cleavage will initiate at any given angle. Erdogan and Sih [29] argued that, in brittle fracture, cracks nucleate in the immediate vicinity of the tip and then propagate radially. Under mixed-mode I/II loading, however, a propagating crack almost always kinks out of its original plane, whereas pure mode-I growth is self-similar. Within a weakest-link picture, failure of the first micro-crack anywhere along the front causes catastrophic fracture of the whole specimen. In the Weibull-stress formalism, that event is governed by the scalar σ_w, so regions where σ_w attains the same value are equiprobable failure sites (assuming the material is locally homogeneous). Gao *et al.* [28] captured this idea for three-dimensional crack fronts with the lineal indicator $(\sigma_w/\sigma_u)^m$ per unit length, where σ_u is the scale parameter and m the Weibull modulus.

Here we introduce an analogous angular Weibull-stress density, $\sigma_{w\text{-}\theta}$, defined as the incremental contribution of σ_w^m per unit angular increment around the crack tip. Peaks in $\sigma_{w\text{-}\theta}$ identify the most probable kink angles, while integration of $\sigma_{w\text{-}\theta}$ furnishes the cumulative probability that cleavage initiates within any prescribed angular sector.

$$\sigma_{w-\theta} = \left[\frac{1}{V_0(2\Delta\theta)} \int_0^B \int_{\theta-\Delta\theta}^{\theta+\Delta\theta} \int_0^{\bar{r}(\theta)} \sigma_{eff}^m r dr d\theta dz \right]^{1/m} \tag{4.32}$$

where B is the specimen thickness, θ is the counter-clockwise polar angle measured from the crack plane at the tip (Figure 4.22a), $\Delta\theta$ represents the angular increment

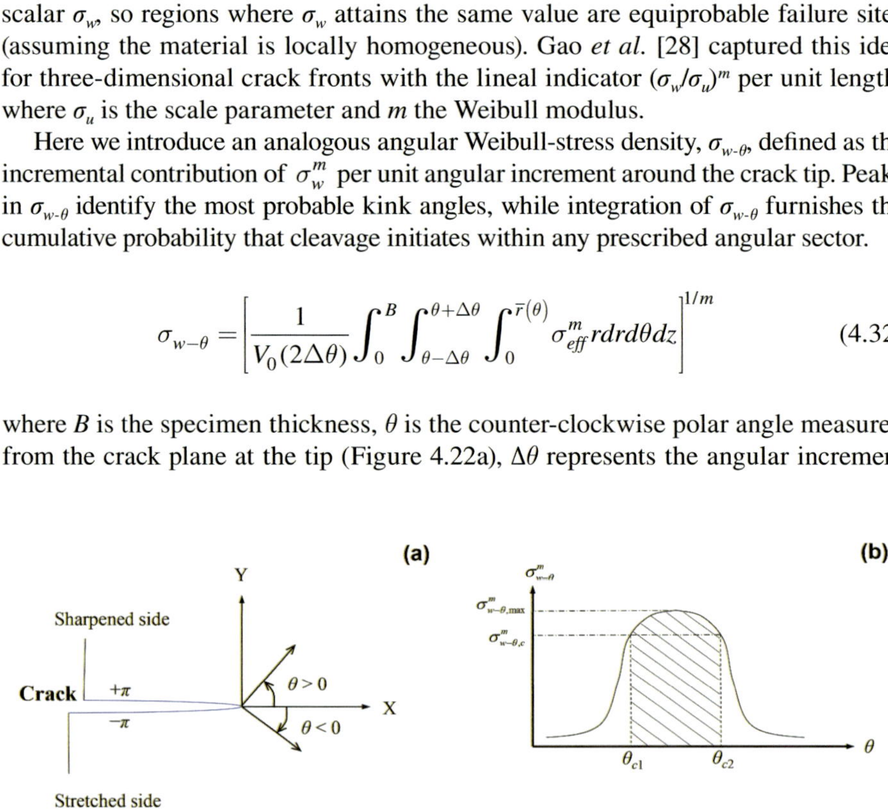

FIGURE 4.22 (a) Schematic illustration of angular position θ near the crack tip; (b) confidence interval of cleavage initiation angles based on the angular Weibull stress density.

set by the number of FE sectors surrounding the tip, and the upper integration limit, $\bar{r}(\theta)$, corresponds to the angular extent of the FPZ.

Following the weakest-link rationale, the probability that cleavage initiates somewhere within the polar sector $-\pi \le \theta \le \theta_c$ is obtained by integrating the angular Weibull-stress density and substituting the result into,

$$P_f(-\pi \le \theta \le \theta_c) = \frac{\int_{-\pi}^{\theta_c} \sigma_{w-\theta}{}^m d\theta}{\int_{-\pi}^{\pi} \sigma_{w-\theta}{}^m d\theta} \tag{4.33}$$

Equation (4.33) defines a probability for cleavage initiation angle in the polar interval $-\pi \le \theta \le \pi$ (Figure 4.22a). By construction, its integral over the entire $\theta = -\pi$ to $\theta = \pi$ domain equals 1, so the cumulative probability that initiation occurs within $-\pi \le \theta \le \theta_c$ is the ratio of the $\sigma_{w-\theta}^m$ calculated from $\theta = -\pi$ to $\theta = \theta_c$ over the total $\sigma_{w-\theta}^m$ computed over the entire fracture process zone from $\theta = -\pi$ to $\theta = \pi$.

Figure 4.22b shows the typical variation of the angular Weibull stress with respect to the crack tip angle θ. The angular location with the highest $\sigma_{w-\theta}^m$ value has the highest probability for the cleavage failure to initiate. If we introduce a critical cutoff value, $\sigma_{w-\theta,c}^m$, the domain covered by $\sigma_{w-\theta}^m \ge \sigma_{w-\theta,c}^m$ corresponds to the angular range $\theta_{c1} \le \theta \le \theta_{c2}$, as shown in Figure 4.22b. The area below the curve from θ_{c1} to θ_{c2}, that is, shaded area in Figure 4.22b, over the entire area under the curve, represents the cumulative probability or the confidence level that cleavage fracture will initiate in the angular range of $\theta_{c1} \le \theta \le \theta_{c2}$.

4.5.2 Fracture Initiation Criteria

This section reviews three distinct failure criteria that govern unstable micro-crack growth in ferritic steels. In the following sections, each criterion is embedded in the Weibull-stress framework to assess how well it predicts cleavage-crack initiation angles under mixed-mode I/II loading.

1) Maximum Principal Stress Criterion: The maximum principal stress criterion assumes that unstable cleavage fracture initiates in a direction perpendicular to the maximum principal stress. The effective stress driving the cleavage fracture, therefore, becomes,

$$\sigma_{eff} = \sigma_1 \tag{4.34}$$

2) Coplanar Energy Release Rate: Evans [30] assumes that the fracture process zone contains randomly distributed, penny-shaped flaws, with their normal orientated along the angular θ direction. The mode I and II stress intensity factors for a penny-shaped flaw with a radius a in a uniformly stressed material equal,

$$K_I = 2\sigma_n \sqrt{\frac{a}{\pi}} \tag{4.35}$$

$$K_{II} = \frac{4\tau_{in}}{2-\upsilon}\sqrt{\frac{a}{\pi}} \tag{4.36}$$

where σ_n indicates the tensile stress normal to the microcrack plane, and τ_{in} denotes the maximum in-plane shear stress. The microcrack failure initiates when the strain-energy release rate, G_c, attains a critical value,

$$G_c = \frac{1-\upsilon^2}{E}K_I^2 + \frac{1-\upsilon^2}{E}K_{II}^2 = \frac{1-\upsilon^2}{E}K_{eff}^2 = \frac{1-\upsilon^2}{E}\left(2\sigma_{eff}\sqrt{\frac{a}{\pi}}\right)^2 \tag{4.37}$$

Substituting Eqs. (4.35) and (4.36) into Eq. (4.37), the effective stress driving the cleavage fracture, under the coplanar energy release rate criterion, becomes,

$$\sigma_{eff} = \sqrt{\sigma_n^2 + \frac{4\tau_{in}^{\ 2}}{(2-\upsilon)^2}} \tag{4.38}$$

Evan's approach was originally proposed to assess statistically the brittle fracture failure in ceramic materials [30]. Subsequent researchers have extended this approach to many other materials including steels [2] and polymer composite materials [31].

3) Modified Coplanar Energy Release Rate: Chang *et al.* [32] assume a kinked crack propagation along the θ direction and propose that the energy release rate G_θ under mixed-mode I and II loadings follows,

$$G_\theta = \frac{\kappa+1}{8G}\frac{1}{2}\cos^2\frac{\theta}{2}\left[(1+\cos\theta)K_I^{\ 2} - 4\sin\theta K_I K_{II} + (5-3\cos\theta)K_{II}^{\ 2}\right] \tag{4.39}$$

where $\kappa = 3-4\nu$ for a plane strain condition and $\kappa = (3-\nu)/(1+\nu)$ for a plane stress condition. Substituting Eqs. (4.35) and (4.36) into Eq. (4.39) leads to,

$$G_\theta = \frac{\kappa+1}{8G}\frac{4a}{\pi}\frac{1}{2}\cos^2\frac{\theta}{2}\left[(1+\cos\theta)\sigma_n^{\ 2} - \frac{8\sin\theta}{2-\nu}\sigma_n\tau_{in} + \frac{4(5-3\cos\theta)}{(2-\nu)^2}\tau_{in}^{\ 2}\right] \tag{4.40}$$

The modified coplanar energy release rate criterion thus yields an effective stress,

$$\sigma_{eff} = \left\{\frac{1}{2}\cos^2\frac{\theta}{2}\left[(1+\cos\theta)\sigma_n^{\ 2} - \frac{8\sin\theta}{2-\upsilon}\sigma_n\tau_{in} + \frac{4(5-3\cos\theta)}{(2-\upsilon)^2}\tau_{in}^{\ 2}\right]\right\}^{1/2} \tag{4.41}$$

Chang *et al.* [32] have verified the above criterion on aluminum alloys. Subsequent researchers have validated and extended the modified co-planar energy release rate in other materials, including steels [33, 34] and polymeric glasses [35, 36].

The three idealized initiation criteria above are formulated for a single, sharp micro-crack whose dimensions are comparable to those of an intermetallic inclusion. They do not account for the number, size, or spatial distribution of micro-cracks evolves ahead of the macroscopic crack tip as plastic deformation increases.

4.5.3 Experimental Validation

An experimental program was designed to measure cleavage-crack initiation angles in S550 steel under mixed-mode I/II loading with an asymmetric four-point-bend configuration. S550 is a fully killed, fine-grain, thermo-mechanically rolled plate; every specimen was cut from the same plate in the T–L orientation. Two specimen sizes were used: 29 "0.75 T" compacts (19 mm thick) and 28 "1.25 T" compacts (32 mm thick). Independent mode-I tests on companion samples [37] established a Weibull modulus of $m = 12$ for this steel. Mechanical properties are 640 MPa yield strength and 734 MPa ultimate strength at room temperature, increasing to 715 MPa and 810 MPa, respectively, at –90 °C. The reference temperature T_0, defined by a median toughness of 100 $\text{MPa}\sqrt{\text{m}}$, is about –150 °C. Accordingly, all mixed-mode tests were run at –90 °C (≈60 °C above T_0), with the specimens cooled by liquid nitrogen in an environmental chamber.

Figure 4.23a presents the geometries of the 0.75 T and 1.25 T compact-bend specimens, respectively. Each plane-sided specimen contains a machined starter notch with a ratio a_n/W of 0.39 for the 0.75 T pieces and 0.44 for the 1.25 T specimens, respectively. Before the low-temperature fracture tests at –90 °C, all specimens were fatigue-pre-cracked in three-point bending at room temperature in accordance with ASTM E1820 [24]. The resulting fatigue-crack lengths were typically 3.2 mm–4.4 mm. Figure 4.23c shows the asymmetric four-point-bend/shear fixture and boundary conditions used in the mixed-mode tests. *CMOD* was monitored throughout with a clip-on crack-opening gauge (COG). The loading was displacement-controlled, following the procedure of Yang and Qian [38]. To characterize the relative proportions of mode I and mode II loading, the equivalent mode-mixity angle $\beta_{eq} = \tan^{-1}\left(K_I/K_{II}\right)$ proposed by Maccagno and Knott [39] was adopted.

Table 4.8 summarizes the specimen dimensions and mode-mixity angles for the two sets examined in this study. He and Hutchinson [40] showed that the position of the loading points has a negligible effect on the stress-intensity factors, provided the distance between the crack tip and the nearest bottom roller exceeds 1.4*W*, that is, $(S + C) > 1.4W$ (see Figure 4.23c).

Following the *J*-integral procedure in ASTM E1820 [24] for assessing fracture resistance under pure Mode I loading, this study generalizes the same *J*-based methodology to determine the mixed-mode (I + II) energy-release rates for asymmetric four-point-bend and shear specimens,

$$J = J_{el} + J_{pl} = \frac{\left(K_I^2 + K_{II}^2\right)(1-\upsilon^2)}{E} + \frac{\eta_{pl,\text{mixed}} A_{pl}}{Bb_0} \tag{4.42}$$

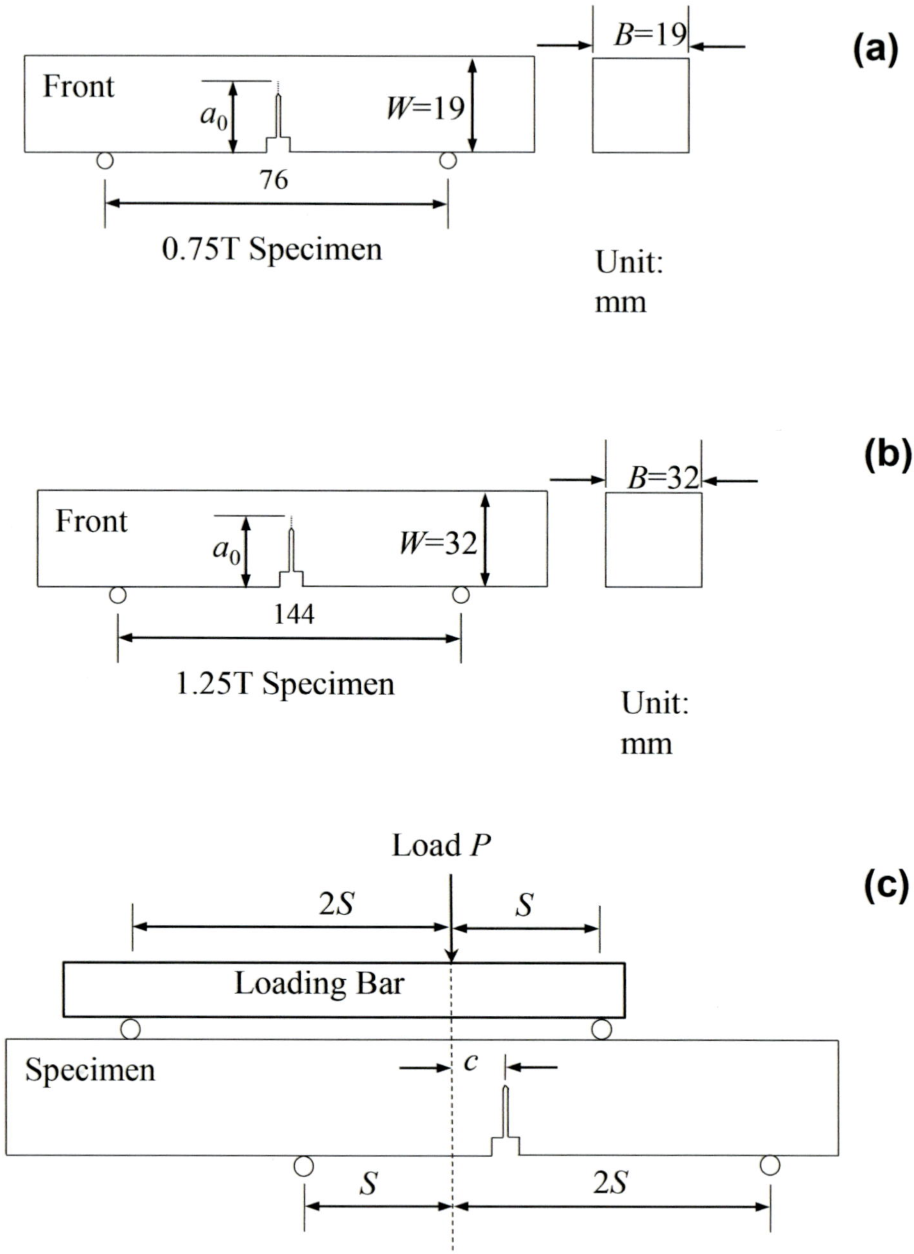

FIGURE 4.23 Configurations of: (a) a 0.75T specimen and (b) a 1.25T specimens; and (c) the schematic setup for the four-point bend and shear test.

TABLE 4.8
Summary of Mixed-Mode Experimental Plan.

Specimen set	β_{eq}(°)	*B* (mm)	*W* (mm)	*S* (mm)	*c* (mm)	a_0/W	No. of specimens
0.75T	78	19	19	32	13	0.581	29
1.25T	78	32	32	45	22	0.566	28

where J_{el} and J_{pl} refer to the elastic and plastic components of energy release rate, respectively. A_{pl} represents the area below the applied force versus the plastic *CMOD* curve, *B* refers to the specimen thickness, b_0 indicates the initial ligament length, and,

$$K_I = \frac{F_I P\sqrt{\pi a}}{WB} \tag{4.43}$$

$$K_{II} = \frac{F_{II} P\sqrt{\pi a}}{WB} \tag{4.44}$$

where *P* refers to the applied load, while F_I and F_{II} denote the non-dimensional geometric factors derived from linear-elastic FE analysis. The $\eta_{pl,mixed}$ values thus derived from Eq. (4.42) from the numerical analysis, with $\eta_{pl,mixed} = 0.552$ for $a/W = 0.56$ and $\eta_{pl,mixed} = 0.539$ for $a/W = 0.59$ [41].

A detailed inspection of the cleaned, corrosion-protected fracture surfaces reveals well-defined river patterns that converge on the cleavage-initiation sites (Figure 4.24a). In a few specimens, a narrow band of stable crack growth, typically < 1 mm, preceded the onset of cleavage.

After testing, each fracture surface was replicated with a high-fidelity silicone compound that faithfully reproduces both the fatigue-crack front and the subsequent cleavage facets. The silicone replica was then sectioned along a plane perpendicular to the local crack front at the initiation point, and the cross-section was examined with an optical microscope. By comparing the orientations of the fatigue-pre-crack surface and the cleavage facet on the micrograph, the cleavage-initiation angle was extracted from a linear fit to the surface profile immediately ahead of the original crack tip.

Figures 4.24b and c define the cleavage-initiation angle, θ_c, as the angle between the cleavage-fracture facet and the original, undeformed fatigue-crack plane in the *x*–*y* coordinate system. The reference angle θ is measured counter-clockwise from the +*x* axis and is taken as positive toward +*y*. Figure 4.24c shows how θ_c is resolved on a half-fractured specimen,

$$\theta_c = \theta_1 + \theta_2/2 \tag{4.45}$$

where θ_1 is the angle between the fatigue pre-crack plane and the cleavage facet at the subsurface crack-front (i.e., not on the free surfaces), and θ_2 is the crack-mouth

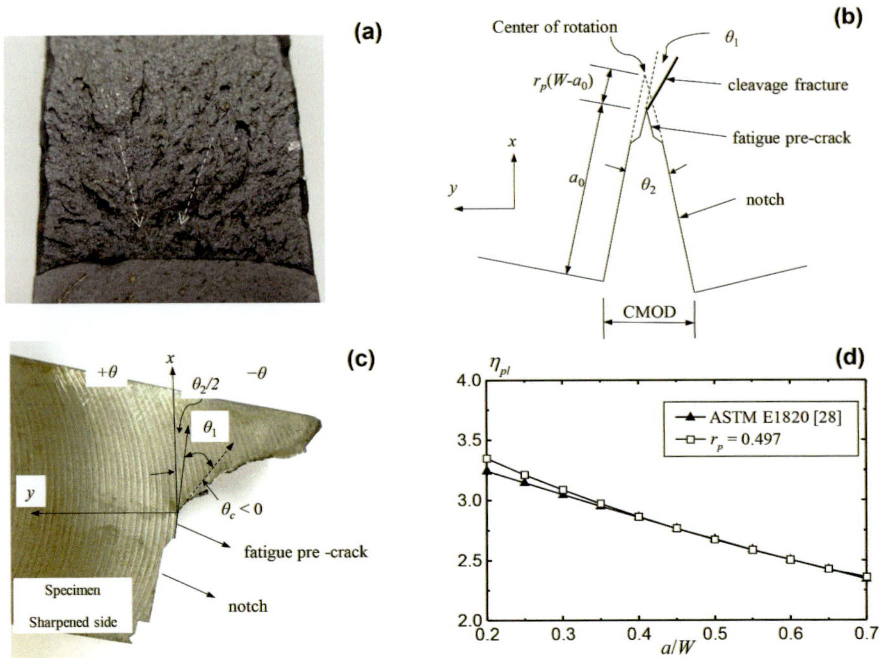

FIGURE 4.24 (a) The fracture surface with river patterns for a typical fractured specimen; (b) definition of cleavage initiation angles; (c) components of cleavage initiation angles illustrated in a half broken specimen; and (d) the comparison of η_{pl} values over different *a*/*W* ratios.

opening angle measured at the instant of cleavage. This decomposition allows θ_c to be determined from two readily measurable quantities on the fracture-surface replica. The angle θ_2 follows,

$$|\theta_2| = \frac{CMOD_{fracture}}{a_0 + r_p(W - a_0)} \tag{4.46}$$

where $CMOD_{fracture}$ indicates the *CMOD* value at the load level causing the cleavage failure of the specimen, a_0 denotes the initial crack length (including the fatigue pre-crack), W is the specimen width, and r_p represents the plastic rotational factor, equal to the ratio between the distance from the crack tip to the center of rotation and the remaining ligament.

Equation (4.46) treats the crack-plane center of rotation as fixed during opening and independent of the initial crack length a_0 and specimen width W. Following the procedure of Keiichiro and Hitoshi [42], a constant pivot ratio of $r_p = 0.497$ is adopted for mixed-mode I/II loading. With this assumption, the *J*-integral obtained from the area beneath the M–θ_2 curve should be equal to the value calculated from the area under the *P*–*CMOD* curve,

$$J_I^{pl} = \frac{2}{Bb_0}\int_0^{\theta_{2,pl}} Md\theta_{pl} = \frac{\eta_{pl}}{Bb_0}\int_0^{CMOD_{fracture,pl}} PdCMOD_{pl} \tag{4.47}$$

Substituting Eq. (4.46) and the bending moment applied on the cracked section ($M = PW$) into Eq. (4.47) and neglecting the elastic *CMOD* component, we have,

$$\eta_{pl} = \frac{2}{a_0 / W + r_p(1 - a_0 / W)} \tag{4.48}$$

Figure 4.24d indicates that the η_{pl} values corresponding to $r_p = 0.497$ in Eq. (4.48) agrees well with the recommended η_{pl} value in the ASTM E1820 [24] for medium to deep cracks ($a_0 / W \geq 0.4$).

Tables 4.9 and 4.10 list the cleavage-initiation angles (θ_c) measured for the 0.75 T and 1.25 T specimens. In both series, the data shows substantial scatter. Because θ_c is controlled mainly by θ_1, the angle between the fatigue pre-crack plane and the subsequent cleavage facet, its value changes across the thickness of each mixed-mode specimen, whereas θ_2 (the crack-mouth opening angle) remains essentially constant for a given test piece. The θ_c values reported here are thickness-averaged results obtained at crack-front positions that exhibit clear river patterns.

4.5.4 Cleavage Angle Assessment

This study computes the Weibull stress in the fracture specimens using the large-deformation, elastic-plastic analysis performed in an open-source FE code WARP3D [20]. Figure 4.25a shows the true stress-true strain relationship for the S550 steel measured at –90 °C, which follows the standard Ramberg-Osgood constitutive model,

$$\varepsilon = \frac{\sigma}{E}\left[1 + \alpha\left(\frac{\sigma}{\sigma_{ys}}\right)^{n-1}\right] \tag{4.49}$$

where the yield stress σ_{ys} equals 715 MPa, the strain hardening component n has a value of 8.0, the Young's modulus E is 210 GPa, and the Poisson's ratio v takes a value of 0.3.

Figure 4.25b illustrates a half-symmetry FE model of the four-point bending–shear specimen. The mesh uses 20-node quadratic brick elements with reduced integration (C3D20R). Nodes on the symmetry plane are restrained in the thickness (out-of-plane) direction. Sixteen graded layers of elements are applied through the thickness, with element size progressively refined toward the free surface. Figure 4.25c zooms in on the crack tip. Concentric rings of focused elements create an initial crack-root radius of 2.5–5 μm, and 40 elements are distributed circumferentially (θ-direction) around the crack front to resolve the steep near-tip gradients.

The specimen's constitutive response follows the true-stress versus true-strain curve in Figure 4.25a. To approximate a rigid support, the loading bar is assigned

TABLE 4.9

Summary of Measured Crack Initiation Angles for 0.75T Specimens.

Specimen no	$\Delta\bar{a}$ (mm)	θ_t (°)	$\bar{\theta}_t$ (°)	θc (°)	θ_1 (°)	$\lvert\theta_2/2\rvert$ (°)	Specimen no	$\Delta\bar{a}$ (mm)	θ_t (°)	$\bar{\theta}_t$ (°)	θc (°)	θ_1 (°)	$\lvert\theta_2/2\rvert$ (°)
1	0.218	–	–	-26.9	-24.2	2.69	16	0	–	–	-31.7	-29.4	2.32
2	0.401	15.39	19.60	14.1	18.3	4.19	17	0.317	–	–	-15.4	-10.9	4.50
3	0.235	–	–	-30.7	-27.3	3.46	18	0	–	–	-29.0	-26.7	2.37
4	0.368	–	–	10.7	15.1	4.41	19	0.157	–	–	-27.1	-26.6	0.43
5	0.448	14.37	19.64	5.9	11.2	5.28	20	0.450	17.27	22.89	-16.2	-10.6	5.60
6	0.443	12.56	17.50	9.0	13.9	4.91	21	0.445	20.60	25.68	11.4	16.4	5.05
7	0	–	–	-27.0	-24.5	2.48	22	0.445	–	–	-13.6	-10.0	3.67
8	0.396	21.02	25.39	7.3	11.6	4.35	23	0.469	12.80	18.09	19.3	24.6	5.27
9	0.134	–	–	-12.2	-9.1	3.03	24	0	–	–	-42.7	-41.0	1.73
10	0.454	18.99	22.63	-16.5	-12.9	3.61	25	0.326	–	–	-18.5	-14.5	4.04
11	0.272	–	–	-32.2	-26.8	5.42	26	0	–	–	-31.3	-29.5	1.81
12	0.506	13.20	18.97	10.2	15.9	5.74	27	0.259	–	–	-19.0	-15.0	3.98
13	0	–	–	-29.7	-29.2	0.54	28	0.473	17.19	22.54	1.8	7.1	5.31
14	0.444	18.28	23.31	2.4	7.5	5.01	29	0.229	–	–	-21.6	-18.0	3.58
15	0	–	–	-51.0	-49.6	1.33	Mean	0.307	16.52	21.48	-13.8	-10.1	3.66

TABLE 4.10

Summary of Measured Crack Initiation Angles for 1.25T Specimens.

Specimen no	$\Delta\bar{a}$ (mm)	θ_t(°)	$\bar{\theta}_t$(°)	θc (°)	θ_1 (°)	$\lvert\theta_2/2\rvert$(°)	Specimen no	$\Delta\bar{a}$ (mm)	θ_t (°)	$\bar{\theta}_t$ (°)	θc (°)	θ_1 (°)	$\lvert\theta_2/2\rvert$ (°)
1	0	–	–	−7.8	−7.3	0.50	16	0.236	–	–	−11.2	−8.5	2.68
2	0	–	–	−36.0	−35.5	0.53	17	0	–	–	−27.3	−25.1	2.12
3	0	–	–	−4.7	−4.4	0.32	18	0	–	–	−29.6	−29.0	0.57
4	0	–	–	−38.5	−36.1	2.36	19	0.206	–	–	−9.5	−7.1	2.37
5	0	–	–	−30.2	−28.9	1.35	20	0	–	–	−37.1	−34.7	2.40
6	0	–	–	−44.7	−41.0	3.66	21	0.425	18.88	22.50	−4.7	−1.1	3.61
7	0.306	–	–	27.4	29.8	2.37	22	0	–	–	−16.8	−14.5	2.32
8	0	–	–	−29.5	−26.6	2.99	23	0.290	–	–	−28.9	−25.9	3.00
9	0.262	–	–	−20.7	−19.2	1.50	24	0	–	–	−39.4	−37.4	2.01
10	0	–	–	−37.4	−34.8	2.57	25	0.410	19.28	23.00	−6.2	−2.5	3.73
11	0.215	–	–	−18.2	−16.2	1.95	26	0.299	–	–	−14.4	−11.6	2.76
12	0	–	–	11.1	15.1	4.01	27	0	–	–	−34.5	−33.1	1.32
13	0	–	–	−18.2	−16.3	1.92	28	0.425	17.05	20.72	−20.1	−16.5	3.67
14	0	–	–	−22.4	−21.4	1.06	Mean	0.354	18.40	22.07	−20.4	−18.2	2.20
15	0	–	–	−21.5	−19.5	1.96							

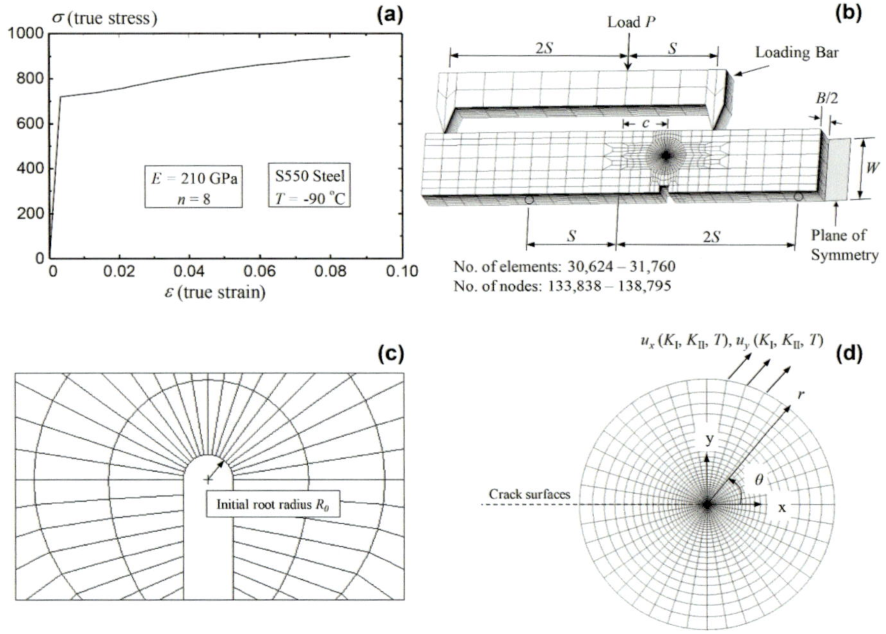

FIGURE 4.25 (a) Uniaxial true stress-strain curve of the S550 steel at −90 °C; (b) one-half symmetric FE model for the four-point bend specimen; (c) a close-up view of the initial root radius formed by focused rings of elements surrounding the crack-tip; and (d) an SSY model.

an elastic material with an artificially high Young's modulus, eliminating any plastic deformation in that component.

To assess the influence of mode-mixity and T-stress on the cleavage-initiation angle under SSY, a plane-strain SSY model (Figure 4.25d) is employed. This model consists of a single layer of roughly 4600 C3D20R elements, with all nodes fixed against out-of-plane motion. A mixed-mode I/II displacement field, supplemented by prescribed T-stresses, is applied along the outer boundary to reproduce the desired remote loading conditions.

$$\begin{Bmatrix} u_x \\ u_y \end{Bmatrix} = \sqrt{\frac{r}{2\pi}} \left\{ \frac{K_I}{2G} \begin{bmatrix} \cos\frac{\theta}{2}\left(\kappa - 1 + 2\sin^2\frac{\theta}{2}\right) \\ \sin\frac{\theta}{2}\left(\kappa + 1 - 2\cos^2\frac{\theta}{2}\right) \end{bmatrix} + \frac{K_{II}}{2G} \begin{bmatrix} \sin\frac{\theta}{2}\left(\kappa + 1 + 2\cos^2\frac{\theta}{2}\right) \\ -\cos\frac{\theta}{2}\left(\kappa - 1 - 2\sin^2\frac{\theta}{2}\right) \end{bmatrix} \right\} + \frac{T}{8G} \begin{bmatrix} (\kappa+1) r\cos\theta \\ -(3-\kappa) r\sin\theta \end{bmatrix} \tag{4.50}$$

where $\kappa = 3 - 4\nu$ for a plane strain condition.

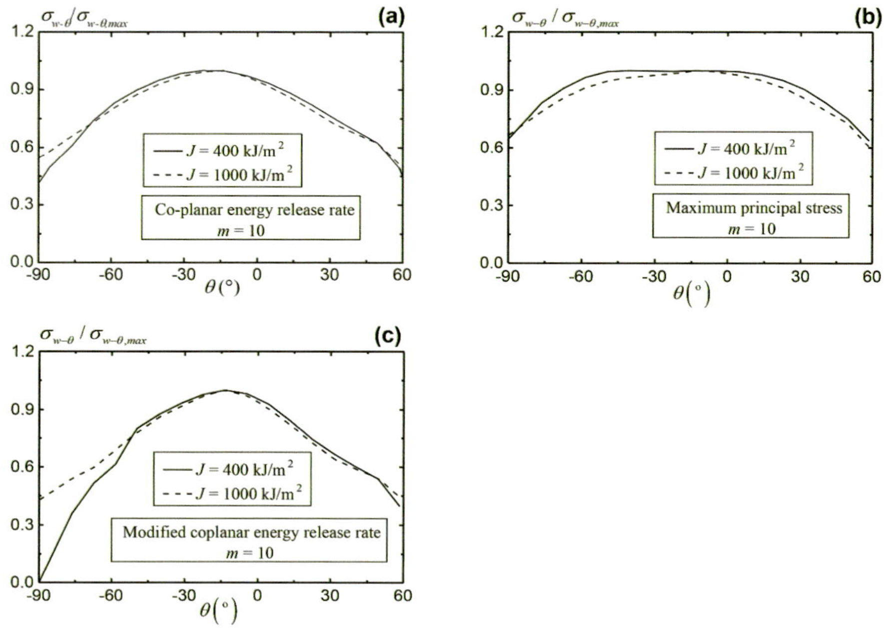

FIGURE 4.26 Variation of the normalized angular Weibull stress $\sigma_{w-\theta}$ along θ based on: (a) the coplanar energy release rate criterion; (b) the maximum principal stress criterion; and (c) the modified coplanar energy release rate criterion.

Figure 4.26 compares the normalized angular Weibull stress, $\sigma_{w\text{-}\theta}$, for two loading levels levels (J = 400 kJ/m² and J = 1000 kJ/m²), predicted with three fracture criteria: (i) the co-planar energy-release-rate-based effective stress (Eq. 4.38), (ii) the maximum principal stress criterion (Eq. 4.34), and (iii) the modified co-planar energy-release-rate criterion (Eq. 4.41). Figure 4.26a shows that $\sigma_{w\text{-}\theta}$ is essentially load-independent and exhibits a distinct peak near the zone at $\theta \approx 15°$. Figure 4.26b, based on the maximum principal stress criterion, displays an extended plateau, indicating nearly equal crack-initiation probabilities across a broad angular range and, consequently, a wide spectrum of possible initiation angles. Figure 4.26c contrasts two loadings for the modified coplanar energy release rate criterion. Noticeable deviations appear only for $\theta < -60°$, but the subsequent analysis demonstrates that these low-stress differences have a negligible influence on the numerically predicted cleavage-initiation angles.

The discussion below compares the cumulative distribution of cleavage-initiation angles predicted by the Weibull-stress model with the empirical rank probabilities obtained from the tests. The empirical rank probability for the ith ranked, measured angle, $\theta_{c,i}$, is calculated following Eq. (4.25). Figure 4.27 compares the cumulative distribution of cleavage-initiation angles predicted by three criteria for four-point-bend specimens of 0.75T and 1.25T thickness.

Figures 4.27a and b contrast the Weibull-stress predictions based on the co-planar energy-release-rate criterion with the experimental data. The model tracks the

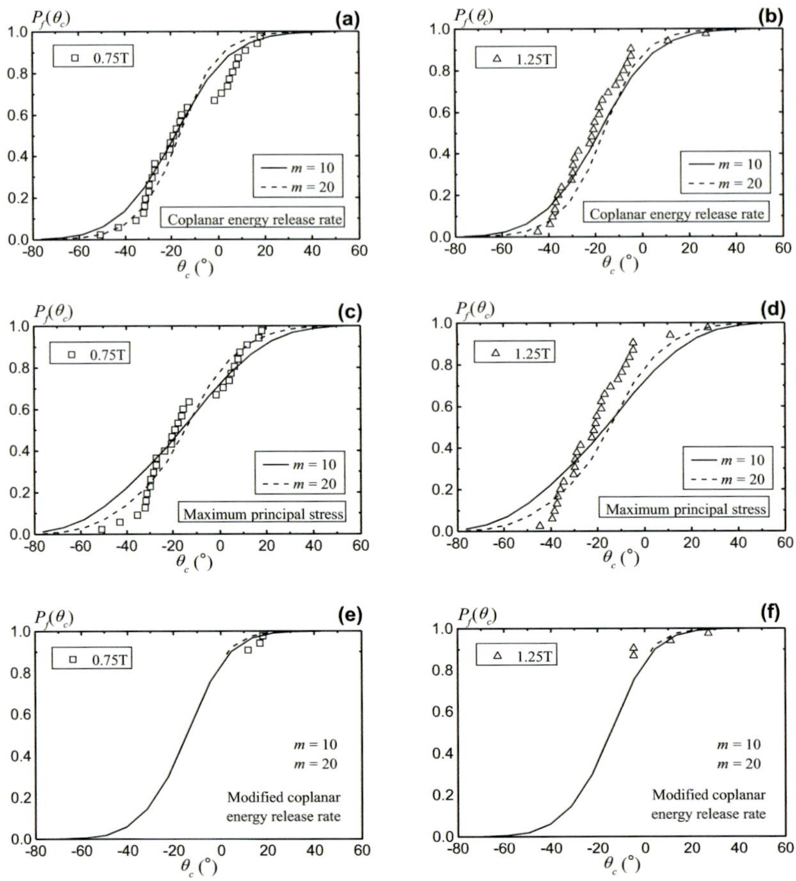

FIGURE 4.27 Cumulative probability of the cleavage initiation angle based on the coplanar energy release rate criterion for: (a) 0.75T and (b) 1.25T; specimens based on the Maximum principal stress criterion for: (c) 0.75T and (d) 1.25T; specimens based on the modified coplanar energy release rate criterion for: (e) 0.75T and (f) 1.25T.

measured failure probabilities closely, except in the upper tail of the 0.75T dataset, where appreciable ductile tearing precedes final cleavage. Over the practical range m = 10 and 20, the Weibull modulus has little effect on the predicted CDF. The calibration, therefore, remains tied to thickness- and constraint-corrected fracture-toughness data [41]. Figures 4.27c and d indicate that the maximum-principal-stress criterion overestimates the scatter in initiation angles. Although this shifts the upper tail in the correct direction for the 0.75T specimens, the overall match is poorer than that of the Weibull-stress approach. Figures 4.27e and f show that the modified co-planar energy-release-rate criterion consistently underpredicts the initiation-angle probabilities, with the discrepancy most pronounced at $m = 20$.

To quantify the discrepancy between model predictions and experiments, we define an error function that, at each cumulative probability P, takes the absolute

TABLE 4.11
Normalized Errors for Different Stress Criteria.

	Normalized errors			
	1.25T Data set		0.75T Data set	
Stress criteria	$m = 10$	$m = 20$	$m = 10$	$m = 20$
Maximum principal stress criterion	2.19	1.71	2.66	1.88
Coplanar energy release rate criterion	1.00	1.19	1.00	1.10
Modified coplanar energy release rate criterion	1.59	2.05	1.75	2.99

difference between the experimentally observed cleavage-initiation angle and the numerically predicted angle,

$$Error = \sum_{i=1}^{N_{1.25T}} \left|\theta_{c-1.25T,i} - \theta_{c-FEM,i}\right| WF_{(i)} + \sum_{j=1}^{N_{0.75T}} \left|\theta_{c-0.75T,j} - \theta_{c-FEM,j}\right| WF_{(j)} \quad (4.51)$$

In Eq. (4.51), the symbols are defined as follows. N_{xT} (with $x = 0.75$ or 1.25) is the total number of data points for specimens of thickness xT. $\theta_{c,xT,i}$ is the experimentally measured cleavage-initiation angle having rank i. $\theta_{c,FEM,I}$ is the numerically predicted initiation angle at the same cumulative probability $P_{i,rank}$. $WF_{(i)}$ is a weighting factor that scales the contribution of each point, following Wasiluk *et al.* [17].

Because the ten largest-angle observations in the 0.75T set showed substantial ductile tearing before final cleavage, they are excluded from the scalar error calculation. Table 4.11 reports the error magnitudes for the 1.25T and 0.75T datasets, each normalized to the baseline error obtained for $m = 10$ with the co-planar energy-release-rate criterion. The table reveals that the maximum principal stress criterion and the modified co-planar energy-release-rate criterion yield markedly higher relative errors than the original co-planar energy-release-rate approach.

Figure 4.28 plots the cumulative distributions predicted with the co-planar energy-release-rate criterion for fracture-process-zone sizes $\lambda = 1.5$, 2.0, and 2.5. For $m = 20$ the three curves are nearly coincident, confirming that λ has a negligible impact on the predicted failure probability. At lower Weibull moduli, however, increasing λ eliminates the more weakly stressed material from the σ_w evaluation, slightly reducing the scatter of the predicted initiation angles.

The remainder of this section investigates how the equivalent mode-mixity angle β_{eq} and the T-stress influence cleavage-initiation angles within a plane-strain SSY framework. Earlier studies [7, 43] show that, under plane-strain SSY, the near-tip stress field retains a self-similar form, providing a convenient basis for isolating the effects of β_{eq} and T-stress on crack-initiation behavior,

$$\sigma = \sigma_{ys} f\left(\frac{r\sigma_{ys}}{J}, \frac{T}{\sigma_{ys}}, n, \upsilon, \frac{E}{\sigma_{ys}}, \beta_{eq}\right) g(n,\theta) \quad (4.52)$$

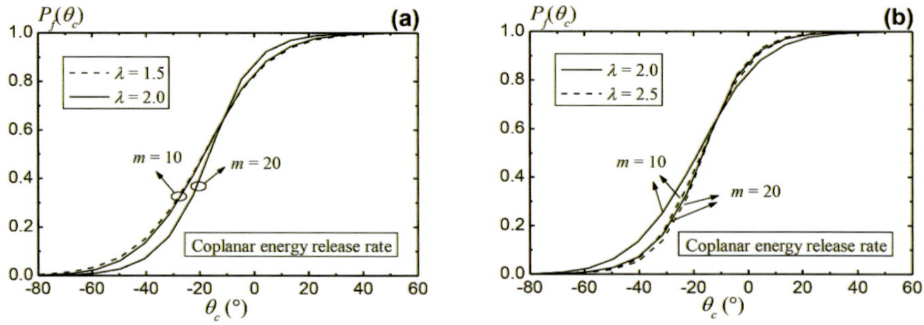

FIGURE 4.28 Cumulative probability of the crack initiation angle based on the coplanar energy release rate criterion corresponding to different fracture process zone sizes with: (a) $\lambda = 1.5$ versus $\lambda = 2.0$; and (b) $\lambda = 2.0$ versus $\lambda = 2.5$.

where *f* and *g* define non-dimensional functions. Substituting Eq. (4.52) into Eq. (4.32), the angular Weibull stress density follows,

$$\sigma_{w-\theta}^{m} = BJ^2 \frac{\sigma_{ys}^{m-2}}{V_0} \int_0^{\bar{\rho}} (f \cdot g)^m \rho d\rho \tag{4.53}$$

where ρ denotes the non-dimensional radius $r\sigma_{ys}/J$, and the upper integration limit $\bar{\rho}$ depends on the size of the fracture process zone defined by λ. Substituting Eq. (4.53) into Eq. (4.33), the cumulative probability of cleavage initiation angles $P(\theta_c)$ becomes independent of *J*-integral, and is only a function of material flow properties, in-plane geometric constraint $T/\sigma ys$, mode-mixity angle β_{eq}, the Weibull modulus *m*, and the effective stress criterion.

Figure 4.29 summarizes the influence of mode-mixity and in-plane constraint on the angular Weibull-stress field and the resulting cleavage-initiation statistics derived with the co-planar energy-release-rate criterion. Figure 4.29a illustrates the normalized angular Weibull stress ($\sigma_{w\text{-}\theta}/\sigma_{w\text{-}\theta,max}$) versus θ curves for mode-mixity angles $\beta_{eq} = 72°$, 78°, and 84°. As β_{eq} decreases (*i.e.*, the Mode II component grows), the peak of $\sigma_{w\text{-}\theta}/\sigma_{w\text{-}\theta,\,max}$ migrates toward negative θ, indicating a greater propensity for cleavage on the negative-θ side of the crack front. Figure 4.29b shows the cumulative probability of initiation angles for varying β_{eq} with $T = 0$. Consistent with the shift observed in Figure 4.29a, lower β_{eq} values translate the cumulative-distribution curve toward negative θ and broaden the overall scatter of predicted initiation angles. Figure 4.29c presents the effect of *T*-stress with $\beta_{eq} = 78°$ and $m = 10$.

A positive T-stress widens the distribution. The upper tail (large positive θ) moves further right, while the lower tail (large negative θ) extends left. The peak of $\sigma_{w\text{-}\theta}/\sigma_{w\text{-}\theta,max}$ simultaneously drifts to positive θ. Conversely, once $T/\sigma_{ys} < -0.5$, further reductions in the (negative) *T*-stress have little impact, as the $\sigma_{w\text{-}\theta}/\sigma_{w\text{-}\theta,max}$ profile effectively stabilizes.

Figure 4.29d compares 1.25T four-point-bend probability at different load levels (J = 120, 400, 1000 kJ/m²) for $\beta_{eq} = 78°$ and $T = 0$. The SSY prediction aligns

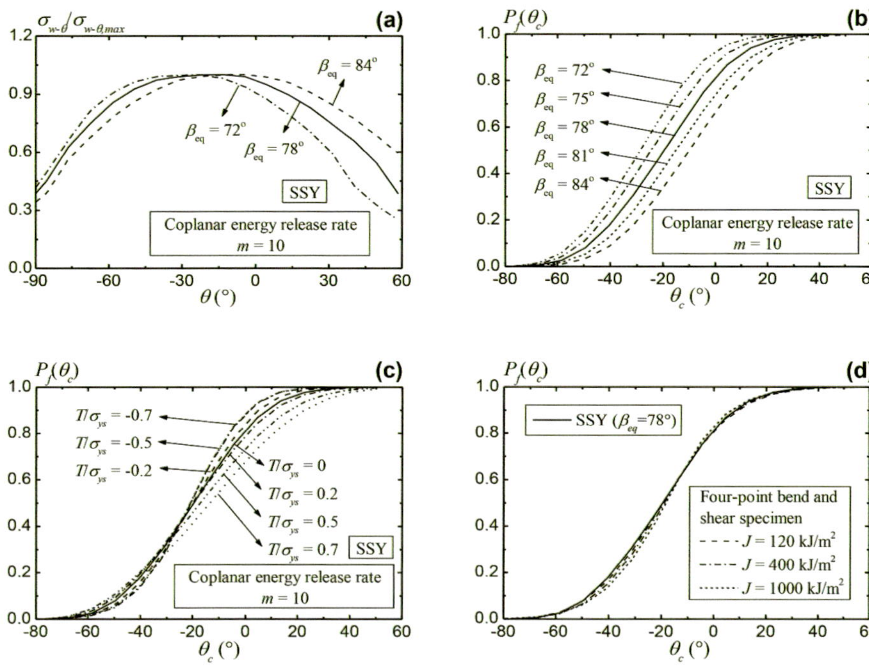

FIGURE 4.29 (a) Variation of the angular Weibull stress along θ in the SSY models for β_{eq} = 72°, 78°, and 84°; cumulative probability of the cleavage initiation angle estimated from the SSY model (b) for β_{eq} = 72°, 75°, 78°, 81°, and 84°, and (c) T/σ_{ys} = −0.5, −0.2, 0.0, 0.2, and 0.5; and (d) comparison of cumulative probability of the cleavage initiation angle between SSY models and four-point bend and shear specimens.

closely with experiments, even though the 1.25T specimens experience a small positive geometric constraint ($T/\sigma_{ys} \approx 0.25$) and through-thickness free-surface effects. This agreement suggests that while large plasticity-induced constraint loss changes the absolute Weibull-stress levels, it has only a marginal influence on the shape of the normalized $\sigma_{w\text{-}\theta}/\sigma_{w\text{-}\theta,max}$ distribution, and hence on the predicted scatter of initiation angles.

This study extends the Weibull-stress framework to mixed-mode (I + II) ductile-to-brittle transitions and proposes a weakest-link statistical method for predicting cleavage-initiation angles. A comprehensive experimental program was conducted on high-strength steel S550 using four-point-bend and mixed-mode shear specimens of two thicknesses (0.75 T and 1.25 T), tested at −90 °C. The experimental observations underpin the numerical evaluation of three stress-based fracture criteria. The key findings are:

Cleavage-initiation angles were determined by combining crack-plane rotation with microscopic assessment of the near-tip fracture-surface tilt. Fracture toughness was calculated via an extended η-approach that is fully compatible with ASTM E1820. Both specimen geometries exhibited wide scatter in initiation angles under mode-I-dominant mixed loading. The cumulative distributions are essentially thickness-independent.

Because the normalized angular Weibull-stress profile is nearly load-invariant, the predicted initiation-angle distribution shows negligible dependence on applied J. This allows direct comparison across specimens with different toughness levels. The co-planar energy-release-rate criterion reproduces the experimental initiation-angle statistics most accurately. The maximum-principal-stress criterion over-predicts scatter, whereas the modified co-planar criterion under-predicts it.

Varying the Weibull exponent between 10 and 20 has minimal effect on the predicted angle distribution, indicating that crack-front constraint influences initiation direction far less than it affects absolute toughness. SSY simulations with modest positive T-stress ($T/\sigma\_y \approx 0.25$) match the experimental distributions for 1.25T specimens. Increased mode-II contribution drives the most probable initiation angle toward negative θ, while T-stress chiefly widens the upper tail of the probability curve.

Together, these results validate the extended Weibull-stress approach and highlight the coplanar energy-release-rate criterion as the most reliable predictor of cleavage-initiation angles in mixed-mode fracture of high-strength steels.

4.6 DUCTILE TEARING INCORPORATED CLEAVAGE ASSESSMENT UNDER MIXED-MODE LOADING

4.6.1 Ductile Tearing Correction

Wallin [44], assuming that ductile tearing is statistically independent of the cleavage fracture toughness K_{Jc}, derived a simplified correction that incorporates tearing into the cumulative failure-probability calculation,

$$\left(\log\frac{1}{1-P_f}\right)^{1/4} = \left(\frac{K_{Jc}-K_{\min}}{K_0-K_{\min}}\right)\left(\frac{B}{B_{1T}}\right)^{1/4}\left(1+\frac{2\Delta a\sigma_{flow}^2}{K_{Jc}^2\beta}\right)^{1/4} \tag{4.54}$$

where Δa denotes the length of ductile tearing, σ_{flow} represents a flow stress measure, and the non-dimensional constant β, multiplied by $\left(K_{Jc}/\sigma_{flow}\right)^2$, defines the size of the active plastic zone. Based on Eq. (4.54), we have,

$$P_f = 1-\exp\left\{-\left[C_1\cdot B\left(\frac{K_{Jc}}{\sigma_y}\right)^4 + C_2\cdot BH_{fp}\Delta a\right]\right\} \tag{4.55}$$

where H_{fp} denotes the height of the fracture process zone and $Bh\Delta a$ refers to the increased, sampled volume of the fracture process zone caused by ductile tearing.

Figure 4.30a depicts the conventional FPZ, whose volume varies with the applied load level, quantified here by the through-thickness, average energy-release rate J_{avg}. Figure 4.30b illustrates the cumulative FPZ volume swept out as the crack advances

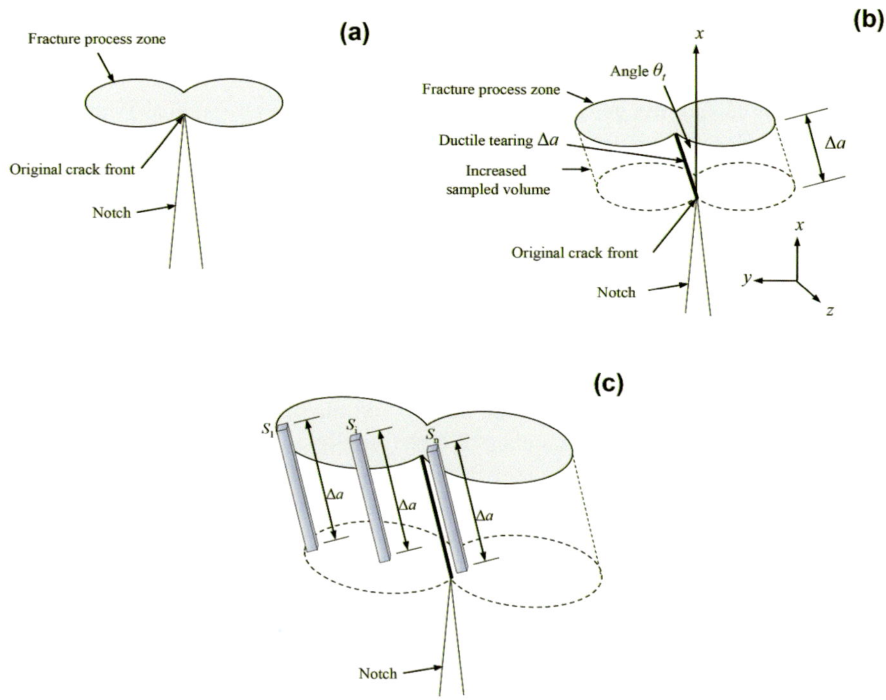

FIGURE 4.30 (a) Conventional definition of the fracture process zone; (b) cumulative sampled volume of the fracture process zone caused by ductile tearing Δa; and (c) projection of the fracture process zone onto a plane perpendicular to the direction of ductile tearing.

by an increment Δa (bold line). Under mixed-mode I/II loading, ductile tearing deviates from self-similar growth and kinks out of the original crack plane by an angle θ_t. As defined in Figure 4.30b, θ_t is the angle between the tearing surface and the undeformed fatigue-crack plane (the x-axis). It is measured counter-clockwise from the positive x-direction and is positive toward the positive y-direction, as further clarified in Figure 4.22a.

In the Weibull-stress framework, the history effect is introduced by recognizing that the material immediately ahead of a propagating crack cannot "heal" during the partial unloading that accompanies ductile tearing. Consequently, the effective stress that drives cleavage, σ_{eff}, should be taken as the largest value attained at each material point over the entire loading history, rather than the value calculated at the current load step.

To evaluate this maximum-stress field, the FPZ is partitioned into slender rectangular blocks whose long axes are aligned with the direction of crack growth (Figure 4.30c). The cross-sectional area of the ith block, measured on a plane that intersects the current crack tip and is normal to the crack-extension direction, is denoted S_i. Under steady-state growth, the maximum effective stress acting over S_i represents the peak stress ever experienced by all materials within that block. As the crack advances an increment Δa, each block contributes an incremental FPZ volume $\Delta V = S \times \Delta a$, where $S = \Sigma S_i$ is the total FPZ area on a plane perpendicular to the crack path.

Following the approach of Sobotka and Dodds [45], the cumulative Weibull stress that governs cleavage initiation is then obtained by replacing the instantaneous σ_{eff} in the standard definition with its history-dependent maximum, yielding the cumulative Weibull stress σcw,

$$\sigma_{cw} = \left[\frac{1}{V_0} \int_V \sigma_{eff}^m dV + \frac{1}{V_0} \int_S \sigma_{eff,max(\theta_t)}^m dS \Delta a \right]^{1/m} \tag{4.56}$$

which is strongly governed by the cumulative volume of the highly stressed FPZ.

To capture the combined influence of constraint relaxation and ductile tearing on the Weibull stress, we introduce a dimensionless correction factor $h(M)$ formulated in terms of the cumulative Weibull stress, σ_{cw}, which exhibits strong dependence on the cumulative volume of the highly stressed FPZ.

A dimensionless correction factor $h(M)$ is introduced to incorporate the combined influence of constraint relaxation and ductile tearing into the Weibull-stress description based on the cumulative Weibull stress σ_{cw},

$$\sigma_{cw}{}^m = CBK_J^4 h(M) \text{ or } \sigma_{cw}{}^m = \bar{C}BJ_{avg}^2 h(M) \tag{4.57}$$

In the special case of zero ductile tearing ($\Delta a = 0$), the cumulative stress field collapses to the instantaneous one, so $h(M) = g(M)$, recovering the earlier, constraint-only correction factor. The fracture toughness measured in a test must likewise be normalized to the 1T high-constraint, no-tearing baseline. Accounting for thickness effects, constraint differences, and any pre-cleavage tearing, the corrected toughness is,

$$K_{Jc}^{1T} = K_{\min} + \left[K_{Jc}^{(xT)} h\left(M^{(xT)}\right)^{1/4} - K_{\min} g\left(M_{\min}^{(xT)}\right)^{1/4} \right] \left(\frac{B_{xT}}{B_{1T}} \right)^{1/4} \tag{4.58}$$

This treatment presumes that no ductile tearing is initiated at the threshold load (K_{min}), that is, at the load level where crack growth first becomes energetically possible; the incremental extension Δa is taken to be zero.

The procedure outlined above does not explicitly model ductile crack advance in the FE analysis. Instead, it represents crack growth indirectly, as an equivalent increase in the volume sampled when computing the Weibull stress. Within this enlarged volume, the effective stress is taken to be the maximum value attained along material lines parallel to the crack-propagation direction, consistent with the Weibull-stress history effect, which assumes that micro-scale flaws cannot "heal" during the local unloading that accompanies ductile tearing.

If a damage-mechanics model is introduced to explicitly simulate ductile extension before the Weibull-stress evaluation, the analysis must first capture the way tearing changes the size, density, and spatial distribution of microflaws near the crack tip.

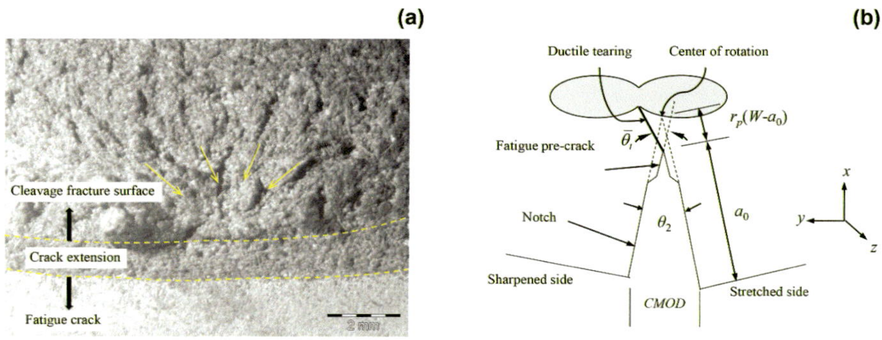

FIGURE 4.31 (a) Fracture surface revealed under an optical microscope; and (b) schematic illustration of the ductile tearing angle.

Local unloading during extension can make the current stress field appear too low because the peak stresses experienced immediately before unloading may already have enlarged critical flaws. To avoid under-estimating the subsequent cleavage-driving force, the post-processing routine should therefore record, for every integration point in the prospective FPZ, the maximum historical stresses and use these maxima in the final Weibull-stress calculation.

Figure 4.31a shows a typical (cleaned and anti-corrosion coated) fracture surface under an optical microscope, which reveals a small amount of ductile crack extension preceding the final cleavage failure characterized by river patterns on the fracture surface. The post-test procedure measures the original crack depth and the ductile tearing length in each fractured specimen, following the recommendations in ASTM E1921 [5], which requires measurements from nine equally spaced points centered about the centerline of the specimen. The measured ductile tearing length in the FPBS specimens equals the projected distance $\Delta\bar{a}$, of the crack extension Δa at the sharpened side of the specimen, on the fatigue pre-crack plane, as recommended by Yang and Qian [38]. This implies,

$$\Delta\bar{a} = \Delta a \cos\theta_t \tag{4.59}$$

where θ_t, as indicated in Figure 4.31b, defines the angle between the ductile tearing surface and the original, undeformed fatigue pre-crack plane along the x-coordinate. This angle remains positive in the sharpened side of the specimen, as indicated in Figure 4.22a. As shown in Figure 4.31b, the ductile tearing angle θ_t decomposes into two measurable components, that is, $\theta_t = \bar{\theta}_t - \theta_2/2$, where $\bar{\theta}_t$ represents the angle between the deformed fatigue pre-crack plane and the ductile tearing surface, and θ_2 denotes the crack mouth opening angle at the cleavage failure, following Eq. (4.46)

The silicone replica materials applied on the fracture surface reproduce the profile of the fracture surface and therefore allow measurement of the angle $\bar{\theta}_t$, following a similar procedure described by Zhang and Qian [46]. To ensure an accurate measurement of the ductile tearing angle, this study only measures $\bar{\theta}_t$ for specimens exhibiting a ductile tearing length $\Delta a \geq 0.4$ mm .

Tables 4.9 and 4.10 list the measured tearing length $\Delta\bar{a}$, and the ductile tearing angle θ_t for the 0.75T and 1.25T fracture specimens, respectively. The estimation of the mean ductile tearing length and angle excludes specimens with negligible ductile tearing. The 0.75T specimen set contains a higher percentage of specimens, which experience ductile tearing prior to the final cleavage fracture, than does the 1.25T specimen set. Tables 4.9 and 4.10 illustrate that 0.75T and 1.25T four-point bend and shear specimens (with $\beta_{eq} = 78^o$) show similar values of the $\bar{\theta}_t$ and θ_t angles, which exhibit very little scatter.

4.6.2 Assessment of Mixed-Mode Brittle Fracture

This study refines the calibration of the Weibull-stress parameters, the Weibull modulus m and the threshold stress-intensity factor, K_{min}, by explicitly accounting for ductile tearing. Calibrating m and K_{min} from a single fracture-toughness dataset can yield multiple admissible solutions, as shown by Ruggieri *et al.* [11]. To resolve this ambiguity, Ruggieri *et al.* [11] proposed using two specimen sets with different crack-front constraints – HC and LC. Wasiluk *et al.* [17] improved the method by introducing a weight factor that down-weights data with large experimental scatter. We further extend Wasiluk's approach by embedding a ductile-tearing correction in the measured fracture-toughness values.

The calibration procedure utilizes two sets of mixed-mode (I + II) specimens at –90 °C, that is, the 1.25T four-point-bend and shear (HC) specimens and the 0.75T four-point-bend and shear (LC) specimens. For each fracture event, the ductile-tearing length $\Delta\bar{a}$ and tearing angle θ_t were recorded.

For the trial Weibull moduli from $m = 1$ ~30, FEA were carried out for the actual four-point-bend/shear (FPBS) geometries, and an idealized SSY model under identical mixed-mode loading (as shown in Figure 4.25). Each FEA produced a relationship between the global driving force K_J and the cumulative Weibull stress $\boldsymbol{\sigma_{cw}}$. The SSY solutions supply the scaling constant C (equivalently $\sigma_{w\text{-}min}$) required by the three-parameter Weibull formulation. Using the measured θ_t and $\Delta\bar{a}$, the σ_{cw} values for the mixed-mode specimens are corrected, yielding the tearing-adjusted function *h(M)* for both HC and LC cases (Eq. 4.57). Finally, the optimal pair (m, K_{min}) is obtained by minimizing the weighted error between the tearing-corrected experimental toughness values and the corresponding FEA predictions, with weights assigned according to Wasiluk *et al.*'s scheme.

This two-constraint, tearing-corrected calibration ensures a unique and physically consistent set of Weibull-stress parameters for high-strength steel S550 under mixed-mode ductile-to-cleavage transition conditions.

The numerical procedure then corrects each measured fracture toughness value, K_{Jc}^{HC} and K_{Jc}^{LC} , to a 1T, plane-strain SSY reference condition, denoted as $K_{Jc}^{\mathrm{HC}-SSY(1\mathrm{T})}$ and $K_{Jc}^{\mathrm{LC}-SSY(1\mathrm{T})}$, through Eq. (4.58), for each trial value of K_{min}. The trial K_{min} ranges from 20 $\mathrm{MPa}\sqrt{\mathrm{m}}$ to 60 $\mathrm{MPa}\sqrt{\mathrm{m}}$ in this study. Other procedures are same to that in Figure 4.1.

The calibrated Weibull modulus m and threshold stress intensity K_{min} are obtained by minimizing the error function so that Eq. (4.58) collapses the HC and

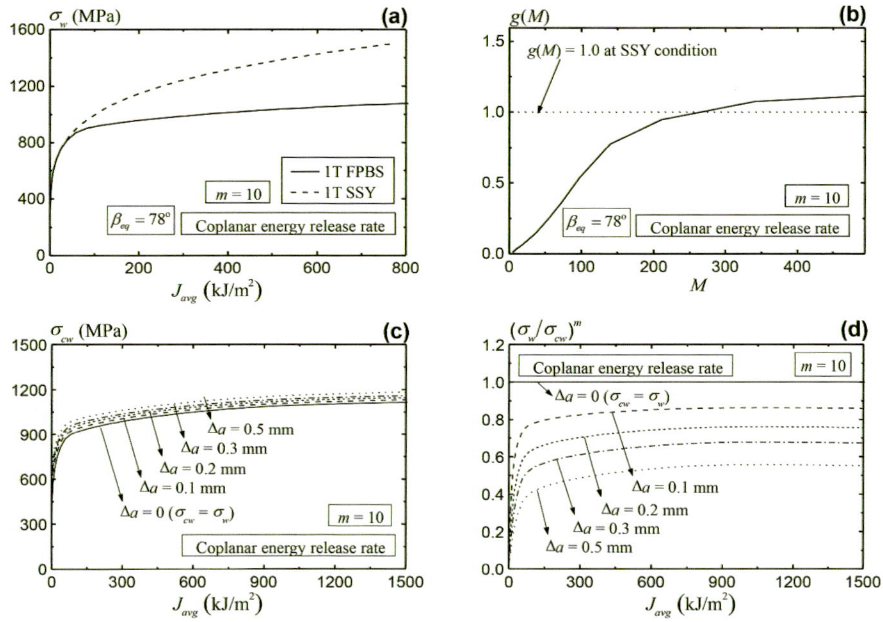

FIGURE 4.32 (a) Weibull stress value σ_w against the through-thickness, average J-integral for 1T FPBS specimen and 1T SSY model with a mode-mixity angle of 78°; (b) the constraint correction function $g(M)$ for m = 10; (c) Cumulative Weibull stress σ_{cw} versus the through-thickness, average J-value for 1T FPBS specimen with different Δa; and (d) $\left(\sigma_w / \sigma_{cw}\right)^m$ for 1T FPBS specimen with different Δa.

LC toughness datasets onto a single, idealized 1T SSY reference curve. Under this reference condition, cleavage is driven exclusively by the local cumulative Weibull stress σ_{cw}, independent of specimen thickness, crack-front constraint, or prior ductile tearing.

To illustrate the behavior of the key quantities, Weibull stress σ_c, cumulative Weibull stress σ_{cw}, and the constraint-correction function $h(M)$, we consider a 1T four-point-bend and shear (FPBS) specimen with a_0/W = 0.56 made from S550 steel and tested at –90 °C, evaluated with the coplanar energy-release-rate criterion. Figure 4.32a plots σ_w (m = 10) versus the through-thickness average J_{avg} for both the 1T FPBS geometry and an equivalent 1T SSY model at β_{eq} = 78°. The widening gap between the two curves as J_{avg} increases quantifies the plasticity-induced loss of crack-front constraint in the FPBS specimen at high load levels. Figure 4.32b shows the resulting constraint-correction function $g(M)$, with $M = b\sigma_{ys}/J_{avg}$. The FPBS specimen remains in a high-constraint regime up to $M \approx 200$, after which $g(M)$ declines as constraint diminishes. Values of $g(M) > 1$ indicate a small positive T-stress. Because $g(M)$ depends only on geometry, its values are identical for all geometrically similar specimens.

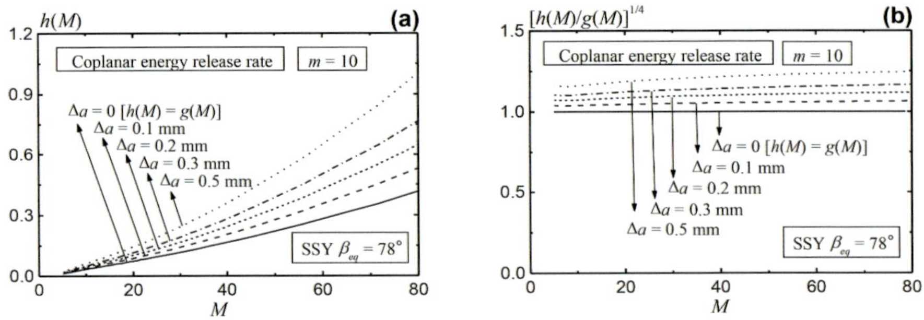

FIGURE 4.33 (a) Correction function $h(M)$ for the constraint loss and ductile tearing; and (b) the effect of ductile tearing on the correction of fracture toughness K_J through a non-dimensional parameter, $\left(h(M)/g(M)\right)^{1/4}$.

Figure 4.32c plots the cumulative Weibull stress, σ_{cw} against the through-thickness average J_{avg} for the 1T FPBS specimen while accounting for ductile-tearing increments Δa measured at the tearing angle θ_t. When $\Delta a = 0$, σ_{cw} reduces to the conventional Weibull stress σ_w. Figure 4.32d isolates the σ_w contribution within σ_{cw}, underscoring how σ_{cw} grows with the expanding, highly stressed FPZ ahead of the advancing crack tip. For the 1T FPBS geometry, stresses in the additional material volume created by $\Delta a = 0.5$ mm supply roughly 45% of the total σ_{cw}.

Figure 4.33a plots the tearing-correction function $h(M)$, for a 1T FPBS specimen with $m = 10$. As expected, $h(M)$ rises with increasing ductile-tearing length Δa but falls progressively as the load level grows (i.e., as J_{avg} increases or the non-dimensional parameter M decreases).

Figure 4.33b illustrates how ductile tearing modifies the apparent fracture toughness K_{Jc} through the dimensionless factor $[h(M)/g(M)]^{1/4}$. For $\beta_{eq} = 78°$, a modest crack extension of $\Delta a = 0.1$ mm raises the corrected toughness by roughly 3% in a 1T FPBS specimen. When $\Delta a < 0.2$ mm, the tearing correction produces only minor changes in the mixed-mode toughness values considered here.

This discussion describes the simultaneous calibration of the Weibull modulus m and the threshold Weibull stress $\sigma_{w\text{-}min}$ (derived from K_{min}) using the mixed-mode fracture data from high-constraint (1.25T) and low-constraint (0.75T) four-point bend and shear (FPBS) specimens.

Figure 4.34a displays the rank-ordered mixed-mode effective toughness K_{Jc} for both specimen sets at –90 °C. Substantial scatter is evident in the fracture toughness data, and many tests show stable ductile tearing before cleavage. A reference transition temperature of $T_0 \approx -170$ °C is obtained from the pooled dataset.

In the calibration procedure, only data with the non-dimensional load parameter M $(= b\,\sigma_{ys}/J_{avg}) > 6$ are retained, thereby eliminating cases with excessive plasticity or tearing. For trial values $m = 1$ to 30 and $K_{min} = 20$ to 60 $\text{MPa}\sqrt{\text{m}}$, we minimize the weighted error defined in Eq. (4.13). Figure 4.34b plots the normalized minimum error for each m across all K_{min}. The global minimum occurs at $m = 17$ and $K_{min} = 30$ $\text{MPa}\sqrt{\text{m}}$. Figure 4.34c traces the K_{min} that yields the minimum error for every m.

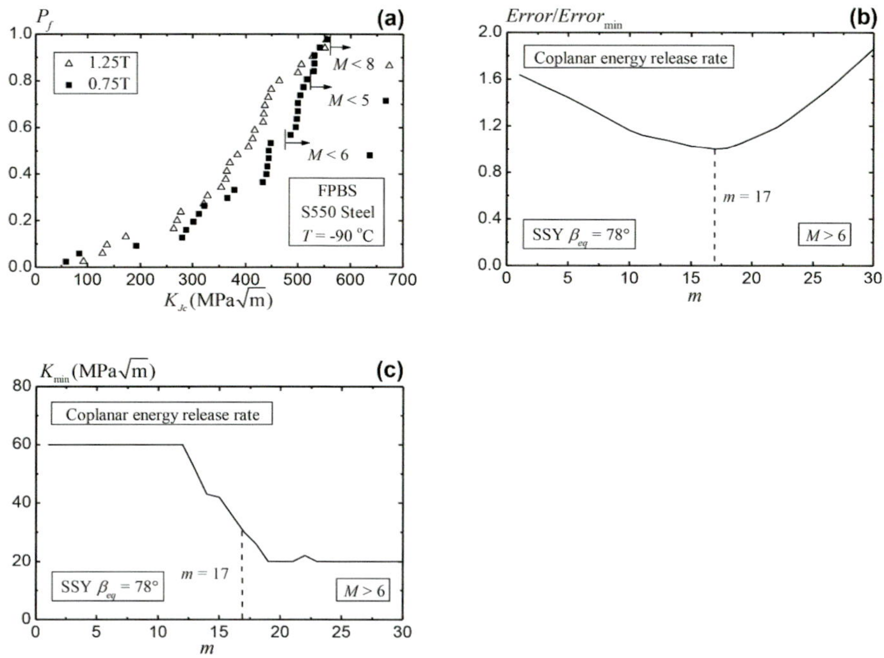

FIGURE 4.34 (a) Rank probability of the 1.25T and 0.75T FPBS fracture toughness data; (b) variation of the normalized error with respect to m based on the coplanar energy release rate criterion; and (c) K_{min} corresponding to the minimum error for each m.

Figure 4.35a rescales the 1.25T and 0.75T toughness values to the idealized 1T plane-strain SSY condition using m = 17 and K_{min} = 30 MPa$\sqrt{\text{m}}$. The two datasets collapse, implying that the local cumulative Weibull stress σ_{cw} is the governing parameter, independent of specimen constraint. The solid curve represents the predicted failure probability for the 1T SSY reference with the scale parameter σ_u taken from the corrected HC data.

Figure 4.35b repeats the scaling with an arbitrary pair (m = 4, K_{min} = 20 MPa$\sqrt{\text{m}}$). The HC and LC datasets no longer overlap, demonstrating misrepresentation of tearing and constraint effects. Figure 4.35c substitutes the conventional constraint function $g(M)$ for the tearing-aware $h(M)$. Toughness values are under-corrected, shifting the scaled data downward. Figure 4.35d compares the measured failure probability for specimens with Δa = 0 against predictions obtained with $g(M)$ (m = 17, K_{min} = 30 MPa$\sqrt{\text{m}}$). Agreement is excellent, confirming that $g(M)$ suffices when ductile tearing is absent. Overall, the calibrated pair (m = 17, K_{min} = 30 MPa$\sqrt{\text{m}}$) successfully reconciles fracture-toughness data from specimens with markedly different constraint levels by accounting for both plasticity-induced constraint loss and ductile tearing through the combined $g(M)$–$h(M)$ corrections.

Figure 4.36 presents calibration with the maximum principal stress criterion. Figure 4.36a shows the normalized error, *Error*/*Error*$_{min}$, which reaches its global

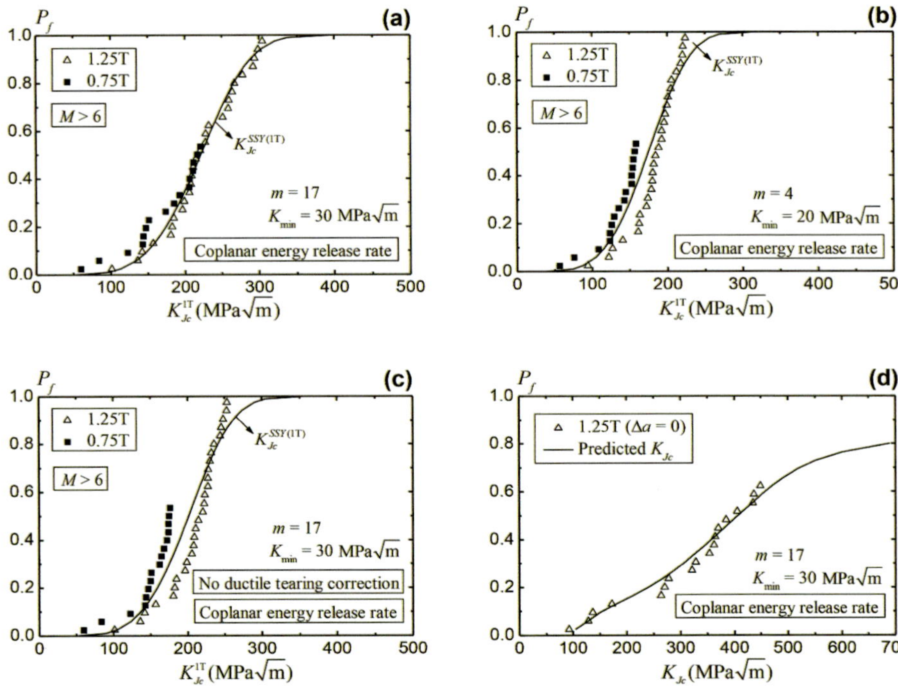

FIGURE 4.35 Comparison between the rank probability for the experimental data and the probability of fracture computed from the Weibull stress approach using the coplanar energy release rate and: (a) the calibrated m and K_{min} value; (b) $m = 4$ and $K_{min} = 20\ \text{MPa}\sqrt{\text{m}}$; (c) calibrated m and K_{min} without considering the ductile tearing effect; and (d) for 1.25T specimens without ductile crack extension.

minimum at $m = 23$ and $K_{min} = 42\ \text{MPa}\sqrt{\text{m}}$ but forms a broad plateau for $15 \leq m \leq 24$, indicating only modest sensitivity to m within this range. Figure 4.36b plots the K_{min} that minimizes the error function for each m, highlighting a slowly varying trend that flattens in the plateau region identified above. As demonstrated in Figure 4.36c, The Weibull-stress model reproduces the mixed-mode fracture-probability distributions for both 1.25T (HC) and 0.75T (LC) datasets with good fidelity, using the optimal pair ($m = 23$, $K_{min} = 42\ \text{MPa}\sqrt{\text{m}}$).

Replacing the tearing-aware correction $h(M)$ with the conventional constraint function $g(M)$, thus neglecting the stressed volume created by prior ductile tearing, shifts the scaled toughness values downward and underpredicts failure probabilities, mirroring the behavior observed for the coplanar-energy-release-rate calibration, as demonstrated in Figure 4.36d. When the Weibull parameters obtained from pure Mode I data [46] are applied to the present mixed-mode dataset, the predicted failure probabilities remain in close agreement with experiment, as illustrated in Figure 4.37. This suggests that, for high-strength steel S550, parameters calibrated

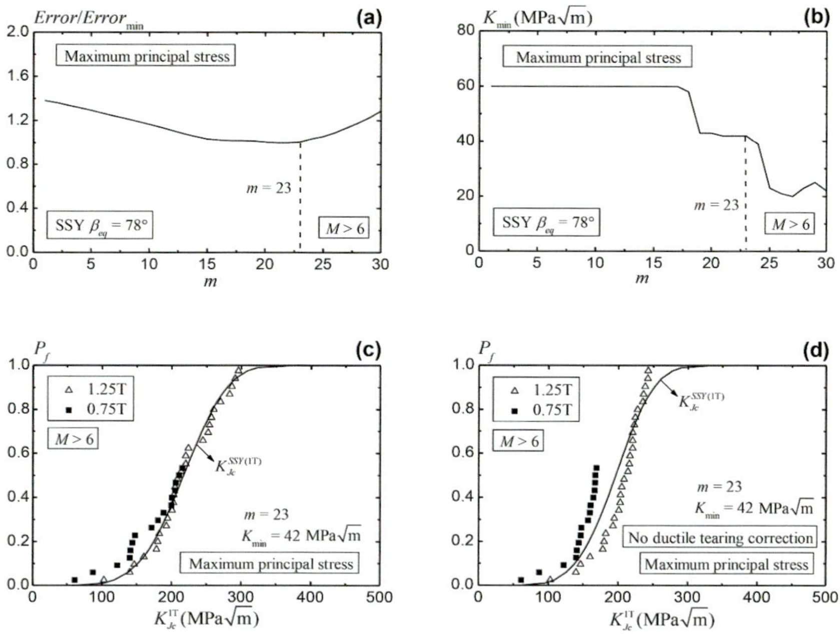

FIGURE 4.36 Calibration based on maximum principal stress: (a) variation of the normalized error with respect to m; (b) K_{min} corresponding to the minimum error for each m; comparison between the rank probability of the fracture specimen and the probability estimated from the Weibull stress approach using the calibrated m and K_{min} values; (c) considering the ductile tearing; and (d) without considering the ductile tearing.

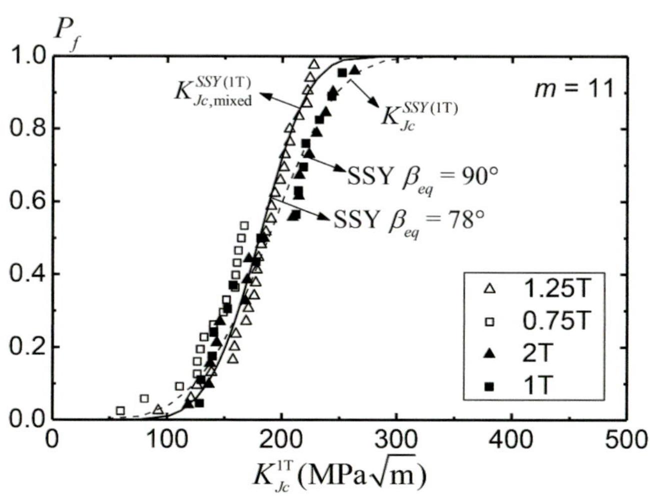

FIGURE 4.37 Comparison of the probability of failure for pure mode I and mixed-mode I and II specimens using the Weibull modulus calibrated from the pure mode condition.

in Mode I provide a reasonable first-order description of mixed-mode cleavage, although the dedicated mixed-mode calibration (m = 23, K_{min} = 42 $\text{MPa}\sqrt{\text{m}}$) delivers the most accurate overall fit.

REFERENCES

[1] K. Ando, K. Mogami, K. Tuji, Probabilistic aspects of cleavage crack initiation sites and fracture toughness, Fatigue & Fracture of Engineering Materials & Structures, 15 (1992) 1171–1184.

[2] W.-S. Lei, A cumulative failure probability model for cleavage fracture in ferritic steels, Mechanics of Materials, 93 (2016) 184–198.

[3] C. McMahon Jr, M. Cohen, Initiation of cleavage in polycrystalline iron, Acta Metallurgica, 13 (1965) 591–604.

[4] G. Qian, V. González-Albuixech, M. Niffenegger, Calibration of Beremin model with the Master Curve, Engineering Fracture Mechanics, 136 (2015) 15–25.

[5] ASTM1921, Standard test method for determination of reference temperature, T0, for ferritic steels in the transition range, in, ASTM, 2023.

[6] F.M. Beremin, A. Pineau, F. Mudry, J.-C. Devaux, Y. D'Escatha, P. Ledermann, A local criterion for cleavage fracture of a nuclear pressure vessel steel, Metallurgical Transactions A, 14 (1983) 2277–2287.

[7] X. Gao, C. Ruggieri, R.H. Dodds Jr, Calibration of Weibull stress parameters using fracture toughness data, International Journal of Fracture, 92 (1998) 175–200.

[8] J.P. Petti, R.H. Dodds Jr, Coupling of the Weibull stress model and macroscale models to predict cleavage fracture, Engineering Fracture Mechanics, 71 (2004) 2079–2103.

[9] F. Minami, M. Ohata, Constraint assessment of brittle fracture of steel components, ISO 27306 vs. FITNET FFS, Engineering Fracture Mechanics, 84 (2012) 67–82.

[10] Gao Tregoning, Joyce Link, A Weibull stress model to predict cleavage fracture in plates containing surface cracks, Fatigue & Fracture of Engineering Materials & Structures, 22 (1999) 481–493.

[11] C. Ruggieri, X. Gao, R.H. Dodds Jr, Transferability of elastic–plastic fracture toughness using the Weibull stress approach: Significance of parameter calibration, Engineering Fracture Mechanics, 67 (2000) 101–117.

[12] K. Wallin, The size effect in KICresults, Engineering Fracture Mechanics, 22 (1985) 149–163.

[13] J.P. Petti, R.H. Dodds Jr, Calibration of the Weibull stress scale parameter, σu, using the master curve, Engineering Fracture Mechanics, 72 (2005) 91–120.

[14] X. Qian, S. Chen, A K0-based calibration procedure for the Weibull stress cleavage model, Fatigue & Fracture of Engineering Materials & Structures, 37 (2014) 391–405.

[15] J. Heerens, D. Hellmann, Development of the Euro fracture toughness dataset, Engineering Fracture Mechanics, 69 (2002) 421–449.

[16] J. Heerens, D. Hellmann, R. Ainsworth, Fracture toughness determination in the ductile-to-brittle transition regime-pre-cracked Charpy specimens compared with standard compact specimens, in: European structural integrity society, Elsevier, 2002, pp. 297–305.

[17] B. Wasiluk, J.P. Petti, R.H. Dodds Jr, Temperature dependence of Weibull stress parameters: Studies using the Euro-material, Engineering Fracture Mechanics, 73 (2006) 1046–1069.

[18] X. Qian, S. Zhang, S. Swaddiwudhipong, Calibration of Weibull parameters using the conventional mechanism-based strain gradient plasticity, Engineering Fracture Mechanics, 78 (2011) 1928–1944.

[19] G. Bernauer, W. Brocks, Micro-mechanical modelling of ductile damage and tearing–results of a European numerical round robin, Fatigue & Fracture of Engineering Materials & Structures, 25 (2002) 363–384.
[20] B. Healy, A. Gullerud, K. Koppenhoefer, A. Roy, S. RoyChowdhury, J. Petti, M. Walters, B. Bichon, K. Cochran, A. Carlyle, WARP3D: 3-D dynamic nonlinear fracture analyses of solids using parallel computers, Structural Research Series (SRS 607) UILU-ENG-95-2012, University of Illinois at Urbana-Champaign, 2015.
[21] K. Wallin, Master curve analysis of the "Euro" fracture toughness dataset, Engineering Fracture Mechanics, 69 (2002) 451–481.
[22] A. Khalili, K. Kromp, Statistical properties of Weibull estimators, Journal of Materials Science, 26 (1991) 6741–6752.
[23] K. Trustrum, A.D.S. Jayatilaka, On estimating the Weibull modulus for a brittle material, Journal of Materials Science, 14 (1979) 1080–1084.
[24] ASTME1820, ASTM E1820–22, Standard test method for measurement of fracture toughness, in, ASTM International, West Conshohocken, PA, 2022.
[25] J. Rice, P. Paris, J. Merkle, Some further results of J-integral analysis and estimates, in: Progress in flaw growth and fracture toughness testing, ASTM International, 1973, pp. 231–245.
[26] Gao Faleskog, F. Shih, Analysis of ductile to cleavage transition in part-through cracks using a cell model incorporating statistics, Fatigue & Fracture of Engineering Materials & Structures, 22 (1999) 239–250.
[27] C. Shih, B. Moran, T. Nakamura, Energy release rate along a three-dimensional crack front in a thermally stressed body, International Journal of Fracture, 30 (1986) 79–102.
[28] Gao Faleskog, F. Shih, Analysis of ductile to cleavage transition in part-through cracks using a cell model incorporating statistics, Fatigue & Fracture of Engineering Materials & Structures, 22 (1999) 239–250.
[29] F. Erdogan, G. Sih, On the crack extension in plates under plane loading and transverse shear, Journal of Basic Engineering, 85 (1963) 519–525.
[30] A.G. Evans, A general approach for the statistical analysis of multiaxial fracture, Journal of the American Ceramic Society, 61 (1978) 302–308.
[31] N.N. Nemeth, Unit-sphere multiaxial stochastic-strength model applied to a composite material, Journal of Composite Materials, 48 (2014) 3395–3424.
[32] J. Chang, J.-q. Xu, Y. Mutoh, A general mixed-mode brittle fracture criterion for cracked materials, Engineering Fracture Mechanics, 73 (2006) 1249–1263.
[33] M. Ayatollahi, B. Saboori, Maximum tangential strain energy density criterion for general mixed mode I/II/III brittle fracture, International Journal of Damage Mechanics, 24 (2015) 263–278.
[34] X. Ju, K. Tateishi, Fatigue crack behavior at rib-to-deck weld bead in orthotropic steel deck, Advances in Structural Engineering, 17 (2014) 1459–1468.
[35] T. Chakherlou, H. Maleki, A. Aghdam, B. Abazadeh, Effect of bolt clamping force on the fracture strength of mixed mode fracture in an edge crack with different sizes: Experimental and numerical investigations, Materials & Design, 45 (2013) 430–439.
[36] M. Todinov, S. Same, A fracture condition incorporating the most unfavourable orientation of the crack, International Journal of Mechanics and Materials in Design, 11 (2015) 243–252.
[37] S. Chen, X. Qian, A. Ahmed, Cleavage fracture assessment for surface-cracked plates fabricated from high strength steels, Engineering Fracture Mechanics, 161 (2016) 1–20.
[38] W. Yang, X. Qian, Fracture resistance curve over the complete mixed-mode I and II range for 5083 aluminum alloy, Engineering Fracture Mechanics, 96 (2012) 209–225.

[39] T. Maccagno, J. Knott, The fracture behaviour of PMMA in mixed modes I and II, Engineering Fracture Mechanics, 34 (1989) 65–86.
[40] M. He, J. Hutchinson, Asymmetric four-point crack specimen, Journal of Applied Mechanics, 67 (2000) 207–209.
[41] Z. Zhang, X. Qian, Effect of prior ductile tearing on cleavage assessment for high-strength steel under mixed-mode I and II conditions, Engineering Fracture Mechanics, 175 (2017) 293–311.
[42] T. Keiichiro, I. Hitoshi, Elastic-plastic fracture toughness test under mixed mode I-II loading, Engineering Fracture Mechanics, 41 (1992) 529–540.
[43] C.-F. Shih, Small-scale yielding analysis of mixed mode plane-strain crack problems, in: Fracture analysis: Proceedings of the 1973 national symposium on fracture mechanics, 1974.
[44] K. Wallin, The effect of ductile tearing on cleavage fracture probability in fracture toughness testing, Engineering Fracture Mechanics, 32 (1989) 523–531.
[45] J. Sobotka, R. Dodds Jr, Effects of steady ductile crack growth on cleavage fracture from three-dimensional, small-scale yielding simulations, Engineering Fracture Mechanics, 127 (2014) 211–225.
[46] Z. Zhang, X. Qian, A local approach to estimate cleavage angles in steel under mixed-mode I and II conditions, Engineering Fracture Mechanics, 166 (2016) 164–181.

5 High-Cycle Fatigue of Welded Joints

5.1 INTRODUCTION

Fatigue cracks originating from surface flaws near the hot-spot weld-toe locations impose significant risks to the safety of offshore structures and onshore bridges during overloading events. The widely accepted assessment of the high-cycle fatigue of welded joints is the *S-N* curves developed through experimental data and stress in either global (Nominal Stress) or local locations (Structural Hot-Spot Stress/ENS) over decades. The determination of the stress concentration factors relies on the numerical FEA at hot-spot locations or weld toe/root. The existing *S-N* curves for tubular connections [1–4], developed through experimental data accumulated over decades of research efforts [5–9], estimate the fatigue life of tubular joints with the through-thickness propagation of weld toe cracks at hot-spot locations. The Structural Hot-Spot Stress method describes the stress-raising effect caused by the global geometric discontinuity of the welded joints without considering the local notch geometries of the welds [4, 10]. The hot-spot stresses have evolved as widely accepted engineering procedures in estimating the high-cycle fatigue life of welded plate connections due to their enhanced accuracy compared to the ENS method. However, the stress gradient caused by different loading conditions on the welded plate connections, similar to the observation reported on the smooth bars under tension and bending. This implies that the fatigue life of the same welded plate connections with equal magnitudes of the above hot-spot indicators may correspond to different *S-N* curves under different loadings. This leads to a separate compendium of *S-N* curves not only for different geometries but also for different loading conditions. The ENS approach, on the other hand, addresses this issue effectively through a single and universal *S-N* curve. However, the calculation of the ENS requires a rounded notch with a fine mesh, which introduces pronounced difficulties in describing the weld toe along the brace to chord intersection using a continuous FE model.

The fabrication of welded components in these structures should therefore incorporate a high-quality and convenient profile to meet fatigue requirements outlined in design codes[2, 11]. The complete joint penetration (CJP) welds, widely implemented in offshore jacket structures, require highly skilled workmanship in delivering quality welds meeting the code requirement[2, 11]. The potential defects near the open root of the CJP welds may lead to detrimental failures of the platform under fatigue loads [12]. Qian *et al.* [12] introduced the enhanced partial joint penetration (PJP+) welds, which include a part of the member wall thickness as the inherent backing plate for the welding material. The enhanced partial joint penetration welds anticipate a controlled weld profile and inconsequential unfused weld root at the limit of API RP 2X [13] level C non-destructive test (NDT) requirement or equivalently class X NDT requirement

DOI: 10.1201/9781003509806-5

in AWS [11]. The understanding on the fatigue performance of this new set of weld details requires experimental efforts to support their applications in realistic engineering projects. To ensure the safety of existing welding structures and save costs, effective maintenance techniques anticipate being well and quantitatively developed. Therefore, this chapter addresses the above-mentioned challenges including convenient determination approach of ENS, the volume-based Hot-Spot Stress, and effects of post-welding treatment in fatigue performance of welded joints (toe grinding on tubular joints and ultrasonic impact treatment (UIT) on welded plate joints).

5.2 EXTRAPOLATION APPROACH TO DETERMINE ENS

5.2.1 The Extrapolation Method

The ENS extrapolation method originates from the comparison of linear-elastic stress fields between FE models with a "sharp" weld toe ($r_{ref} = 0$ mm) and those with a "rounded" weld toe ($r_{ref} = 1$ mm). The stress at the sharp notch is theoretically infinite, and in the context of FE analysis, the stress is non-converging. In contrast, the stress at a rounded notch converges to a certain value, with peak stress residing typically at the midpoint of the curved notch surface. Figure 5.1a illustrates this phenomenon by plotting the typical variation of the normalized maximum principal stress against the normalized distance from the sharp weld toe in a log-log scale for a cruciform joint. The log-log scale prohibits the plotting of the sharp notch peak stress at $x/t = 0$. Nevertheless, as shown in Figure 5.1a, the sharp notch shows a significantly higher stress at the point where the rounded notch reaches the maximum stress.

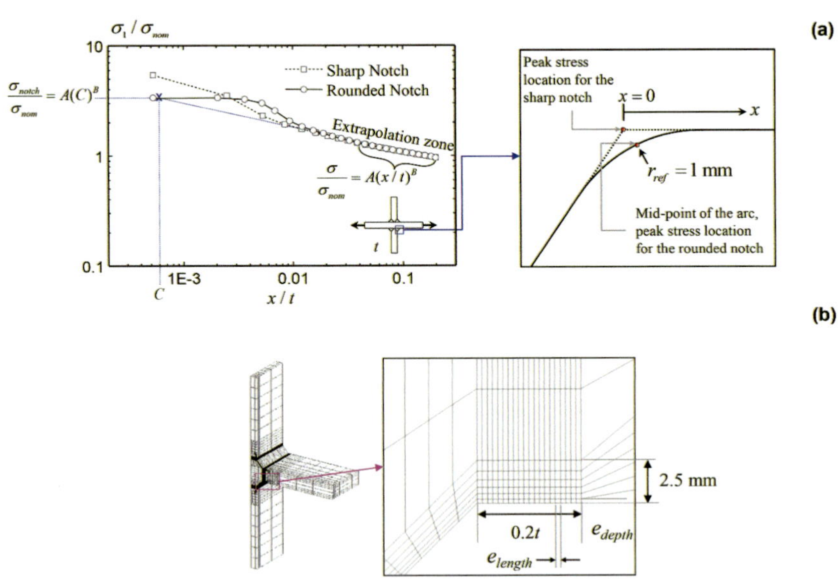

FIGURE 5.1 (a) The proposed effective notch stress (ENS) extrapolation method; (b) the typical weld toe mesh required for the extrapolation method according to Pradana *et al.* [14].

Previous work [14] has shown that for full-penetration welded cruciform joints, the stress fields from the sharp and rounded notch ($r_{ref} = 1$ mm) converge at a distance about 2% of the plate thickness from the weld toe, as illustrated in Figure 5.1a. The converging stress fields appear as a linear segment, and thus provide a potential extrapolating curve which follows,

$$\frac{\sigma_1}{\sigma_{nom}} = A(x/t)^B \quad (0.04 \leq x/t \leq 0.2) \tag{5.1}$$

where A and B represent the dimensionless scaling constant and exponent, respectively, while x/t denotes the normalized distance measured from the sharp weld toe. The nominal stress (σ_{nom}) corresponds to the remotely applied stress without the presence of the weld geometry. The common practices for CHS joints often define σ_{nom} as the remote fiber stress applied on the brace member under axial loading or the maximum fiber stress in the brace under in-plane bending or out-of-plane bending conditions [1]. The critical point ($x/t = C$) represents the location at which the value of the extrapolation curve equals the ENS (σ_{notch}) calculated according to the procedure described in IIW [4]. With both the extrapolation curve and the critical point known, the ENS becomes tractable via extrapolation:

$$\frac{\sigma_{notch}}{\sigma_{nom}} = A(C)^B \tag{5.2}$$

With a parametric expression for the variable C, and the values of A and B, the extrapolation in Eq. (5.2) presents a significantly simplified method to estimate the ENS, eliminating the detailed modeling requirement as prescribed in the IIW [4] procedure.

Pradana *et al.* [14] reported that the critical point of extrapolation depends on the joint type, dimensions and loading modes. Furthermore, they reported that consistent extrapolation requires a prescribed extrapolation zone with strictly controlled mesh sizes. Pradana *et al.* [14] proposed an extrapolation zone of $0.04t - 0.2t$ from the weld toe, with mesh density defined by the element depth e_{depth} and the element length e_{length} (see Figure 5.1b). The element length should not exceed $0.01t$, while the element depth depends on the plate thickness. When $t \geq 20$ mm, follows the IIW [4] recommended size of 0.25 mm (for quadratic element), while cases with $t < 20$ mm require a smaller, thickness-scaled e_{depth},

$$e_{depth} = \left(\frac{t}{20 \text{ mm}}\right) \times 0.25 \text{ mm} \quad (\text{for } t < 20 \text{ mm}) \tag{5.3}$$

The complex geometry of the CHS X-joint naturally creates a more complex stress variation near the weld toe, which requires additional verification for the ENS extrapolation approach. This section presents a preliminary numerical study aiming to verify: 1) the existence of the extrapolation curve, 2) the required extrapolation zone and mesh sizes to ensure consistent A and B values, and 3) the convergence of the ENS (σ_{notch}) computed using the proposed mesh sizes. This preliminary study utilizes a CHS X-joint subjected to a unit tensile stress on both brace ends, with

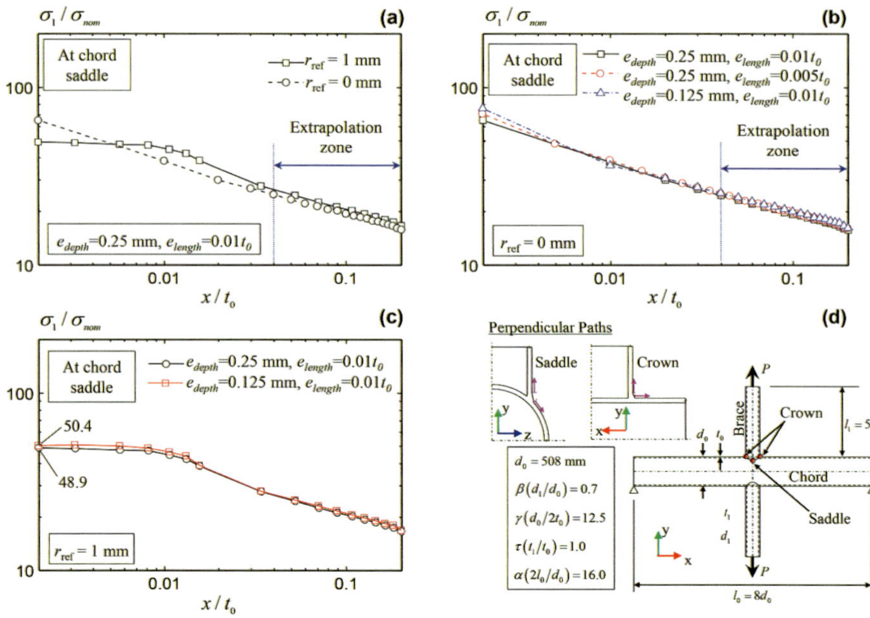

FIGURE 5.2 Results from the preliminary numerical study showing: (a) the effect of the weld toe radius on the near-toe stress fields; (b) the effect of the near-toe mesh sizes on the stress fields within the extrapolation zone; (c) the effect of near-toe mesh sizes on the computed ENS; and (d) the X-joint considered.

dimensions as shown in Figure 5.2d. The definition of the X-joint adopts the commonly used non-dimensional parameters, which include: the brace-to-chord outer diameter ratio, β (d_1 / d_0), the chord outer radius to the wall-thickness ratio, γ ($d_0 / 2t_0$), the brace-to-chord wall thickness ratio, τ (t_1 / t_0), and the chord length to the chord radius ratio, α ($2l_0 / d_0$).

Figures 5.2a–c summarize the results of the preliminary study by plotting the stress fields along a path perpendicular to the weld line at the chord saddle (see Figure 5.2d), which is the critical position for CHS X-joints subjected to brace axial tension. Figure 5.2a shows that the stresses computed from the "sharp" and the "rounded" weld toe converge to a linear segment, therefore verifying the existence of the extrapolation curve. Figure 5.2a also implies that the proposed extrapolation zone ($0.04t - 0.2t$) remains valid for CHS X-joints since the converging segment stays linear within this zone. The comparison in Figure 5.2b shows that further mesh refinement from the recommended e_{depth} and e_{length} does not lead to significant changes in the stress field within the extrapolation zone. Similarly, as shown in Figure 5.2c, the overall near-toe stress variation and the ENS (the peak stress) remain relatively constant as e_{depth} becomes smaller. The subsequent numerical investigation, therefore, adopts the same extrapolation zone and mesh sizes as recommended by Pradana *et al.* [14].

Table 5.1 compares the *A* and *B* values obtained from the original extrapolation zone against those computed from the enlarged extrapolation zone ($0.04t - 0.4t$)

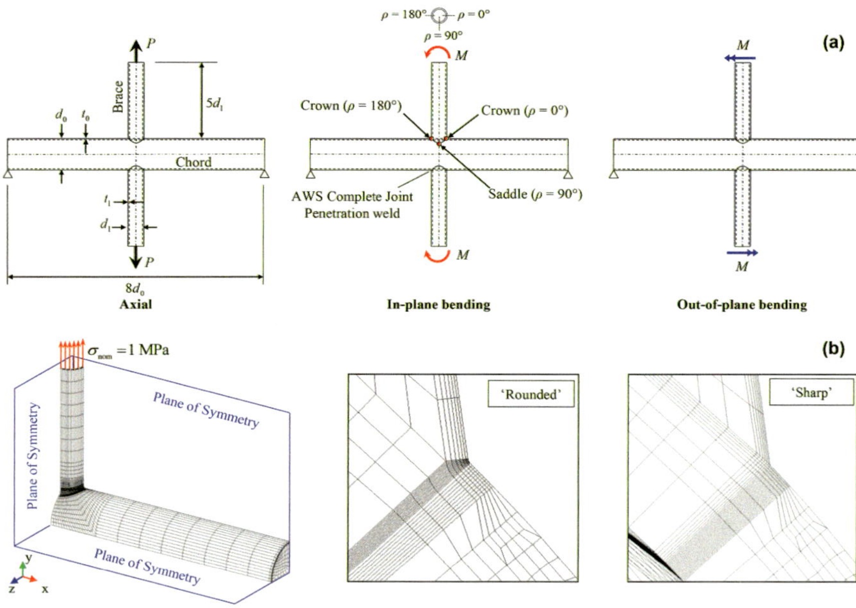

FIGURE 5.3 The numerical investigations on the CHS X-joints: (a) the configuration of the X-joint and the load cases considered; and (b) the FE model and the typical mesh at the sharp and the rounded weld toe.

TABLE 5.1
Numerical *A* and *B* Values.

	Finite element (FE) Extrapolation zone			
	0.04*t*–0.2*t*		0.04*t*–0.4*t*	
Location	***A***	***B***	***A***	***B***
Chord Crown	2.05	−0.36	2.25	−0.32
Brace Crown	0.93	−0.25	0.92	−0.26

based on the numerical analysis [14]. The development of parametric relationships is to estimate the proper extrapolation critical point, *C*, for a given geometry and loading condition. The development involves extensive parametric studies covering the geometric range of commonly encountered joints, under three loading modes, namely the brace axial tension, the brace in-plane bending, and the brace out-of-plane bending. Table 5.2 lists the range of dimensional and non-dimensional parameters considered, while Figure 5.3a illustrates the configuration of the X-joints analyzed in this parametric study. The parametric study includes a total of 110 cases for each loading mode. In all cases, the brace-to-chord intersection angle equals 90°, and the

TABLE 5.2

Parametric Ranges for the CHS X-Joint.

Parameter	Description	Values
β	Brace-to-chord outer diameter ratio ($\boldsymbol{d_1/d_0}$)	0.18, 0.23, 0.33, 0.38, 0.48, 0.64, 0.7
γ	Chord outer radius to thickness ratio ($\boldsymbol{d_0/2t_0}$)	8, 10, 12.5, 18.45, 25.4, 29, 32
τ	Brace-to-chord wall-thickness ratio ($\boldsymbol{t_1/t_0}$)	0.2, 0.5, 1.0
α	Chord length to outer radius ratio ($\boldsymbol{2l_0/d_0}$)	16.0
θ	Brace-to-chord intersection angle	90°
$\boldsymbol{d_0}$	Chord outer diameter	508 mm

chord outer diameter (d_0) remains constant at 508 mm. Moreover, the chord length factor α remains fixed at 16 to eliminate the chord end effect. This study avoids X-joints with $\beta = 1.0$ due to the strong variation observed in the weld profile (e.g., variation in the weld leg length and convexity), which leads to significant scatter in the measured near-toe stress fields [15, 16]. Without valid reference on the appropriate weld profile, the inclusion of $\beta = 1.0$ in the parametric study may produce misleading and unrealistic results.

The computation of a C value for a specific X-joint requires: 1) FE model with a "sharp" toe that provides A and B values, and 2) FE model with a "rounded" toe that computes the ENS according to the IIW [4] procedure. Figure 5.3b illustrates the one-eighth FE model used for the brace axial tension cases, as well as the typical mesh at both "rounded" and "sharp" weld toe, where the element sizes follow the recommendation by Pradana *et al.* [14]. The model simulates the X-joint using 20-node brick elements with reduced integration (C3D20R) in the ABAQUS [17] element library. Both the base metal and the weld (CJP profile AWS [11]) adopt the same linear-elastic material model with a Young's modulus of 205 GPa and a Poisson ratio of 0.3. In all three load cases, the applied load corresponds to a unit nominal stress in the brace member ($\sigma_{nom} = 1$ MPa). The nominal stress for the brace in-plane bending and out-of-plane bending cases refers to the maximum stress at the most extreme fiber in the brace member. Using the correlation between C, the extrapolation curve (A and B), and the normalized ENS ($\sigma_{notch} / \sigma_{nom}$) shown in Eq. (5.2), this study computes the C values in all cases within the considered parametric range. Regression analysis on the C values then allows for the determination of the parametric formulae.

This study adopts the following parametric equation to predict the value of C for the CHS X-joints under brace axial load,

$$C = \lambda(\beta)^{X_1}(\tau)^{X_2}(\gamma)^{X_3} \tag{5.4}$$

where λ, X_1, X_2, and X_3 are constants derived from the regression analyses performed on the computed C values.

5.2.1.1 X-Joints under Brace Axial Load

This study proposes extrapolation only for the saddle point, with two recommended extrapolation critical points, each for the chord saddle and the brace saddle. Figures 5.4a and b illustrate the effect of β on the calculated ENS at the saddle point. As illustrated in Figures 5.4a and b, the ENS increases initially as the β ratio increases from a small value ($\approx$ 0.2) to about 0.4, reaches a peak value over the β range from 0.4 to 0.7, and decreases in magnitude when β further increases. Overall, the ENSs at the chord saddle and those at the brace saddle exhibit a similar trend, although the chord saddle shows consistently higher ENS. The trend observed in

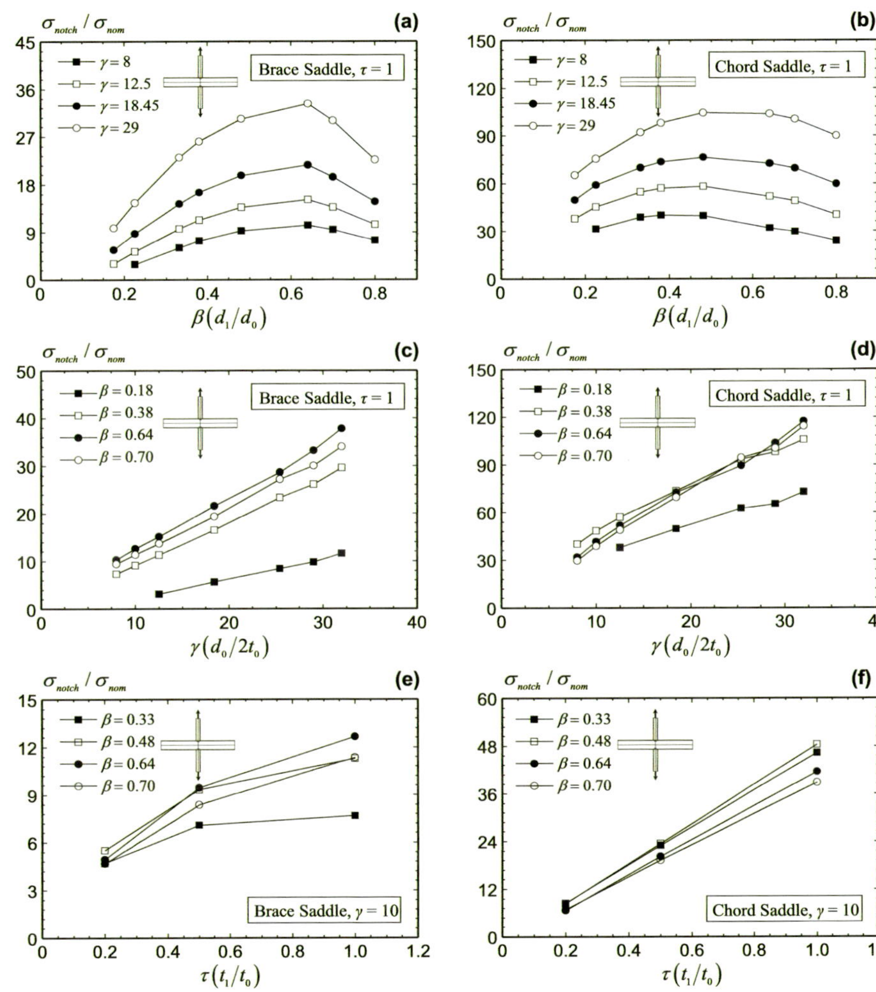

FIGURE 5.4 The variation of computed normalized ENS under brace axial loading: (a) at the brace side with respect to β; (b) at the chord side with respect to β; (c) at the brace side with respect to γ; and (d) at the chord side with respect to γ; (e) at the brace side with respect to τ; and (f) at the chord side with respect to τ.

Figures 5.4a and b are results of the complex interaction between the shell bending and the membrane action. At small β values, the dihedral angle along the intersection is close to 90°, causing the load from the brace to generate predominantly shell bending in the chord wall. However, as β increases, the increase in dihedral angle increases the load transfer gradually involves more membrane action. The changes in proportion between the shell bending and the membrane action at different β ratios lead to the observed trend in Figures 5.4a and b. Figures 5.4c and d show that the ENS at the chord and brace saddle increases approximately linearly as γ increases. A large value of γ (a thin chord wall) causes a more significant chord shell bending than does a thick chord wall under the same brace load. This leads to an increase in the ENS as observed in Figures 5.4c and d. For thin chord walls, the significant shell bending at the saddle point forms the primary mechanism in resisting the applied load, as inferred by the comparisons in Figures 5.4c and d.

Figures 5.4e and f illustrate the typical variation in the saddle ENS with respect to the brace-to-chord wall thickness ratio τ. Both the brace saddle (Figure 5.4e) and the chord saddle (Figure 5.4f) show an increasing ENS trend as τ becomes larger. Nevertheless, as shown in Figure 5.4e, the ENS at the brace saddle increases with a weaker gradient. In general, since the axial load is applied as a unit stress over the cross section of the brace, the increase in τ (therefore increase in the brace cross section) introduces a higher resultant force, which in turn causes an increase in the computed ENS. At the same time, the increase in τ also strengthen the brace member, leading to a less steep increase in the ENS at the brace saddle, as illustrated in Figure 5.4e.

Table 5.3 summarizes the constants obtained from the regression analyses and their accuracy in terms of the ratio between the extrapolated ENS and the ENS computed according to the IIW [4]. Table 5.3 demonstrates that the extrapolation critical point proposed in Eq. (5.4) leads to a reasonably accurate ENS prediction, with the coefficient of variance (COV) lower than 6% for extrapolation at both brace and chord saddle.

5.2.1.2 X-Joints under Brace In-Plane Bending

Under the brace in-plane bending load, the material near the crown point experiences a high notch stress due to the significant shell bending, while the saddle point

TABLE 5.3

Constants and Accuracy of the Proposed Parametric Equation for the X-Joints under Brace Axial Load.

						$\sigma_{notch,Eq.\ (5.4)}/\sigma_{notch,\ IIW}$	
Location	**No. of data**	λ	X_1	X_2	X_3	**Mean**	**COV**
Brace saddle	110	3.04E-04	−0.8994	−0.4179	0.6592	1.004	0.025
Chord saddle	110	4.89E-04	0.2743	−0.0235	0.8393	1.004	0.057

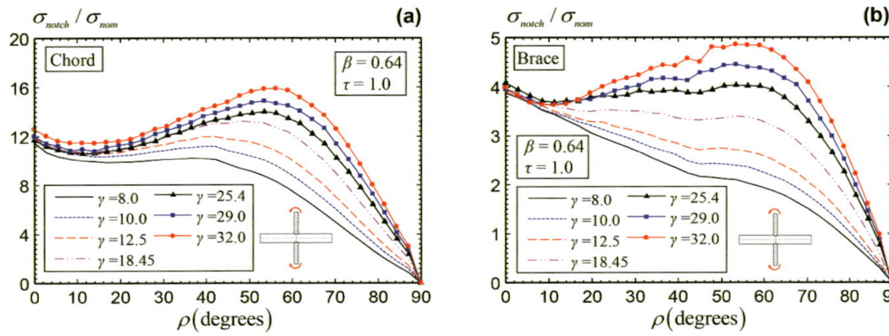

FIGURE 5.5 The variation of the computed ENS along the weld line of CHS X-joints under brace in-plane bending: (a) at the chord side; and (b) at the brace side.

experiences a zero notch stress as it coincides with the neutral axis of the brace. However, current numerical results indicate that the maximum ENS may not locate at the crown point. Figure 5.5 illustrates the ENS variation along the weld line for selected geometries. The angle ρ in Figure 5.5 measures the position along the brace-to-chord intersection as illustrated in the top view of the brace cross-section in Figure 5.3a. Figure 5.5 demonstrates that the location of the maximum ENS in the chord and the brace shifts from the crown towards the saddle as γ increases. The critical location may shift up to 60° towards the saddle, with a magnitude up to 30% higher than the ENS at the crown. This suggests that a special treatment is required to estimate the maximum notch stress in the proposed extrapolation.

Compared to the ENS in the brace axial tension case, the ENS computed under the brace in-plane bending load displays a similar trend with respect to β, γ, and τ, as illustrated in Figure 5.6. Figures 5.6a and b illustrate the effect of β on the maximum ENS in the brace member and that in the chord member. The ENS increases with β at small β values, and decreases after reaching a maximum value at approximately $\beta = 0.4$. The ENS at the chord (Figure 5.6b), however, displays a significantly slower decrease after reaching the peak value, implying that the effect of β on the chord member diminishes beyond a certain β value. Figures 5.6c and d show the variation of the ENS with respect to γ , which demonstrates an approximately linear relationship, as also observed in the brace axial load case. Again, the thin chord wall (large γ value) induces a significant shell bending in the chord, therefore causing an increase in the computed ENS as increases. Figures 5.6e and f demonstrate the effect of τ on the computed ENS, which shows that the ENS increases with the τ ratio. Similar to X-joints under brace axial load, the increase in τ (thus increase in the brace cross section) introduces a larger resultant force when the brace nominal stress remains fixed at 1 MPa. Consequently, the increase in the resultant force leads to higher ENS. Overall, the brace in-plane bending load generates considerably less critical ENS compared to the brace axial load. Under a combined load scenario, the ENS from the axial load case will likely govern the fatigue life of the joint, unless the nominal in-plane bending stress is significantly larger than the nominal axial stress.

TABLE 5.4
Constants and Accuracy of Proposed Parametric Equation for the X-Joints under Brace In-plane Bending Load.

Location	No. of data	λ	X_1	X_2	X_3	$\sigma_{notch,Eq.(5.4)}/\sigma_{notch,IIW}$	
						Mean	COV
Brace	110	4.16E-04	−0.2672	−0.7329	0.7368	1.0030	0.050
Chord	110	3.68E-04	−0.4812	0.304	0.6765	1.0004	0.044

The critical ENS for the in-plane bending may not locate at the crown point, and could exceed the ENS at the crown in Figure 5.5. Therefore, the extrapolation of stresses at the crown point, with an amplification factor embedded in the recommended extrapolation point, enables extrapolation at a well-defined location. The process of deriving the parametric formulae remains the same as in the brace axial load case, except that the computation of C values now involves the maximum ENS instead of the ENS at the crown position. Table 5.4 summarizes the constants and the accuracy of the proposed parametric formulae for the extrapolation critical point at the brace and the chord side. The COV values listed in Table 5.4, which are less than 6%, confirms the accuracy in the proposed approach.

5.2.1.3 X-Joints under Brace Out-of-Plane Bending

The numerical procedure for the X-joints under brace out-of-plane bending remains the same as that under brace in-plane bending. However, the saddle point now becomes a potential critical location as it is located along the extreme fiber of the brace. In contrast to the in-plane bending case, the numerical results show that the maximum ENS always occurs at the saddle point for all geometric parameters. The ENS in the out-of-plane bending case shows generally higher magnitudes than that in the in-plane bending case, but lower than the ENS in the brace axial load case. The trend of ENS under the three load cases shows similarity, especially between the out-of-plane bending case (Figure 5.7) and the axial load case (Figure 5.6). As illustrated in Figure 5.7, the magnitude of ENS in the out-of-plane bending case is approximately half of that in the axial load case since the total brace tension applied on an X-joint model under brace out-of-plane bending is half of the brace tension applied in the axial load case.

The development of the recommended C values for the current load case simply repeats the procedure described earlier, resulting in constants summarized in Table 5.5. Table 5.5 also shows the accuracy of the extrapolation through the COV values at the brace and chord saddle, which remains lower than 4%.

5.2.2 Fatigue Assessment Using the Proposed Extrapolation

The ENS section demonstrates the reliability of the proposed ENS extrapolation in assessing the fatigue life of 23 CHS X-joint specimens extracted from the literature

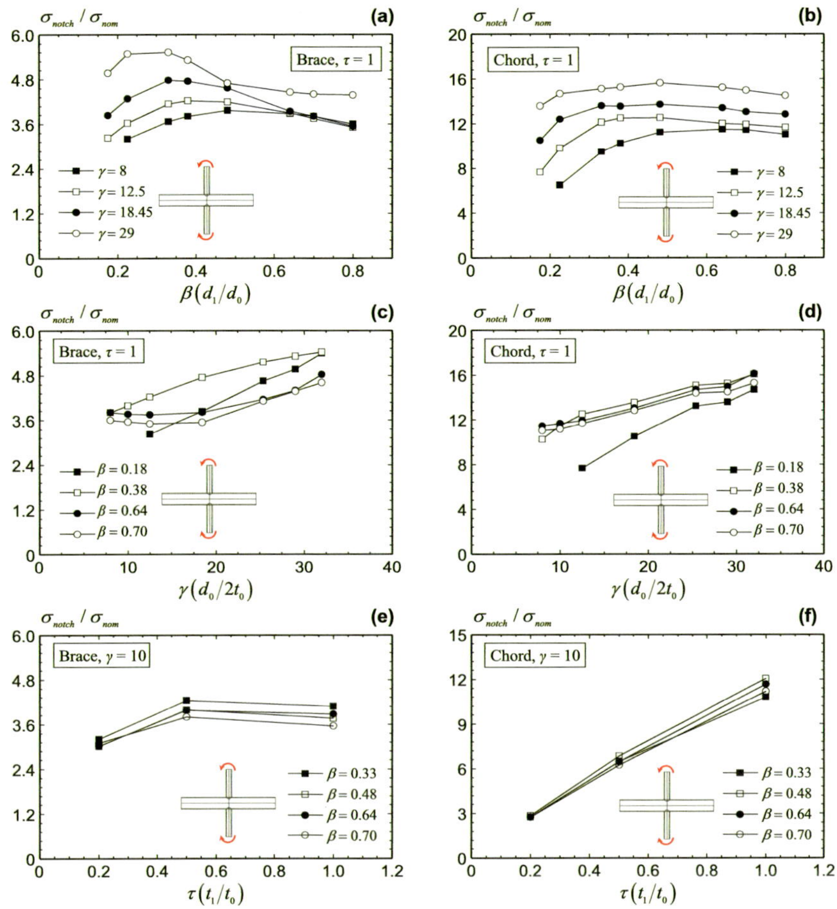

FIGURE 5.6 The variation of computed normalized ENS under brace in-plane bending: (a) at the brace side with respect to β; (b) at the chord side with respect to β; (c) at the brace side with respect to γ; and (d) at the chord side with respect to γ; (e) at the brace side with respect to τ; and (f) at the chord side with respect to τ.

TABLE 5.5

Constants and Accuracy of the Proposed Parametric Equation for the X-Joints under Brace Out-of-plane Bending Load.

Location	No. of data	λ	X_1	X_2	X_3	$\sigma_{notch,Eq.(5.4)}/\sigma_{notch,IIW}$ Mean	COV
Brace saddle	110	1.81E-04	–0.5865	–0.5033	0.8568	1.009	0.036
Chord saddle	110	4.55E-04	0.2096	0.0254	0.8305	1.001	0.039

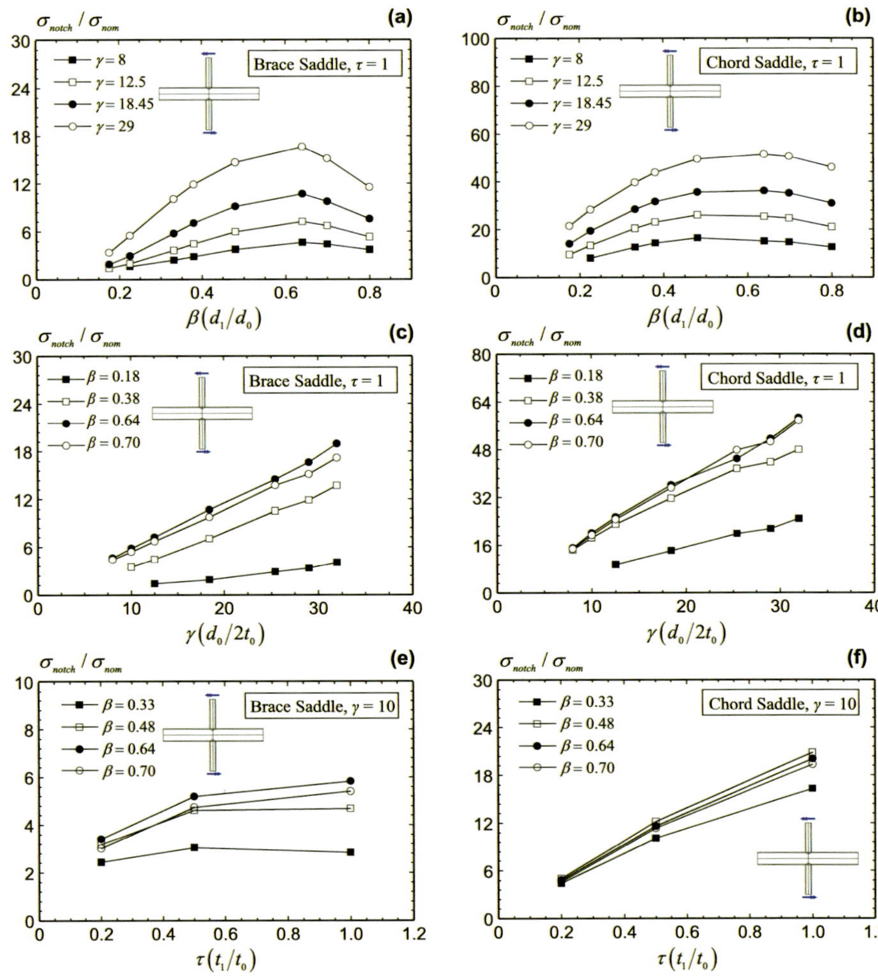

FIGURE 5.7 The variation of computed normalized ENS under brace out-plane bending: (a) at the brace side with respect to β; (b) at the chord side with respect to β; (c) at the brace side with respect to γ; and (d) at the chord side with respect to γ; (e) at the brace side with respect to τ; and (f) at the chord side with respect to τ.

[18–20], all tested under the constant-amplitude load in a laboratory air environment. The selected specimens consist of isolated X-joints in as-welded condition, subjected to either brace axial tension or brace in-plane bending. The reported fatigue failure in all specimens occurred at the weld toe in the chord.

This study generates FE models of the specimens with zero weld toe radius and mesh sizes according to Section 5.2.1. The weld geometry assumes the AWS [11] CJP weld profile, while the material follows the linear-elastic model described in the previous section. The extrapolation then provides an estimation of the most critical ENS for each specimen following the procedure described previously. The numerical

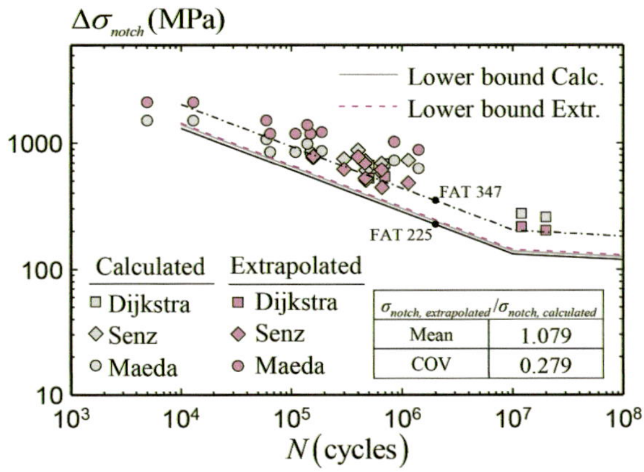

FIGURE 5.8 Fatigue assessment of the selected CHS X-joint tests under the ENS system.

calculation reveals that the most critical stress in all specimens resides in the chord member, consistent with the observed failure location in the actual tests.

Figure 5.8 compares the fatigue data in terms of the extrapolated ENS against the FAT 225 and FAT 347 *S-N* curves. The FAT 225 is the IIW recommended design curve for the ENS approach, while the FAT 347 represents the mean curve ($P_s = 50\%$) from the pioneering experimental data on welded plate joints reported by Olivier *et al.* [21]. The fatigue lives shown in Figure 5.8 correspond to the reported number of cycles at which a through-thickness crack was first detected. In addition, Figure 5.8 includes the same fatigue data in terms of the ENS calculated using FE meshes compliant with the IIW procedure. Compared to the traditionally calculated ENS, the extrapolated ENS shows reasonable agreement as indicated by the COV in Figure 5.8. Comparison against the FAT 347 curve reveals that a number of data points fall below the curve, which is expected since the FAT 347 represents a mean curve and therefore does not include any safety margin. The data points of the CHS X-joints agree well with the FAT 347 curve, despite that the FAT 347 curve being based on welded plate joints. Moreover, the lower bound curves ($P_s = 97.7\%$) from the experimental data show excellent agreement with the FAT225 curve, as indicated in Figure 5.8. These findings confirm the reliability of the proposed ENS extrapolation for fatigue assessment of CHS X-joints, while also supporting the claim that the ENS system allows for fatigue assessment using a single and universal *S-N* curve.

5.3 VOLUME-BASED HOT-SPOT FATIGUE INDICATOR

5.3.1 Volume-Based Hot-Spot Energy Indicator

Esin [22] (reproduced in Figure 5.9a) shows that the fatigue life of a smooth bar under axial loading is always shorter than that subjected to bending actions. Similarly, the fatigue test data of welded plate connections presented by Hyundai Heavy Industries

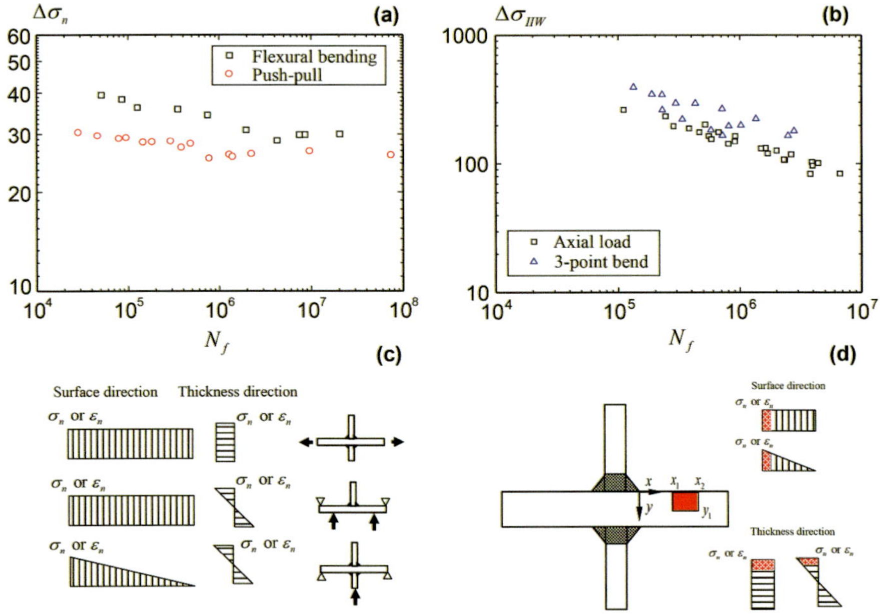

FIGURE 5.9 (a) Fatigue test results for smooth bars under axial loading and bending [22]; (b) fatigue test data for welded plate joints under three-point bending and axial loading; (c) the stress distribution in surface and thickness direction under different loadings; and (d) the schematic hot-spot volume to incorporate the stress gradient.

(HHI) [23], as shown in Figure 5.9b, also illustrates that the stress gradient effect caused by different loading conditions on the fatigue life of welded plate joints indexed by the point-based hot-spot stress in IIW recommendations [4]. The limitation of the point-based hot-spot stress indicators lies in their inability to consider the stress gradient effect, which causes different fatigue lives under the same nominal or hot-spot stress range. Other researchers [24, 25] have reported a similar phenomenon for welded tubular joints and introduced a parameter named degree of bending (DoB) besides the hot-spot stress, to reflect the stress gradient effect on the fatigue life. The volume-based concept, originally proposed by Kuguel [26–28], provides a convenient approach to integrate the stress and strain fields over the finite volume near the weld toe and demonstrates apparent advantages over the point-based stress indicators.

Welded plate connections, the stress gradient effect exists in both the surface and thickness directions as shown in Figure 5.9c, for three different loading conditions (the three-point bending, four-point bending, and the axial loading). The hot-spot volume-based fatigue indicator, as illustrated in Figure 5.9d, should cover the zone, over which the stress and strain variations near the toe of welds reflect the effect of stress gradients caused by the changes in the global structural dimensions and the applied loading. Meanwhile, the stresses and strains in this volume should be independent of the local weld geometry, in line with the hot-spot definition, so as to facilitate its engineering applications. Figure 5.9d indicates the location of the proposed hot-spot volume. The coordinates x_1 and x_2 in Figure 5.9d refer to the starting and ending positions of the hot-spot volume, from the weld toe, and y_1 denotes the depth of the hot-spot volume. The value of x_1 denotes the material point beyond which the stresses become insensitive to the local weld geometry, in line with the hot-spot concept. The magnitude of x_2 aims to incorporate the stress gradient in the surface direction caused by the loading conditions and specimen sizes. The depth y_1 covers the highly stressed material in the thickness direction over which the stress variation reflects the size effect of the plate thickness and imposes a critical effect on the initiation and propagation of small fatigue cracks under different DoBs. The following sections present the procedure to determine the x_1, x_2, and y_1 for the welded plate connections. In determining the location and size of the hot-spot zone, this study has made use of the reference stress solution computed from a FE mesh required in IIW [6] in calculating the ENS. The IIW procedure [4] requires a 1 mm notch radius to be simulated at the weld toe in computing the ENS.

5.3.2 Effect of Local Weld Notch Geometry and Determination of x_1

In line with the hot-spot concept, the value of x_1 should locate far enough to be unaffected by the local weld notch geometry (local notch angle and local notch tip radius). Figure 5.10a presents the geometry configuration, loading, and boundary information of the transverse fillet welded joints. The main plate thickness (t)

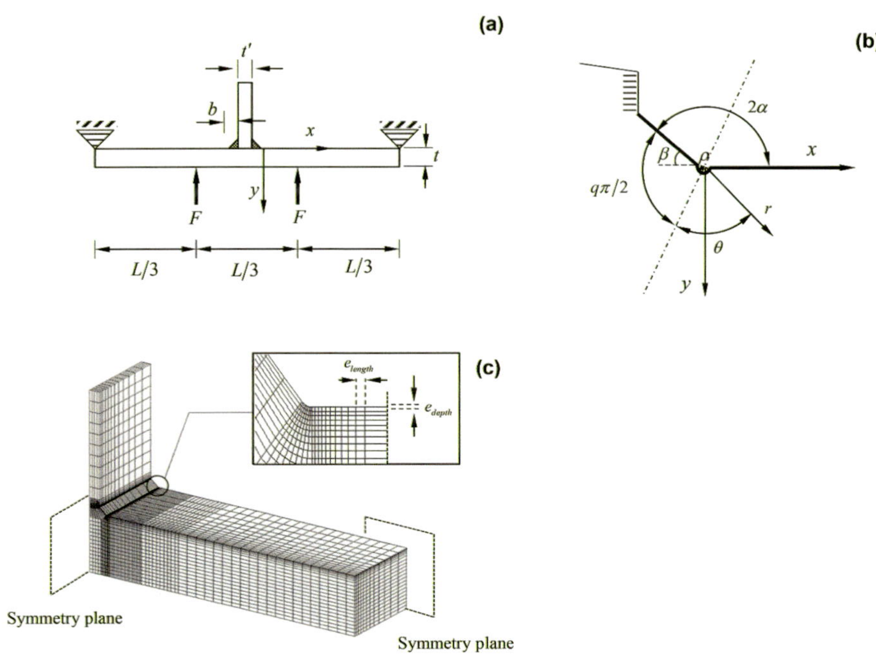

FIGURE 5.10 (a) Geometric configuration, loading, and boundary conditions of four-point bending specimens; (b) Cartesian coordinate system and polar coordinate system at the weld toe; and (c) a quarter-symmetric FE model of the specimen.

TABLE 5.6
Six Types of Local Notch Angle and Notch Tip Radius in the Numerical Study.

	I	II	III	IV	V	VI
ρ(mm)	0	1	0	1	0	1
β	30°	30°	45°	45°	60°	60°

of these welded connections ranges from 5 mm to 80 mm. The local weld notch geometry in Figure 5.10b considered in this study includes the local notch angle (β) and the local notch tip radius (ρ). Linear elastic analysis for specimens with Young's modulus of 200 GPa and Poisson's ratio of 0.3 is conducted in ABAQUS [17]. Figure 5.10c shows the quarter-symmetric FE model, with 20 node hexagonal elements (C3D20 in ABAQUS). The local mesh sizes e_{depth} and e_{length} follow the recommended value by Paradana *et al.* [14].

This study investigated six types of local notch angle and tip radius in the model, listed in Table 5.6. The weld toe with $\rho = 0$ mm represents a sharp weld toe in the plate connection. Figure 5.11a presents, based on the FE analysis, a significant effect of the local weld notch angle and notch tip radius on the σ_{xx}/σ_n distribution near the weld toe of the connection with a main plate thickness (t) equal to 15 mm. In Figure 5.11a, σ_{xx} denotes the stress perpendicular to the weld line and σ_n represents the nominal stress. The effect of the local weld notch angle geometry gradually diminishes at approximately $x = 1$ mm (with $x = 0$ denoting the position of the weld toe). Figure 5.11b compares the difference in the normalized stress among different local geometries with respect to case IV in Table 5.6 at $x = 1$ mm. For most of the 13 welded specimens considered, the difference in the normalized stress at $x = 1$ mm remains within $\pm 5\%$. For a hot-spot zone to reflect the global geometry effect and to remain independent of the local weld notch geometry, this chapter selects $x_1 = 1$ mm.

5.3.3 Determination of x_2 Based on Stress Gradient Along Surface

The value of x_2 determines the zone of the material in the surface direction, which reflects the stress gradient along the surface direction, caused by either the size effect or loading conditions. The linearly extrapolated structural stress, as recommended in IIW [4], does not capture the local stress concentration in some types of joints [29], as reflected by the significant deviation in the extrapolated hot-spot stress from the local stress at the weld toe in Figure 5.11c. Table 5.7 shows the dimensions of the test specimens under four-point bending for different main plate thicknesses reported by Miki *et al.* [30], with the main plate thickness varying from 9 mm to 50 mm. As shown in Figure 5.11d, the contrast difference diminishes in the extrapolation zone ($0.4t$ to $1.0t$ from the weld toe, where t denotes the plate thickness) recommended in IIW [4]. Consequently, the hot-spot stress requires a thickness scaling factor to

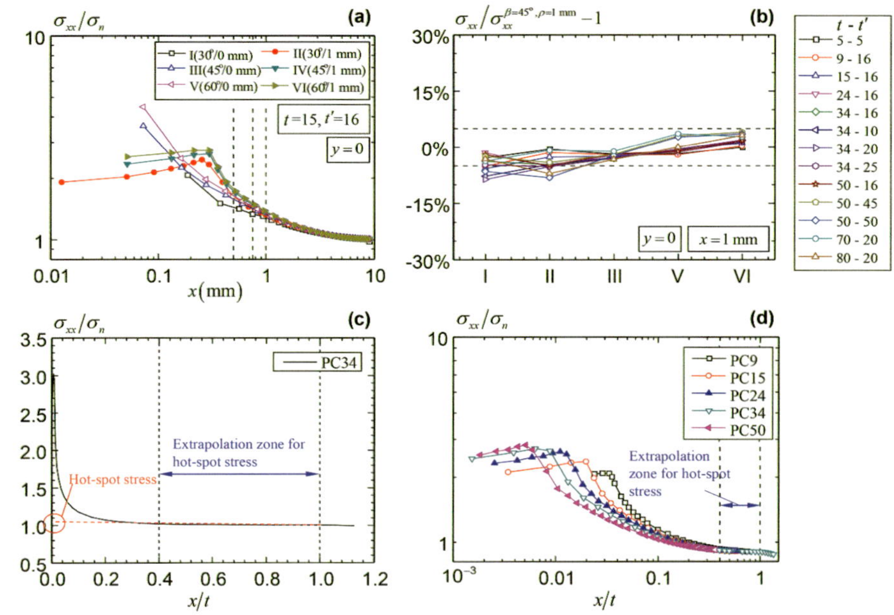

FIGURE 5.11 (a) σ_{xx}/σ_n distribution along the surface direction in logarithmic scale for different local weld geometries; and (b) σ_{xx} in joints with different local weld geometries normalized against σ_{xx} in the joint with a weld toe of $\beta = 45^{\circ}$ and $\rho = 1$ mm at $x = 1.0$ mm : (c) σ_{xx}/σ_n variation near the weld toe in the specimen PC34; (d) σ_{xx}/σ_n variation of different welded plate joints near the weld toe in the logarithmic scale.

TABLE 5.7

Geometries of Different Main Plate Thickness Specimens [30].

Specimen	Load type	*L* (mm)	*t* (mm)	*b* (mm)	*B* (mm)	t'(mm)
PC9	Four-point bend	150	9	6	50	16
PC15	Four-point bend	210	15	6	50	16
PC24	Four-point bend	270	24	6	65	16
PC34	Four-point bend	360	34	6	90	16
PC50	Four-point bend	450	50	6	130	16

compensate for the thickness insensitive hot-spot stress value. This implies that a proper value of x_2 remains essential to enclose the material, which exhibits the size effect and the stress gradient effect near the weld toe, in the hot-spot volume.

Figure 5.12a presents the normalized stress σ_{xx}/σ_n distribution near the weld toe with the sharp notch and that with a rounded notch of $\rho = 1$ mm. Beyond 1 mm from the weld toe, the stress becomes independent of the presence of the

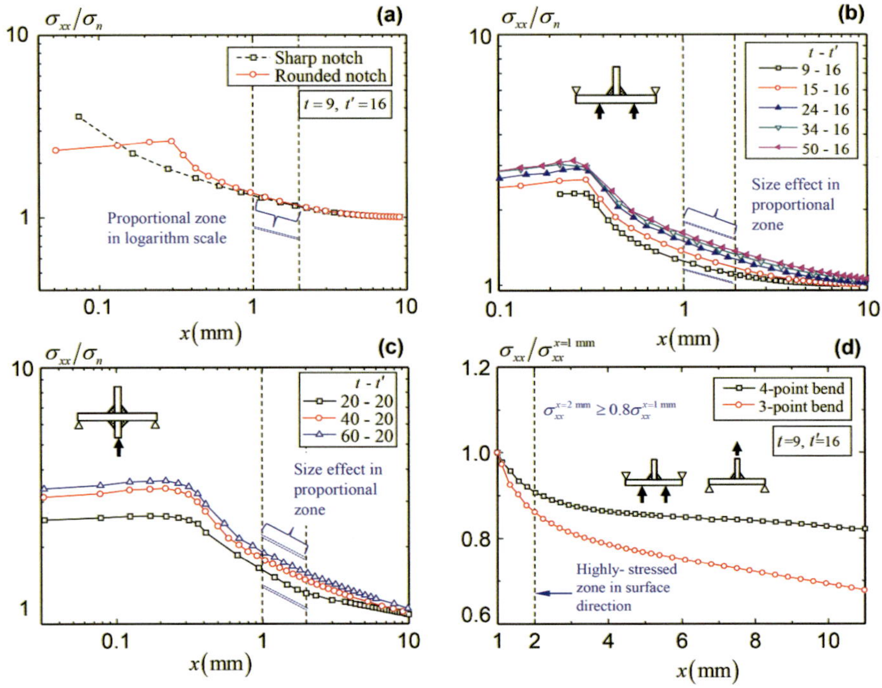

FIGURE 5.12 σ_{xx}/σ_n distribution: (a) near a sharp weld toe and near a rounded weld toe with a notch radius of 1 mm; (b) in five welded specimens under four-point bending; (c) in three welded plate specimens under three-point bending; and (d) $\sigma_{xx}/\sigma_{xx}^{x=1\text{ mm}}$ variation near the weld toe for the same welded plate joint under four-point bending and three-point bending.

rounded notch at the weld toe. Figures 5.12b and c illustrate the normalized near-toe stress distribution along the surface of the welded specimens under four-point bending and three-point bending, respectively. The near-toe stresses maintain a similar level of difference among different main plate thicknesses over the 1~2 mm distance from the weld toe, denoted as the proportional zone in this chapter. Figures 5.12b and c indicate similar stress gradients over this proportional zone for different main plate thicknesses. The stress field beyond 2 mm from the weld toe does not reflect a consistent effect of the main plate thickness. Figure 5.12d presents the near-toe stress variation for a welded plate joint under four-point bending and three-point bending, normalized by the stress at $x = 1$ mm. Figure 5.12d demonstrates the significant difference in the stress gradient caused by different loading conditions over the proportional zone. In addition, the proportional zone covers the highly stressed material over which the stress exceeds 80% of the stress at 1 mm from the weld toe. Beyond 2 mm, the stress magnitude decays rapidly along the surface direction. Therefore, the hot-spot volume in this chapter terminates at $x = 2$ mm along the surface.

5.3.4 Fatigue Crack Propagation and Determination of y_1

BS7910 [31] presents the numerical analysis for the semi-elliptic crack growth in the hopper corner joints (see Figure 5.13a) with the same hot-spot stress, but under different DoBs using fracture mechanics. The DoB follows,

$$DoB = \frac{\sigma_b}{\sigma_b + \sigma_m} \tag{5.5}$$

where σ_b denotes the bending stress and σ_m represents the membrane stress. The effect of the stress gradient corresponding to different DoBs leads to different fatigue propagation lives of welded plate joints with the same hot-spot stress, as illustrated in Figure 5.13b. The hot-spot volume should, therefore, cover the stress gradient in the thickness direction and extend to a depth over which the near-toe stresses influence strongly the fatigue life of welded plate connections. Previous researchers [32–34] have reported that the initiation and propagation of small cracks (with a depth within a couple of mm) contribute to a significant portion of the total fatigue life. Figure 5.13b elucidates a similar observation for the hopper corner joints under constant amplitude loading and different DoBs. Figure 5.13c shows a consistent difference in the fatigue propagation life of hopper corner welded joints under different DoBs within a small crack depth up to around 1 ~ 2 mm. Xiao and Yamada [34] have shown that the fatigue life of a welded joint N_p depends on the structural stress concentration of $K_s(y)$ and the fatigue life N_{pr} of a reference joint with the same local weld geometry.

$$N_p = \frac{1}{\left[K_s(y)\right]^m} N_{pr} \left(0 < y < a_f\right) \tag{5.6}$$

where a_f refers to the crack length (1 mm to 2 mm in this chapter) and $K_s(y)$ denotes the stress concentration factor calculated at a depth of y below the weld toe. This implies that fatigue crack propagation in the welded joint relies essentially on the structural stress concentration of $K_s(y)$ within a small depth (1 mm to 2 mm), and the fatigue crack propagation of a reference joint N_{pr}.

Figures 5.14a and b show the stress variation along the thickness direction of the welded plate specimens with the thickness of the main plate equal to 9 mm and 80 mm at $x = 1$ mm and $x = 2$ mm, under four-point bending. Figures 5.14c and d illustrate the stress along the thickness direction at $x = 1$ mm and $x = 2$ mm for the same set of welded plate joints subjected to pure tension. The converging stress value below 1 mm from the surface reflects that the stress concentration due to macro-structural geometry discontinuity occurs primarily within 1 mm from the surface for different main plate thicknesses. Within 1 mm depth from the surface, the stress at 1 mm depth is above 80% of the maximum surface stress for both four-point bending and pure tension. Therefore, combining the fatigue crack propagation under different loading conditions and the stress concentration effect in the thickness direction, the highly stressed hot-spot volume in this chapter extends to a depth of 1 mm from the surface of the plate for all specimens.

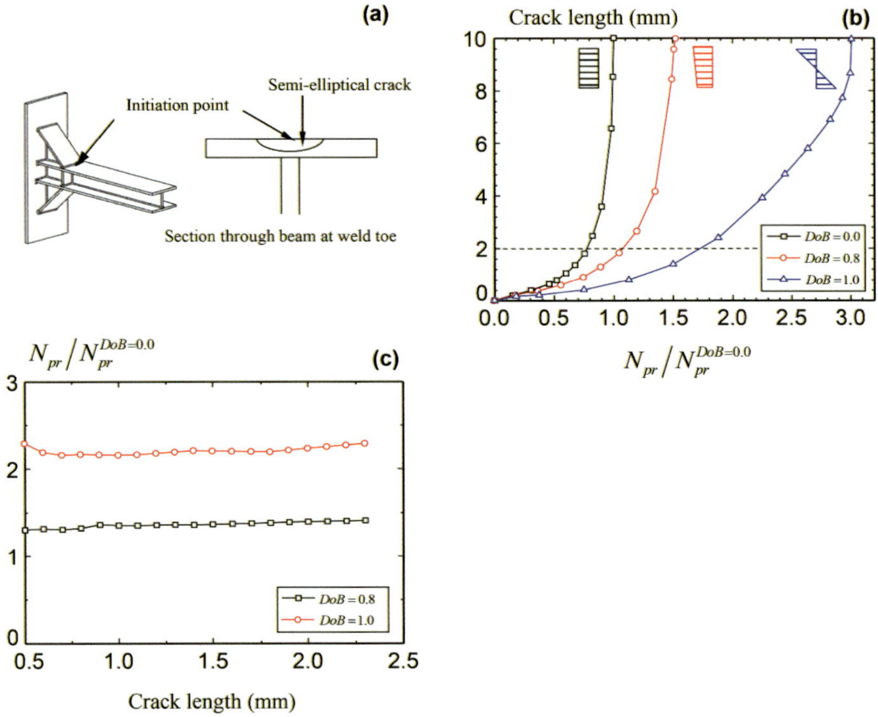

FIGURE 5.13 (a) Hopper corner joints in BS7910; (b) the crack growth under different degrees of bending with the fatigue life normalized by that under pure membrane stress; (c) the fatigue life normalized by that under pure membrane action versus the fatigue crack length.

5.3.5 Energy in the Hot Spot Volume

Based on a close examination of the near-toe stress field in welded connections with various geometries and loading conditions, this study proposes a hot-spot volume, located at 1 mm from the weld toe, with a length of 1 mm along the plate surface and a depth of 1 mm into the thickness of the plate, as shown in Figure 5.15a, to estimate the hot-spot energy as the fatigue driving force. In contrast to the stress-based approaches, the proposed energy-based hot-spot indicator is independent of the orientation of reference axes and incorporates the stress gradient effect in both the surface and thickness directions. The average strain energy density ΔW_V, to quantify the hot-spot fatigue driving force for welded plate connections, follows,

$$\Delta W_V = \frac{1}{V}\int_1^2\int_{-1}^0\int_0^{\varepsilon_{ij}} \sigma_{ij} d\varepsilon_{ij} dxdy \cdot B_V \tag{5.7}$$

where B_V and V represent the width and volume of the hot-spot zone, respectively ($V = 1\times 1\times B_V$ mm^3). The summation convention applies to the repeated index in Eq. (5.7). For linear-elastic materials with a unit width, Eq. (5.7) becomes,

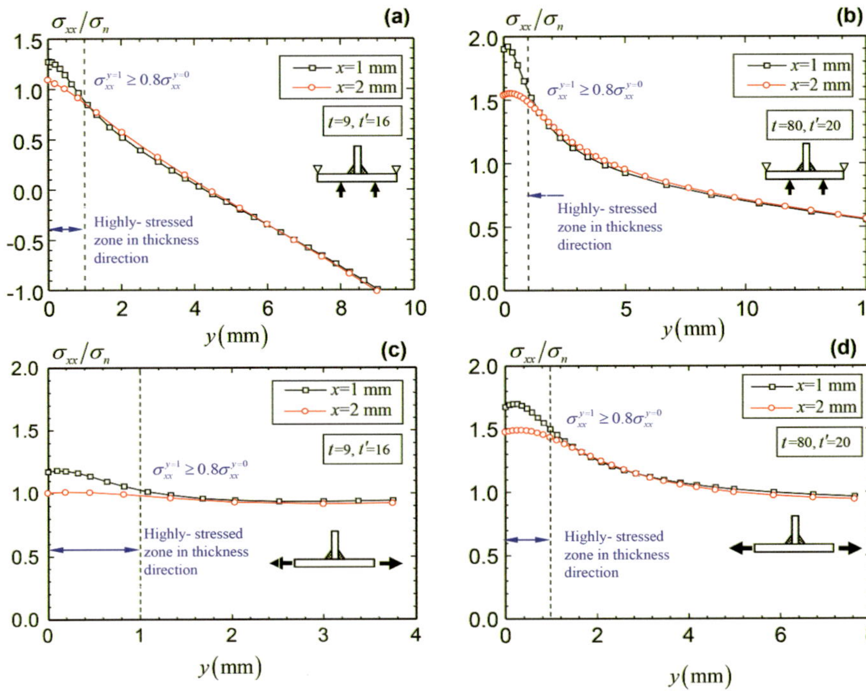

FIGURE 5.14 The normalized stress distribution in the thickness at $x = 1$ mm and $x = 2$ mm for specimens with a main plate thickness of (a) 9 mm under pure bending; (b) 80 mm under pure bending; (c) 9 mm under axial loading; and (d) 80 mm under axial loading.

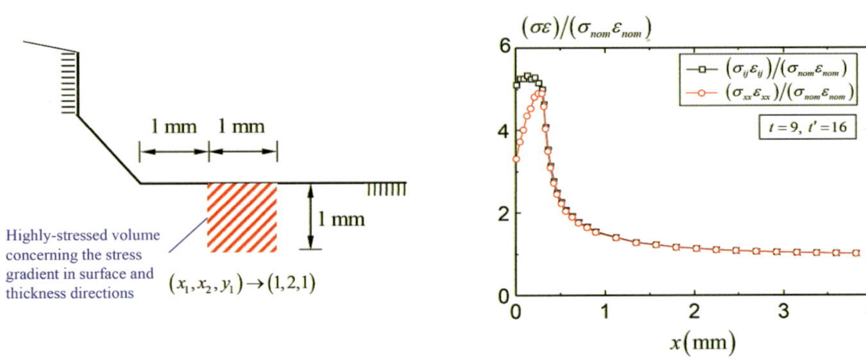

FIGURE 5.15 (a) The proposed highly stressed hot-spot volume to incorporate the stress gradient effect in both the surface and thickness directions; (b) The comparison between the distribution of $\sigma_{xx}\varepsilon_{xx}$ and $\sigma_{ij}\varepsilon_{ij}$ near the weld toe.

$$\Delta W_V = \int_1^2 \int_{-1}^0 \frac{1}{2}\sigma_{ij}\varepsilon_{ij}\,dxdy \tag{5.8a}$$

Figure 5.15b shows the close agreement of the value of $\sigma_{xx}\varepsilon_{xx}$ with the value of $\sigma_{ij}\varepsilon_{ij}$ after about $x = 0.3$ mm. Therefore, this chapter utilizes the strain energy density integrated using the stresses and strains perpendicular to the weld line to represent the primary fatigue driving force.

$$\Delta W_V = \int_1^2 \int_{-1}^0 \frac{1}{2}\sigma_{xx}\varepsilon_{xx}\,dxdy \tag{5.8b}$$

This study calculates the strain energy density in the hot-spot volume, through a FORTRAN code using the nodal stresses and strains computed in ABAQUS [35].

5.3.6 Validation against Fatigue Test Data

Figure 5.16 shows the geometric configuration of the welded plate connections under three-point bending, four-point bending, and pure axial loading. The experimental data covers welded plate joints with a large range of the main plate thickness (t),

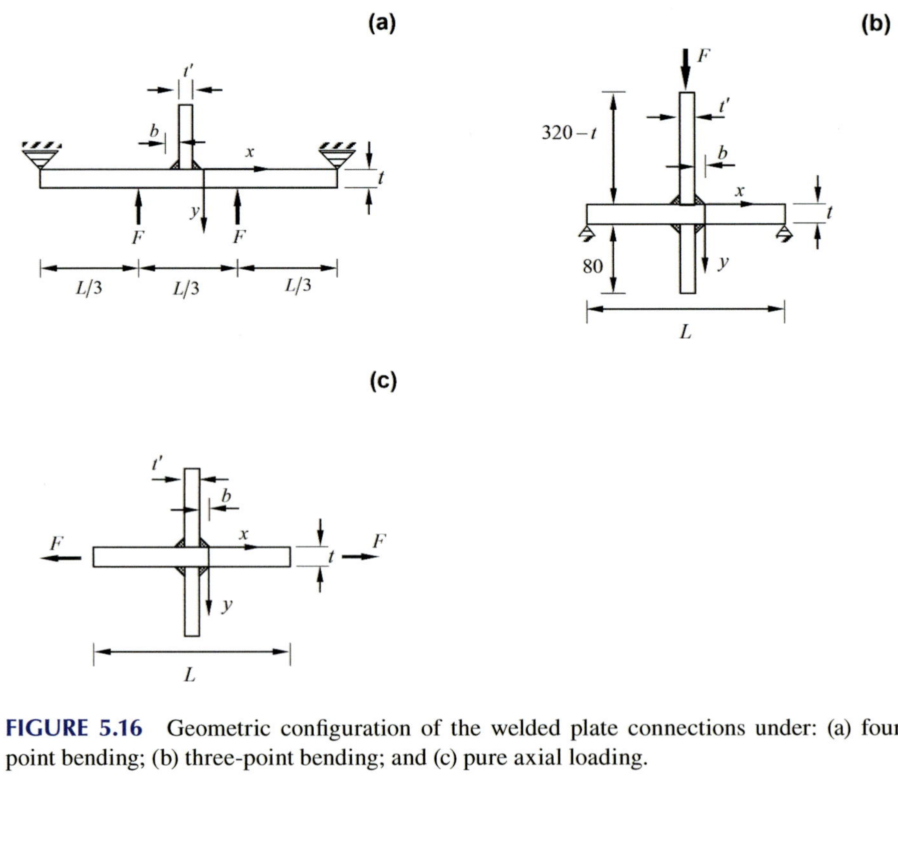

FIGURE 5.16 Geometric configuration of the welded plate connections under: (a) four-point bending; (b) three-point bending; and (c) pure axial loading.

including the four-point bending of transverse fillet welded joints with t ranging from 9 mm to 50 mm, reported by Miki *et al.* [30], the cruciform welded joints under pure axial loading from literatures [36–38] with t from 4.8 mm to 100 mm, and cruciform welded plate joints under three-point bending [14, 39] with t from 20 mm to 60 mm. All fatigue cracks initiate at the weld toe.

Figure 5.17 compares the fatigue life assessment of all test data using different hot-spot fatigue-driving indicators. Figures 5.17a–d show the S-N data for all the fatigue tests using the linearly extrapolated hot-spot stress [4], the 1 mm

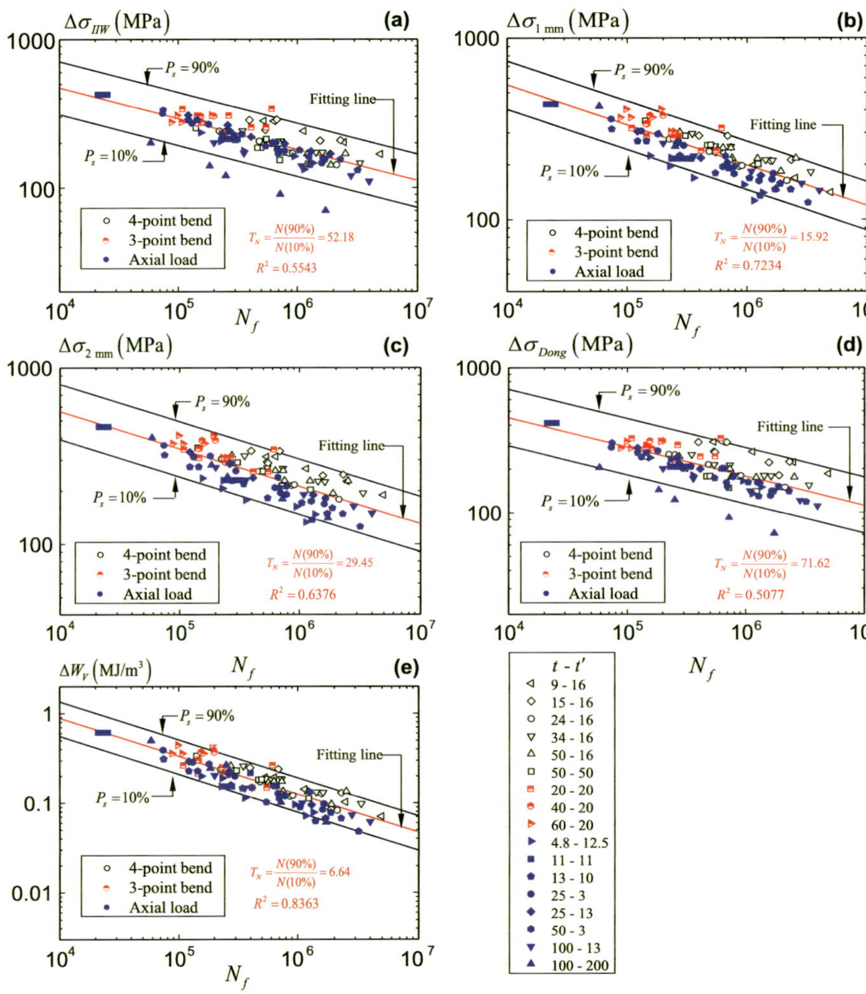

FIGURE 5.17 The comparison of the experimentally measured fatigue life of welded plate joints using: (a) hot-spot stress by IIW [4]; (b) 1 mm stress by Xiao and Yamada [34]; (c) 2 mm stress by Haibach [40]; (d) Dong's [29] structural stress; and (e) the energy in the hot-spot volume.

stress [34], the 2 mm surface stress [40] and Dong's [29] structural stress method. Figure 5.17e presents the fatigue test data using the hot-spot energy as the fatigue driving force. The coefficient of determination R^2, as shown in Figure 5.17, indicates the degree of variation of one variable predicted from the other variable and represents the percentage of the data that is closest to the line of best fit. The solid line (red), dashed line (black), and dash-dot line (black) represent the best fitting curve and the curves with the survival probability of 90% and 10%, respectively. The scatter T_N defines the ratio of the number of cycles to failure with 90% survival probability to that with 10% survival probability. The symbols in hollow (black), half solid (red), and solid (blue) represent the welded plate connections under four-point bending, three-point bending, and axial loading, respectively. Under the same loading conditions, the stress gradient caused by different specimen sizes leads to significant scatters in the fatigue assessment by the hot-spot stress in IIW [4], the 2 mm stress by Haibach [40] and Dong's [29] structural stress. In contrast, the 1 mm stress by Xiao and Yamada [34] and the proposed hot-spot energy reduce the scatter in the fatigue life caused by the size effects of the welded plate joints. However, the 1 mm stress beneath the weld toe by Xiao and Yamada [34] does not reflect the stress gradient in the surface direction. The welded plate joints under different loading conditions thus indicate a slightly larger scatter when assessed using the 1 mm stress than that assessed by the proposed hot-spot energy,

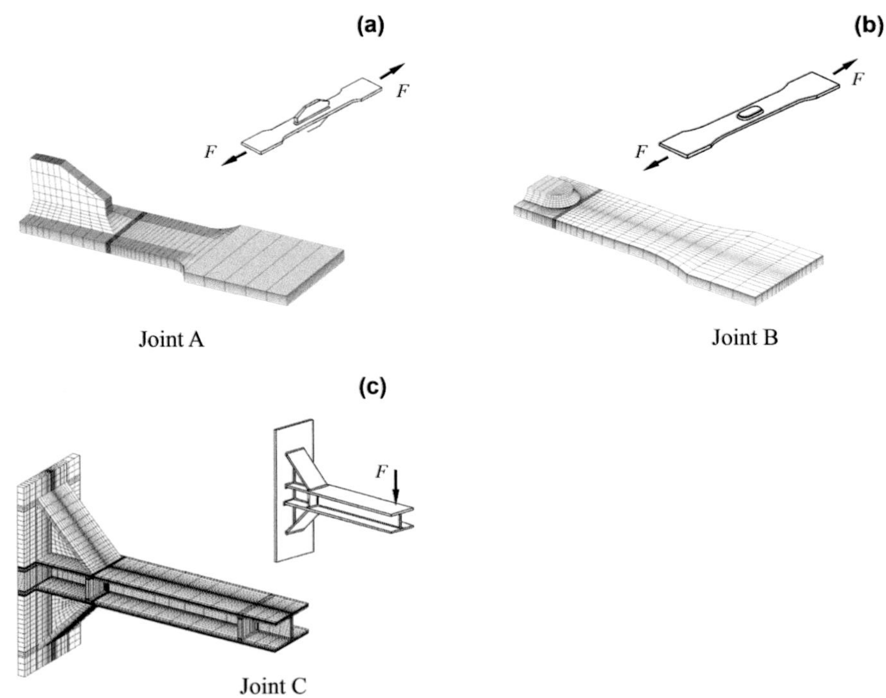

FIGURE 5.18 FE models and configurations for the joint specimens in [41] for: (a) Joint A; (b) Joint B; and (c) Joint C.

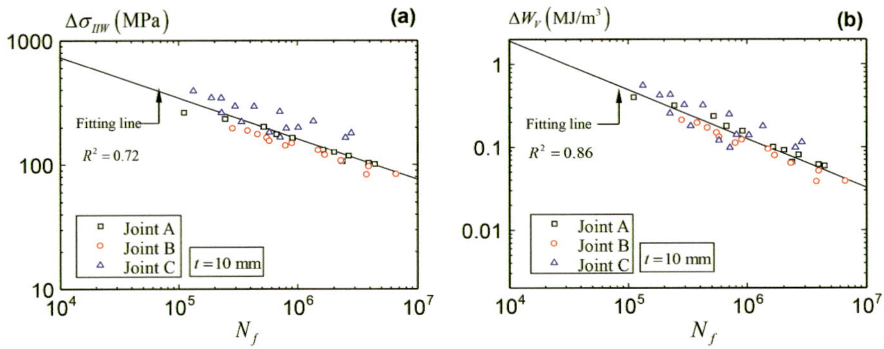

FIGURE 5.19 Assessment of the fatigue test data in [41] using: (a) the hot-spot stress by IIW [44]; and (b) the energy in the hot-spot volume.

as revealed by the comparison between Figures 5.17b and e. The fatigue life of joints with a main plate thickness of 4.8 mm deviates furthest from the best fitting line in Figure 5.17b, which implies that the 1 mm stress by Xiao and Yamada [34] may overestimate the fatigue life for a welded plate joint with a thin plate thickness, despite its consideration of the stress gradient in the thickness direction. The proposed volume-based hot-spot energy fatigue indicator reflects the stress gradient effects in both surface and thickness directions caused by both the size effect and loading conditions and leads to a reduced scatter.

This study further validates the feasibility and accuracy of the volume-based structural geometry indicator in welded plate joints under different loadings, using the fatigue test data reported by Hyundai Heavy Industries (HHI) [41]. Figure 5.18 shows the geometric configuration and FE mesh for the three different joint types [42, 43]. Figure 5.18a shows the fatigue tests assessed using the hot-spot stress without correction for bending and axial loading. Joint C shows a higher fatigue strength under bending than the other two types of joints under axial loading due to the stress gradient effects. Figure 5.18b shows the fatigue test data using the volume-based hot-spot energy fatigue indicator, which leads to reduced scatters in the fatigue life assessment than does the hot-spot stress method.

5.4 FATIGUE OF CONCRETE-FILLED TUBULAR X JOINTS WITH FULL-CAPACITY WELDS

5.4.1 Welding Profile

This section will present experimental and numerical investigations on the fatigue performance of the CHS X-joints with a concrete-filled chord. The welding profile along the brace-chord intersection employs a recently proposed set of welding details, namely the full capacity tubular joint welds. Marshall *et al.* [45] present a new type of welding details for the CHS connections, namely the full capacity tubular joint welds, also known as the welder-optimized CJP-equivalency welds.

Figure 5.20 illustrates the typical details of the full capacity tubular joint welds for various ranges of the local dihedral angle, ψ, along the brace-to-chord intersection. The local dihedral angle, ψ, refers to the angle formed by the tangent to the outer surface of the brace and that to the outer surface of the chord, intersecting at the weld line, as illustrated in Figure 5.20. The full capacity tubular joint welds make use a part of the brace wall as the inherent backing plate to enhance the quality control at the root of the welds, in contrast to the conventional CJP welds, which require high-quality workmanship of the welder. The strength requirement on the welds and, therefore, on the weld volume leads to a large groove angle so as to accommodate sufficient welding deposits in the full capacity tubular joint welds. This large groove angle facilitates the welding procedure, especially for brace-to-chord intersections with a small local dihedral angle (ψ in Figure 5.20). Previous experimental programs [8, 46] have reported the fatigue performance of CHS X-joints with different post-weld treatments subjected to constant-amplitude cyclic actions and overloading conditions. The previous experimental efforts confirm that the toe cracking near the hot-spot locations prevails in the CHS X-joints fabricated with the full capacity tubular joint welds. Root cracking does not initiate unless the post-weld treatment has significantly delayed the initiation of the weld toe cracks.

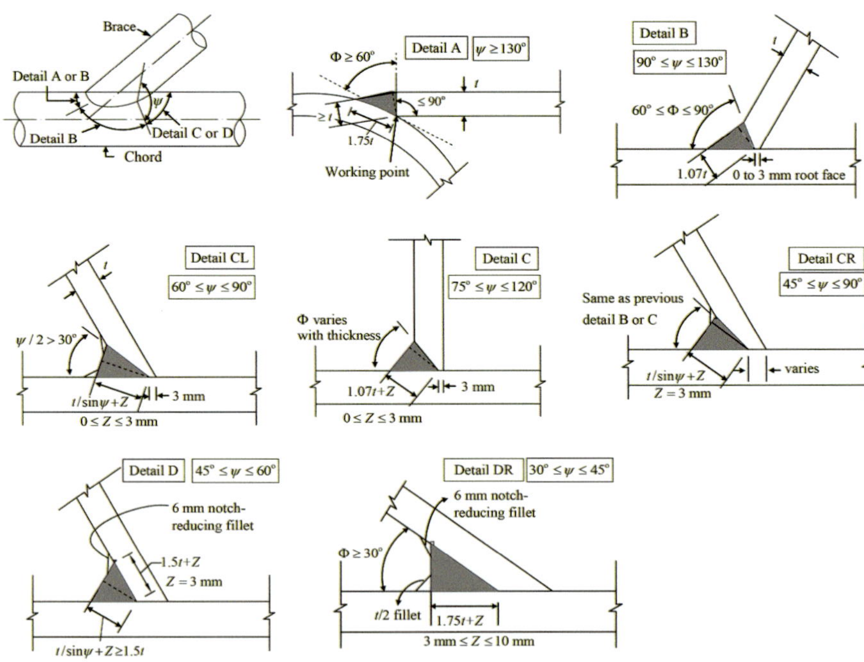

FIGURE 5.20 Typical details of the full capacity tubular joint welds for various ranges of the local dihedral angle, ψ, along the brace-to-chord intersection.

5.4.2 Experimental Study and Numerical Procedure

The entire experimental program to examine the fatigue performance of the full-capacity tubular joint welds consists of five specimens, as indicated in Table 5.8. All five specimens have the same chord length and brace length, with the brace-to-chord intersection angle fixed at 60°, as illustrated in Figure 5.21a. All five specimens have a 13 mm thick end plate welded inside the brace member directly above the end supports, which locate at 100 mm away from the end of the brace (see Figure 5. 21a). The main member (chord) of the X-joint has an outer diameter (d_0) of 750 mm, with the wall thickness (t_0) equal to 25 mm. The secondary member (brace), which is profiled and welded to the chord, has an outer diameter (d_1) of 406 mm. The non-dimensional parameters used in defining the CHS joint geometry include: the brace-to-chord outer diameter ratio, $\beta\left(=d_1/d_0\right)$, the chord radius to wall thickness ratio, $\gamma\left(=d_0/2t_0\right)$, and the brace-to-chord wall thickness ratio, $\tau\left(=t_1/t_0\right)$. Table 5.8 lists these three non-dimensional parameters for all five specimens. The first three specimens, denoted as J1-1, J1-X and J1-2, have a brace thickness of 12.5 mm and a hollow chord member, while the other two specimens, denoted as J2-1 and J2-2, have a 25 mm thick brace member and a concrete-filled chord member. Figure 5.21 shows the typical configuration of the concrete-filled X-joint specimens. The two diaphragm plates within the chord member (shown in Figure 5.21a), each of 12.5 mm thick,

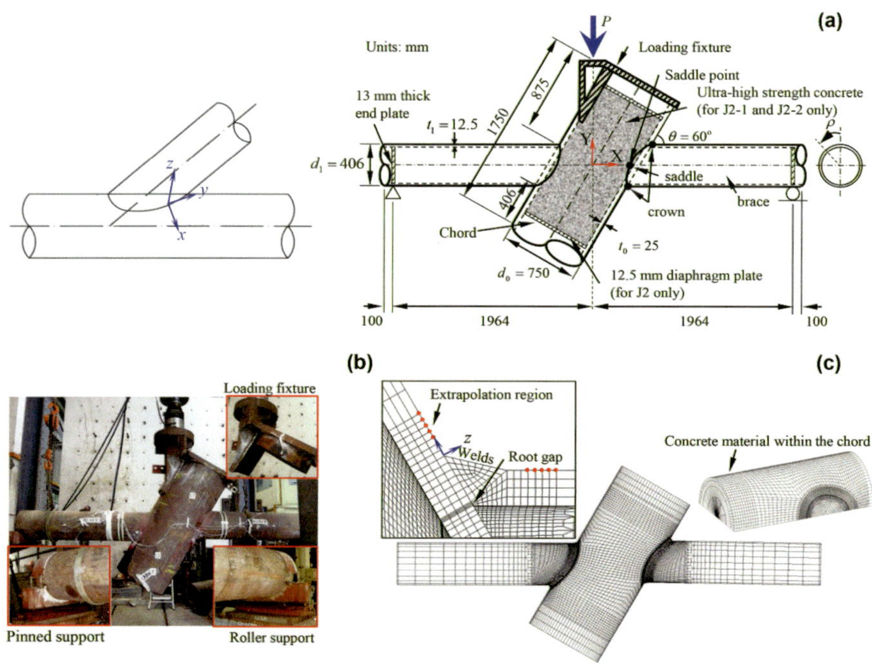

FIGURE 5.21 (a) Geometry and experimental set-up of X-joints; (b) loading and boundary conditions in the experiments; and (c) FE model of the X-joint.

connect to the inner surface of the chord member via fillet welds. These two diaphragm plates facilitate the concrete filling process and hold the weight of the concrete during the cyclic tests. The parameter ρ in Figure 5.21a measures the position along the brace-to-chord intersection in the counter-clockwise direction when viewed from the right side of the specimen. An angle $\rho < 180^{\circ}$ denotes the brace-to-chord intersection facing the observer in Figure 5.21, while an angle $\rho > 180^{\circ}$ refers to the brace-to-chord intersection at the back of the specimen in Figure 5.21. The boundary conditions on the test specimens include a pin support at the end of the left brace and a roller support on the end of the right brace. All joints experience an in-plane bending action generated by an applied vertical load at the top end of the chord member, as indicated in Figure 5.21b, which shows the experimental set-up of the X-joint specimens.

The 200-ton capacity loading rig applies a constant-amplitude cyclic load to the top end of the chord member through a detachable loading fixture, as indicated in Figure 5.21a. The vertical load applied at the top of the chord generates an in-plane bending moment on the X-joint, which experiences a tensile stress near the bottom of the brace-to-chord intersection and a compressive action near the top of the brace-to-chord intersection. The load ratio of the cyclic load, $R\ (= P_{min} / P_{max})$, remains fixed at 0.1 for all fatigue tests. The experimental program includes two constant-amplitude cyclic load tests for each specimen in a non-corrosive laboratory air environment at room temperature. The first cyclic test stops when the maximum measured crack size near the bottom of the brace-to-chord intersection reaches about 80% of the wall thickness of the brace or chord member. The same specimen will then undergo the second cyclic test after being rotated by 180° in the plane of the joint, that is, the X-Y plane in Figure 5.21a. This 180° rotation relocates the fatigue crack generated in the first fatigue test to the top of the brace-to-chord intersection, which anticipates compressive stresses in the second cyclic test. A letter "F" appended to the specimen name denotes the flipped configuration of the specimen. The second cyclic test terminates when the deepest crack front reaches about 80% of the wall thickness. Table 5.8 shows the load ranges for all five specimens.

The specimen preparation applies post-weld grinding treatment near the weld toe to the hollow section specimen, J1-X, and the concrete-filled specimen J2-2. The other three specimens (*i.e.*, J1-1, J1-2, and J2-1) do not undergo the post-weld grinding treatment. The post-weld grinding applied to the specimen J1-X includes surface grinding on the weld surface and burr grinding at weld toes in both the chord and brace members. The concrete-filled specimen J2-2 undergoes burr grinding for the weld toes along the tension side of the brace-to-chord intersection in the chord member prior to the fatigue test. The radius of the burr grinding equals 3 mm for both the J1-X and J2-2 specimens.

The instrumentation during the experiment includes the hot-spot strain measurement through the five-component strip-type strain gauges near the weld toes in both the chord and brace members. The spacing between two adjacent strain measurements equals 2 mm. The nearest strain measurement is located at $0.4t$ away from the weld toe, where t refers to the member (brace or chord) thickness. For the concrete-filled specimens (J2-1 and J2-2), the experimental procedure measures the SCF

TABLE 5.8
Dimensions and Loading Ranges for the X-Joint Specimens.

Specimen	d_0 (mm)	t_0 (mm)	d_1 (mm)	t_1 (mm)	β	γ	τ	ΔP (kN)	Remarks
J1-1	750	25	406	12.5	0.54	15	0.5	210	Without post-weld grinding, hollow chord
J1-X	750	25	406	12.5	0.54	15	0.5	210	With post-weld grinding, hollow chord
J1-2	750	25	406	12.5	0.54	15	0.5	160	Without post-weld treatment, hollow chord
J2-1	750	25	406	25	0.54	15	1.0	180	With a concrete-filled chord and without post-weld grinding
J2-2	750	25	406	25	0.54	15	1.0	210	With post-weld grinding and a concrete-filled chord

values before and after the concrete is filled in the chord member. For the specimen J2-2 with burr-grinding treatment, the experimental procedure has included multiple measurements of the SCF values corresponding to different stages of the specimen preparation: 1) the welded specimen prior to any post-weld burr grinding; 2) the specimen with burr grinding at the weld toe (denoted by a letter "X" appended to the specimen name); 3) the specimen with diaphragm plates welded to the inner surface of the chord (denoted by a letter "P" appended to the name of the specimen) after burr grinding; and 4) the specimen with filled concrete in the chord (denoted by a letter "G" appended to the specimen name) after burr grinding and welding of the internal diaphragm plates.

In addition to the SCF measurement by the strip-gauges, the experimental procedure has utilized single gauges placed at multiple weld toes along the brace-to-chord intersection to measure the notch stress concentration factor (NSCF) in each fatigue test. These post-yield single gauges have a standard gauge length of 2 mm. The monitoring of the crack growth utilizes the alternating current potential difference (ACPD) technique.

All X-joints make use of S355 steel for the brace and the chord members. The coupon test indicates a Young's modulus of 200 GPa for the chord member and 203 GPa for the brace member. The Poisson's ratio equals 0.3. Figure 5.22 illustrates the uniaxial true stress-true strain curves for the brace and chord materials. The steel pipes used in fabricating the joint specimens represent the realistic pipes used in constructing tubular connections in an offshore fabrication yard. The primary member of the joint, that is, the chord, exhibits both superior material strength (see Figure 5.22) and enhanced fracture toughness [47] compared to the brace material. The ultra-high strength concrete has a measured compressive strength of 185 MPa after 28 days of curing, an elastic modulus of 70 GPa and a Poisson ratio of 0.19.

In addition to the experimental investigation, this chapter computes the stress-concentration factors through a linear-elastic FE procedure. Figure 5.21c shows a typical FE mesh for the concrete-filled specimens, which employs about 49,000

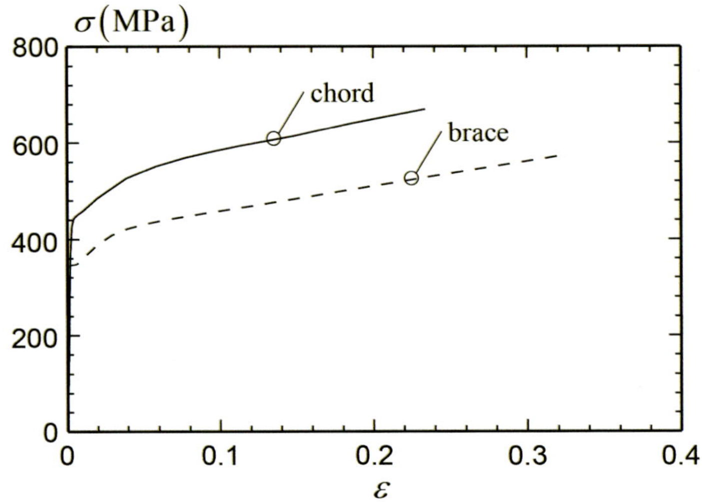

FIGURE 5.22 True stress-strain curve for chord and brace material.

20-node brick elements with 230,000 nodes. The extrapolation zone from $0.4t$ to $1.4t$ [4] includes five elements near the weld toe in both the chord and brace members, as illustrated in Figure 5.21c. Both the chord and brace members contain four elements over the thickness. For concrete-filled specimens, the numerical representation of the contact between the concrete surface and the inner chord surface utilizes the finite sliding, surface-to-surface contact, with the inner chord surface as the master surface and the concrete surface as the slave surface, as illustrated in Figure 5.21c. The inner surface of the chord and the concrete surface follow exactly the same mesh design, as shown in Figure 5.21c. The surface interaction employs the penalty method to enforce the contact constraint between the concrete and steel surfaces. The coefficient of friction between the concrete surface and the steel surface equals 0.6 based on the separately performed push-out tests.

5.4.3 Fatigue Performance

This study derives the stress concentration factor (SCF) at the weld toe from the measured strain concentration factor (SNCF) at the corresponding weld toe. The hot-spot strain, ε_{hs}, at the weld toe follows a linear extrapolation of the five measured strain values over the extrapolation zone from $0.4t$ to $0.4t + 10$ mm away from the weld toe. The relationship between the hot-spot stress, σ_{hs}, and the hot-spot strain, ε_{hs}, adheres to the constitutive relationship for an isotropic, linear-elastic material. The stress concentration factor at the weld toe, therefore, becomes,

$$SCF = \frac{E\varepsilon_{xx-hs}}{\sigma_{nom}}\left[\frac{1}{1+\upsilon}+\frac{\upsilon}{(1+\upsilon)(1-2\upsilon)}\left(1+\frac{\varepsilon_{yy-hs}}{\varepsilon_{xx-hs}}+\frac{\varepsilon_{zz-hs}}{\varepsilon_{xx-hs}}\right)\right] \tag{5.9}$$

where υ denotes the Poisson's ratio. The x - y - z coordinate defines a local coordinate system at the weld toe (see Figure 5.21a). The x-axis remains perpendicular to the weld line and parallel to the outer surface of the base metal. The y-axis aligns along the tangent of the weld line and the outer surface of the base metal. The z-axis points in a direction perpendicular to the outer surface of the base metal, as illustrated in Figure 5.21a. The three hot-spot strain components, ε_{xx-hs}, ε_{yy-hs}, and ε_{zz-hs}, therefore, refer to the hot-spot strain perpendicular to the weld line, that parallel to the weld line and the hot-spot strain in the direction perpendicular to the outer surface of the chord at the weld line, respectively. The parameter σ_{nom} in Eq. (5.9) denotes the nominal stress in the brace member, or the maximum bending stress in the brace wall calculated from the elastic bending theory for the X-joint specimens considered. A plane stress assumption, that is, $\sigma_{zz-hs} = 0$, simplifies Eq. (5.9) into,

$$SCF = \frac{E\varepsilon_{xx-hs}}{\sigma_{nom}} \frac{1}{1-\upsilon^2} \left(1 + \upsilon \frac{\varepsilon_{yy-hs}}{\varepsilon_{xx-hs}} \right) \tag{5.10}$$

Equation (5.10) requires the measurement of both the strain components perpendicular to the weld line and those parallel to the weld line. Engineering design guidelines [8] have proposed a widely accepted relationship to determine the stress concentration factors from the measured hot-spot strain,

$$SCF = 1.15 \frac{E\varepsilon_{hs-xx}}{\sigma_{nom}} \tag{5.11}$$

Equation (5.11) simplifies the strain measurement in the experiment, which now requires only the strain values perpendicular to the weld line.

Figure 5.23 compares the computed SCF values using Eqs. (5.9)–(5.11) with the SCF determined via extrapolating the computed stresses over the extrapolation zone in linear-elastic FEA performed in ABAQUS [17]. The FE models used in calculating the hot-spot stresses and hot-spot strains follow a similar mesh refinement as in the previous study [8]. To compare the SCF calculated using Eqs. (5.9)–(5.11), the numerical procedure first computes the strain fields over the extrapolation zone from 0.4t to 1.4t [4] from a linear-elastic analysis. The hot-spot strains ε_{xx-hs}, ε_{yy-hs}, and ε_{zz-hs} then derive from a linear extrapolation of the strain values over the extrapolation zone to the weld toe. Substituting these hot-spot strains into Eqs. (5.9)–(5.11) leads to the SCF values at the weld toe along the brace-to-chord intersection. The linear-elastic FEA also allows direct calculations of the SCF values based on the extrapolated hot-spot stresses at the weld toe using the stresses computed over the extrapolation zone, that is, $SCF = \sigma_{hs} / \sigma_{nom}$. Figure 5.23 shows that Eqs. (5.9)–(5.11) lead to similar estimations of the stress concentration factors for both the concrete-filled specimen J2-1G and the hollow section specimen J2-1. Subsequent calculations reported in this chapter therefore follow the simple relationship in Eq. (5.9).

Figure 5.24 compares the effect of the brace-wall thickness on the measured SCF values for the hollow section specimens (J1-1 and J2-1), the concrete-filled specimen (J2-1G) and the concrete-filled specimen in its flipped configuration (J2-1GF) along

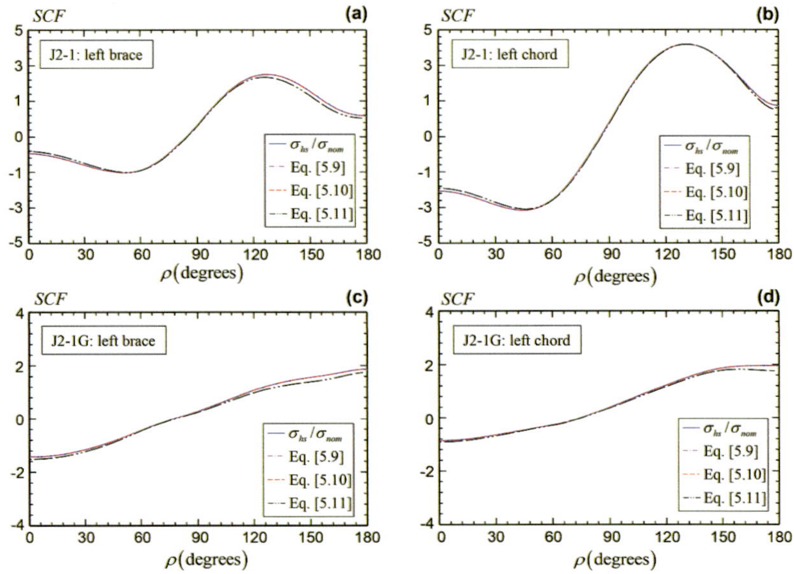

FIGURE 5.23 Comparison in SCF from Eqs. (5.9)–(5.11) and the extrapolation approach in FE analysis.

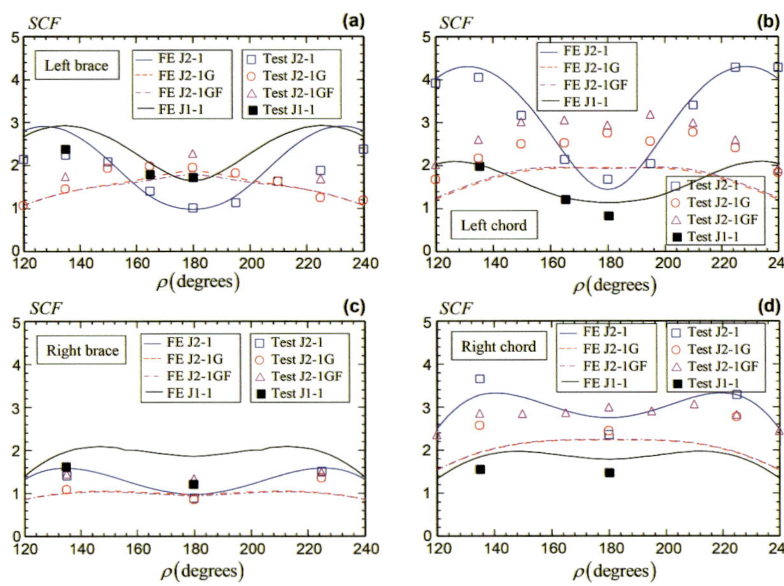

FIGURE 5.24 (a) The effect of the brace-wall thickness on the measured SCF values for the hollow section specimens (J1-1 and J2-1), the concrete-filled specimen (J2-1G) and the concrete-filled specimen in its flipped configuration (J2-1GF) at (a) left brace; (b) left chord; (c) right brace; and (d) right chord.

the tension side of the brace-to-chord intersection ($120^{\circ} \leq \rho \leq 240^{\circ}$). A thick brace slightly decreases the SCF values of J1-1 and J2-1 near the weld toe in the brace member, as reflected in Figures 5.24a and c. However, a stronger brace significantly increases the SCFs at the weld toes in the chord member, due to strong punching actions impinged on the chord wall by the thick brace wall. For both specimens J1-1 and J2-1, the maximum SCF occurs at the weld toe (around $\rho \approx 120^{\circ} \sim 135^{\circ}$ and $\rho \approx 225^{\circ} \sim 240^{\circ}$) in the left side of the chord member. The numerically calculated SCFs (based on $SCF = \sigma_{hs} / \sigma_{nom}$) agree reasonably well with the experimentally measured values. For the concrete-filled specimen J2-1G and J2-1GF, the numerical analysis indicates reasonable agreement with the SCFs measured at the weld toe on the brace members. However, the calculated SCFs significantly underestimates the measured values at the weld toe on the chord member of the concrete-filled specimens, as shown in Figures 5.24b and d. Such discrepancies lie mainly in the certain amount of autogenous shrinkage during the curing stage [48]. The presence of a finite gap between the concrete and steel surfaces leads to strong variations in the SCFs calculated along the weld toe in the chord member.

Figure 5.25 compares the SCF values measured corresponding to four different preparation stages for the specimen J2-2 before the first cyclic test: 1) the hollow section specimen before burr grinding, J2-2; 2) the hollow section specimen with burr grinding applied to the weld toe in the chord member along the tension side of the brace-to-chord intersection, J2-2X; 3) the burr-ground hollow section specimen

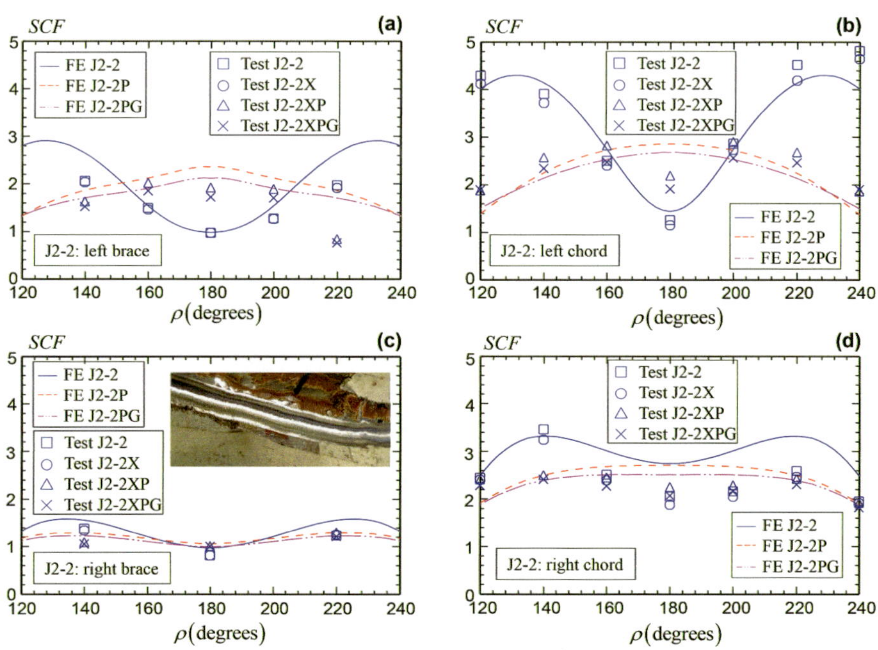

FIGURE 5.25 The SCF values measured corresponding to four different preparations for the specimens J2-2.

with diaphragm plates in the chord member, J2-2XP; and 4) the concrete-filled specimen with burr-grinding and internal diaphragm plates in the chord, J2-2XPG. Figure 5.25c illustrates the burr-ground weld toe in the chord member. The comparison of SCF values in Figure 5.25 indicates that post-weld burr grinding applied to a limited and localized region near the weld toe does not introduce significant changes to the measured hot-spot strains. The internal stiffening of the chord member by the two diaphragm plates, in contrast, introduces significant variations in the SCF values near the weld toe of the chord member. Further stiffening of the chord member by the internal ultra-high-strength concrete causes only marginal decreases in the SCF values. The internal stiffening of the chord member relocates the maximum hot-spot stress from $\rho \approx 120^\circ \sim 135^\circ$ and $\rho \approx 225^\circ \sim 240^\circ$ towards a position close to the bottom crown point ($\rho \approx 160^\circ$ and $\rho \approx 200^\circ$). This implies that the fatigue crack initiation may first occur at the weld toes in the chord member during the fatigue test. However, the presence of the internal concrete and diaphragm plates does lead to more significant reductions in the SCF values near weld toes in the chord member than the reductions in the SCF values measured in the brace.

5.4.4 Crack Initiation and Propagation

Table 5.9 compares the percentage of the fatigue crack initiation life (N_i) over the total fatigue life (N_t) for all ten fatigue tests performed. The crack initiation in the current study corresponds to a measurable crack depth of 0.5 mm through the ACPD probes. The fatigue crack initiation life (N_i), therefore, refers to the number of load cycles before the maximum crack depth measured by ACPD reaches 0.5 mm. Table 5.9 confirms that the fatigue crack propagation life for the welded X-joint specimens contributes to a significant portion of the total fatigue life of the specimens. This advocates indirectly the importance of the fracture-mechanics approach in accurately estimating the remaining fatigue crack propagation life.

TABLE 5.9
Fatigue Crack Initiation and Propagation Life in the X-Joint Specimens.

Test	Crack initiation location	Total fatigue life (N_t)	Percentage of crack initiation life ($N_i/N_t \times 100\%$)	Percentage of crack propagation life ($N_p/N_t \times 100\%$)
J1-1	$\rho = 245^\circ$, chord left	203,000	34.4%	65.6%
J1-1F	$\rho = 235^\circ$, chord left	159,000	16.4%	83.6%
J1X	$\rho = 225^\circ$, chord left	485,000	36.1%	63.9%
J1X-F	$\rho = 225^\circ$, chord left	600,000	49.2%	50.8%
J1-2	$\rho = 240^\circ$, chord left	430,000	23.7%	76.2%
J1-2F	$\rho = 120^\circ$, chord left	478,000	35.1%	64.9%
J2-1G	$\rho = 175^\circ$, chord left	442,700	7.2%	92.8%
J2-1GF	$\rho = 190^\circ$, chord left	410,000	9.0%	90.0%
J2-2XPG	$\rho = 190^\circ$, chord left	760,000	23.9%	76.1%
J2-2XPGF	$\rho = 190^\circ$, chord left	980,000	10.6%	89.4%

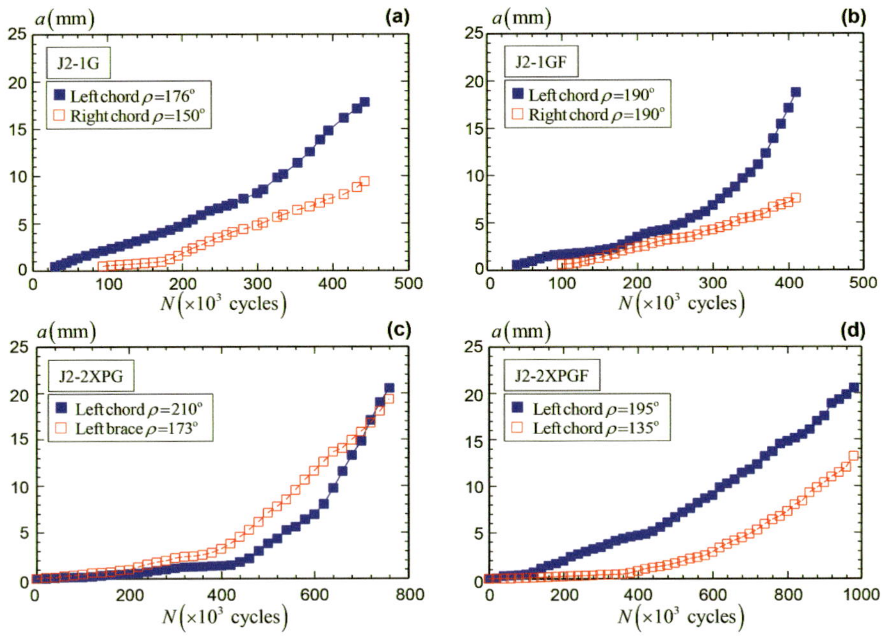

FIGURE 5.26 The evolution of the deepest crack-front location for the four different fatigue tests.

Figure 5.26 shows the evolution of the deepest crack-front location for the four different fatigue tests performed on the concrete-filled joint specimens. In most of the fatigue tests, the fatigue crack on the left side of the chord member has a faster growing rate than the rest of the cracks observed during the test, especially near the final stage of the fatigue test. The faster crack propagation rate derives essentially from the higher stresses driving the fatigue crack initiation and propagation on the left side of the brace, as evidenced by the large SCF values on the left side of the joint.

5.4.5 Comparison with *S-N* Estimations

Figure 5.27 compares the fatigue life of the concrete-filled X-joint specimens with three types of design *S-N* curves: the notch stress-based *S-N* curve FAT 225 SSN curve in IIW [4], the hot-spot stress-based *S-N* curve in API [1], and the nominal stress-based *S-N* curve for tubular joints in AWS [11]. These *S-N* curves develop from experimental databases for hollow section tubular joints fabricated using CJP welds. The hot-spot stresses and notch stresses for different tests follow Eq. (5.11) based on the measured hot-spot strain or notch strain. The filled symbols in Figure 5.27 represent the experimental data corresponding to the fatigue crack initiation at a crack depth of 0.5 mm, while the hollow symbols refer to the fatigue life of specimens at the end of the fatigue test (with the maximum crack depth equal to about 70~80% of the wall thickness). The fatigue crack initiation life for all specimens remains lower

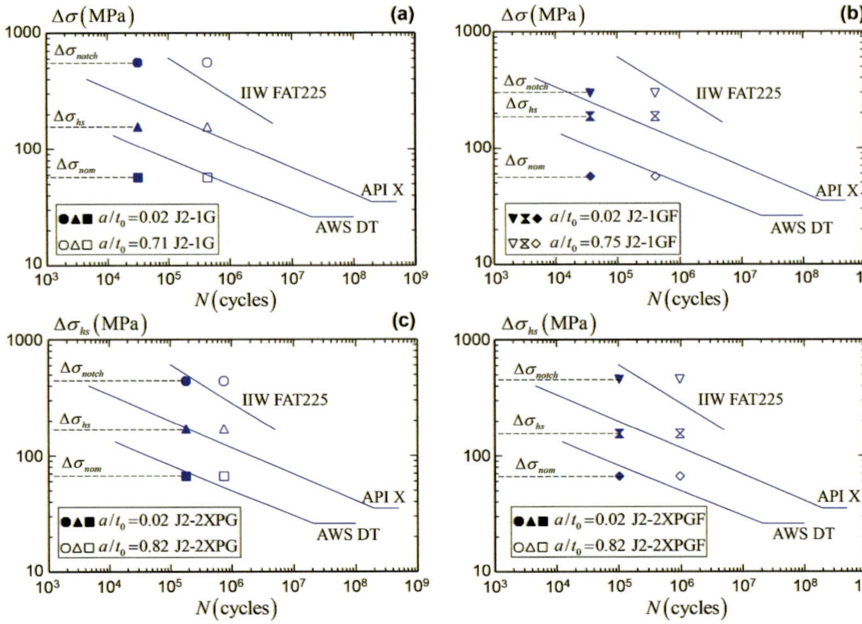

FIGURE 5.27 Comparison in fatigue life of the concrete-filled X-joint specimens with three types of design *S-N* curves.

than that estimated using all three types of the design *S-N* curve, as a considerable portion of the fatigue life (from 76~93% for the concrete-filled specimens) is consumed by the fatigue crack propagation. Compared to the nominal stress *S-N* curve, the fatigue life corresponding to the end of the fatigue test of the concrete-filled joints with weld-toe burr-grinding (J2-2XPG and J2-2XPGF) exceeds the life estimation by the nominal stress *S-N* curve. The fatigue life of the concrete-filled joints without weld-toe burr grinding (corresponding to a slightly smaller maximum crack depth than that for the burr-ground specimen) matches the fatigue life estimated using the nominal stress *S-N* curve.

All fatigue tests indicate a final fatigue life well exceeding the fatigue life calculated based on the hot-spot stress *S-N* curve. The notch stress *S-N* curve also conservatively predicts the fatigue life for most of the fatigue tests, except for the test J2-1GF. Figure 5.28 compares the fatigue life for both the hollow section X-joint specimens and the concrete-filled specimens. The solid symbols correspond to the fatigue crack initiation at a maximum crack depth of about 0.5 mm, while the hollow symbols refer to the fatigue life at the end of the fatigue test with the maximum crack depth equal to 80% of the wall thickness for all hollow section joints and 70~80% for concrete-filled specimens. All hollow section specimens indicate a fatigue life at the end of the test (with the maximum crack depth equal to about 80% of the wall thickness) that exceeds the fatigue life estimated using the nominal stress *S-N* curve.

The comparison based on the hot-spot stress *S-N* estimations includes two hot-spot *S-N* curves, the API X curve [1] and the ISO19902 TJ curve [2]. All fatigue test

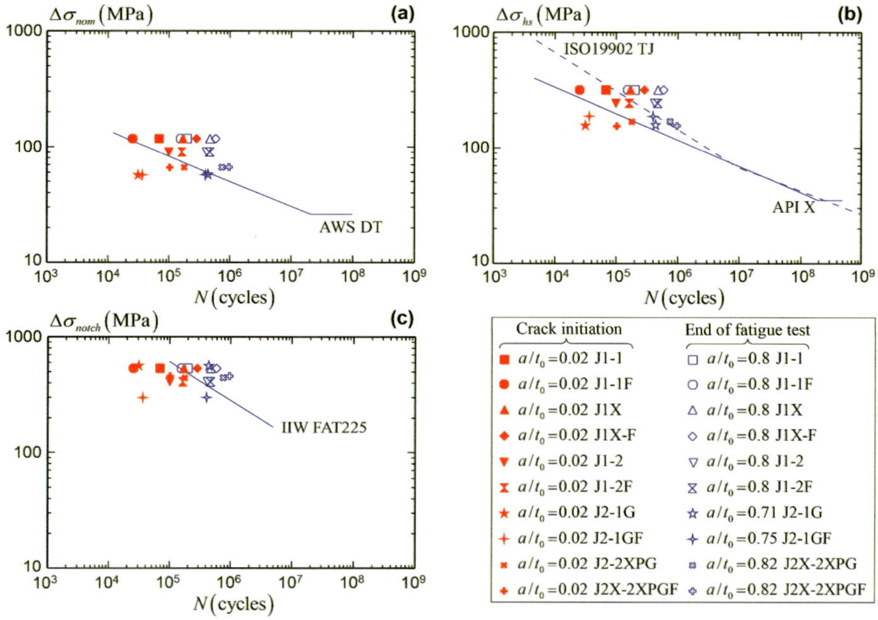

FIGURE 5.28 The fatigue life for both the hollow section X-joint specimens and the concrete-filled specimens.

data at the end of the fatigue test (with the maximum crack depth around 80% of the wall thickness) exceed the life estimation by the API X curve. Compared to the ISO19902 TJ curve, the fatigue life for the two fatigue tests of the concrete-filled specimen (J2-1G and J2-1GF) remains lower than that estimated from the ISO19902 TJ curve. The fatigue life for all specimens indicates reduced conservatism when compared to the ISO19902 TJ curve.

Compared to the notch stress *S-N* curve, all fatigue life data at the end of the fatigue tests, as indicated by the hollow symbols in Figure 5.28, exceed the life estimation by the *S-N* curve, except for the fatigue test J2-1GF on the concrete-filled specimen without weld-toe burr-grinding. Figure 5.28 implies that the fatigue life data for both the hollow section specimens and concrete-filled joints with full capacity tubular joint welds comply with the same *S-N* curves, originally developed for hollow section tubular joints with CJP welds.

5.5 EFFECT OF ULTRASONIC IMPACT TREATMENT (UIT) ON PREFATIGUED WELDED JOINTS

5.5.1 UIT and Experimental Scheme

The demand for existing steel infrastructures' life extension motivates authors to investigate the effect of UIT on fatigue life extension of in-service Q345 welded cruciform joints experimentally. This study examines the welding residual stress, local weld geometry, fatigue strain, and failure surface to quantify the beneficial mechanism of UIT and proposes the two reduction factors to quantify the decreasingly beneficial effect with increasing prefatigued cycles. Through utilizing the nonlinear damage accumulation law and Paris' law to quantify fatigue crack initiation and propagation, respectively, the theoretical relationship of the two reduction factors has been developed.

UIT, one of the high-frequency mechanical impact treatment (HFMI) methods, improves the fatigue life of welded joints by peening the potential fatigue crack initiation regions, which are typically the weld toe and root. Existing studies demonstrate that UIT can significantly improve the fatigue life of welded joints [4, 49–51]. First, UIT reduces the tensile residual stress at the welded zones, enhancing the fatigue resistance to cracks. Second, UIT smoothens the local geometry at the weld toe and reduces the stress concentration at the treated regions. Third, UIT eliminates surface welding defects and refines the microscopic grain at a certain depth of the impacted area.

In contrast, for in-service welded steel structures, subjected to a certain number of cycles of fatigue loads, the effectiveness of the UIT in extending the remaining fatigue life is rarely investigated, but of great importance for industry stakeholders. However, pre-fatigue loading experienced by the aged steels was not discussed adequately, and hence, it is unable to evaluate the effect of the pre-fatigue loading on the fatigue life enhancement of welded joints. Therefore, the effectiveness of the UIT in extending the remaining fatigue life still needs to be further investigated.

Figure 5.29 presents the geometry of the experimental full penetration non-load carrying cruciform welded joints. There are 28 in total welded joints to facilitate the occurrence of fatigue cracks at the weld toe. The K-shaped groove is adopted to fully

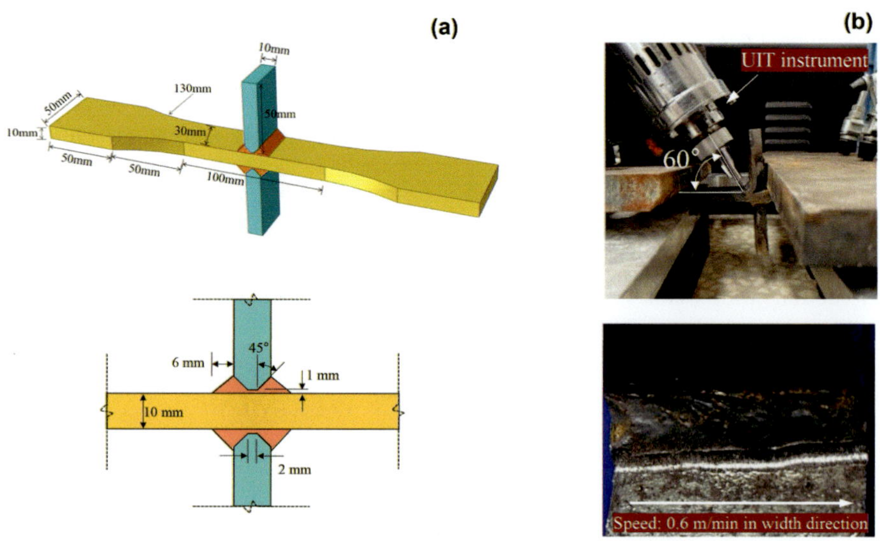

FIGURE 5.29 (a) Geometry of the non-load carrying cruciform welded joints; and (b) UIT instrument set-up and the local geometry profile appearance after UIT.

penetrate the welds. The upper and lower vertical attachments were processed into a 45° K-shaped groove by wire cutting. The welding clearance, blunt edge, and groove depth are 1 mm, 2 mm, and 4 mm, respectively.

A single impact needle with a diameter of 3 mm was used in this study. The operating frequency of the UIT process is 17.7 kHz, the excitation current is 1.9 A, and the amplitude of the impact needle is 26.7 μm. During the UIT process, the ultrasonic impact gun was aimed at the weld toe of the specimen as shown in Figure 5.29b, and the angle between the axis of the impact needle and the surface of the horizontal plate was about 60°. Each weld toe on the horizontal plate of the specimen was processed 8 times at an impact treatment speed of 0.6 m/min till a continuous, uniform and bright impact groove as shown in Figure 5.29b was observed [52]. Twenty-eight specimens are axially subjected to high-cycle fatigue loadings with a stress range ($\Delta\sigma$) of 256 MPa and load ratio (*R*) of 0.1. This study adopts the SDS 100 electro-hydraulic fatigue testing machine to achieve the 10 Hz loading frequency. All fatigue specimens were divided into four groups with 7 specimens in each group. The first group (group A) contains the as-welded specimens, of which the statistical fatigue life is defined as N_{f1}. Specimens in the second group (group B) were treated by UIT after welding. Specimens in the third (group C) and fourth (group D) groups were treated by UIT after prefatigued loading cycles, equal to 50% N_{f1} and 80% N_{f1}, respectively.

Table 5.10 summarizes the experimental fatigue life and failure locations for different groups. Through the statistical fatigue life explained in [53], the statistical fatigue life in Table 5.10 is determined with 50% confidence and 20% failure probability.

5.5.2 Effect of UIT

The section examines the effects of UIT experimentally on the local welding geometry and residual stress. Figure 5.30 illustrates the main geometrical features of the local weld toe zones with and without the UIT process. The main features of as-welded joints involve the weld toe angle (θ), and notch radius (ρ), while the main features of the welded components with UIT consist of the notch radius (ρ), groove depth (h), and groove width (w). This study examines weld toes of the as-welded specimens A9, and subsequently conducts UIT on A9. The local geometrical values are listed in Table 5.11 The average value of the local weld toe angle (θ) and notch radius (ρ) is 0.30 mm and 37.5°, respectively. After the UIT process, the average value of the notch radius (ρ) increases to 1.78 mm. The groove depth (h) and groove width (w) are 0.11 mm and 1.97 mm, meeting the requirement of IIW in minimum groove depth [4] from 0.1 mm to 0.2 mm. The above local geometrical change helps to reduce the stress concentration factor at the weld toe.

Change of the residual stress due to UIT is the observed phenomenon. This study quantifies the residual stress perpendicular to the welding line for different groups. The $\mu - X360S$ residual stress analyzer was utilized to measure the surface residual stress of each group. The target material was Cr, and the spot diameter was 1 mm. Before the X-ray diffraction measurement, the targeted area was cleaned by using electrolytic polishing. The electrolytic solution was saturated sodium chloride

TABLE 5.10
Summary of the Statistical Fatigue Life and Fracture Location for Different Groups.

Group	No.	Fracture location	Prefatigued cycles (N_f^P)	Total fatigue life cycles (N_f)	Total statistical fatigue life ($N_f(p,1-\alpha,v)$)	Residual statistical fatigue life ($N_f^R(p,1-\alpha,v)$)
Group A	A1	WT	0	72686	$N_{f1}(0.5,0.8,6)=$ 81422	/
	A2	WT		113205		
	A3	WT		121378		
	A4	WT		112221		
	A5	WT		78838		
	A6	WT		105079		
	A7	WT		109340		
Group B	B1	BM	0	827086	$N_{f2}(0.5,0.8,6)=$ 676814	/
	B2	WT		773291		
	B3	HAZ		1270891		
	B4	BM		1271282		
	B5	BM		715796		
	B6	Unbroken		1600000		
	B7	HAZ		717388		
Group C	C1	WT	$N_{f3}^P = 40711$	186279	$N_{f3}(0.5,0.8,6)=$ 223465	$N_{f3}^R(0.5,0.8,6)=$ 182478
	C2	WT		561348		
	C3	WT		276346		
	C4	Unbroken		1000000		
	C5	WT		260020		
	C6	WT		738346		
	C7	BM		595844		
Group D	D1	WT	$N_{f4}^P = 65138$	115161	$N_{f4}(0.5,0.8,6)=$ 129523	$N_{f4}^R(0.5,0.8,6)=$ 67251
	D2	WT		168490		
	D3	WT		150846		
	D4	WT		336188		
	D5	WT		596278		
	D6	Unbroken		1000000		
	D7	WT		769628		

**WT:* Weld Toe; BM: Base Material; HAZ: Heated Affected Zone

solution, the working voltage was 13 V, and the electrolytic time of each weld was 2 minutes. There are 7 measurement points adopted along the width direction of welded plates, with an equal space of 4 mm (Figure 5.30b). The distance from point 1 or point 7 to the edge of the base material is 3 mm.

Figure 5.30c presents the effect of UIT process on reducing the residual stress at the weld toe zone. The residual stress perpendicular to the welding line (black line)

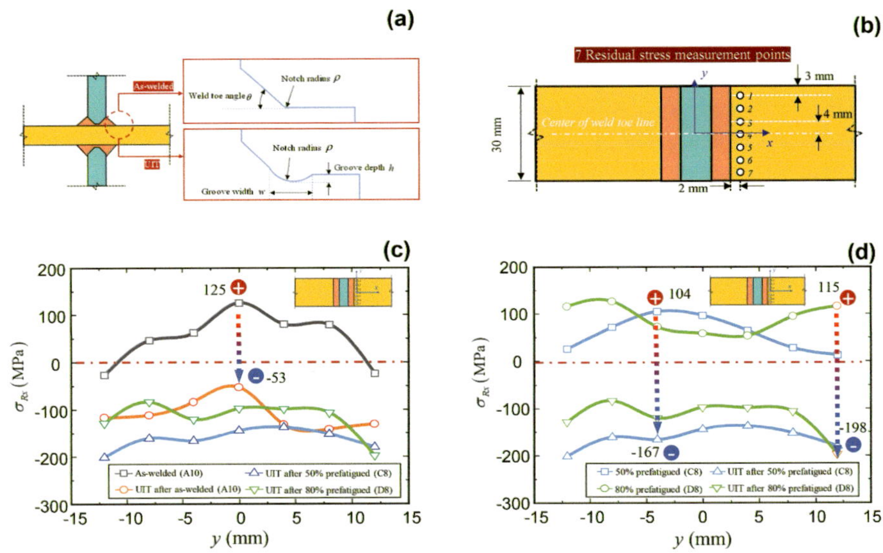

FIGURE 5.30 (a) Local geometry information of the weld toe; (b) residual stress measurement points; (c) the effect of the UIT process on reducing the original residual stress of the as-welded; and (d) stress reduction by UIT on prefatigued specimens.

TABLE 5.11

The Summary of the Local Weld Toe Geometry for As-Welded and UIT Specimens.

		As-welded		After UIT		
Name of specimens	Section number	Notch radius (mm)	Weld toe angle (°)	Notch radius (mm)	Groove depth (mm)	Groove width (mm)
A9	1	0.27	40	1.47	0.10	1.43
	2	0.37	30	1.92	0.10	2.00
	3	0.22	30	1.82	0.10	1.82
	4	0.16	60	1.48	0.12	1.77
	5	0.34	30	1.75	0.12	2.29
	6	0.38	30	2.09	0.08	1.95
	7	0.36	40	1.92	0.11	2.22
	8	0.28	40	1.75	0.12	2.33
Average Value		0.30	37.5	1.78	0.11	1.97

for the as-welded specimens (A10) is almost positive along the width direction, with the maximum stress of 125 MPa. After the UIT process, the peak positive stress decreases to –53 MPa. All the residual stress transforms to negative values (orange line). The residual stress for 50% prefatigued with the UIT process is around –170 MPa (blue line), while the longitudinal residual stress for 80% prefatigued with UIT is a little lower, around −140 MPa (green line). All three specimens entail negative residual stress after the UIT process. Figure 5.30d clearly quantifies the stress reduction for prefatigued specimens. For 50% and 80% prefatigued specimens, the maximum peak residual stress was reduced by 271 MPa and 313 MPa, respectively. In addition, all the residual stress before UIT is smaller than the yield strength of the material.

Based on the fatigue life and failure locations for different groups in Table 5.10, Figure 5.31a depicts the scatter distribution of fatigue life under the stress range of 256 MPa for different groups. The solid curves with different colors represent the statistical normal fitting curves. Based on the fitted normal distribution, Group B entails the longest fatigue life while Group A entails the shortest fatigue life. The UIT extends the fatigue life even after the fatigue cycles (solid orange curve in Figure 5.31a). The fitted standard deviation of Groups A, B, C, and D is 18549, 267922, 224273, and 269987, respectively. The fitted mean fatigue life value of Groups A, B, C, and D is 101821, 929289, 436363, and 356098, respectively. Correspondingly, the COV is 0.182, 0.288, 0.513, and 0.758. Therefore, the scatter of the fatigue life increased due to the UIT process according to the COV of Groups A and B. The scatter of fatigue life for prefatigued-UIT specimens (Groups C and D) is relatively large.

Figure 5.31b shows the two life extension factors K_f for different groups. The orange line represents one life extension factor (K_f), defined as the ratio of the total statistical fatigue life of different groups to the statistical fatigue life of as-welded specimens. The blue line is another life extension factor (K_f), denoting the ratio of the statistical residual fatigue life $N_{fi}(0.5,0.8,6)$ to the original remaining fatigue life $N_{f1}(0.5,0.8,6)$, where i is 1, 2, 3, 4 for Groups A, B, C, and D, respectively. Detailed values and definitions of different statistical fatigue lives can be found in Table 5.10. The UIT process can extend the fatigue life to 8.3 times longer (Group B), compared to the as-welded specimens (Group A). The effect of UIT can still greatly enhance the remaining fatigue life as long as 4.48 and 4.13 for Groups C and D. However, the effect of UIT on the total statistical fatigue life of prefatigued joints decreases from 8.31 to 1.59, with increasing prefatigued cycles from 0 to $0.8N_f$.

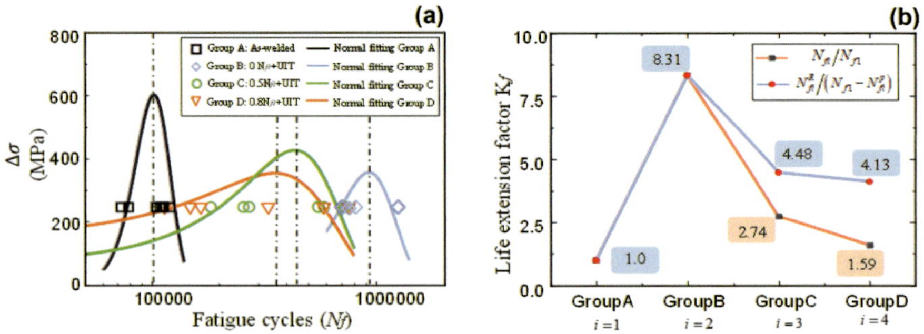

FIGURE 5.31 (a) The statistical fatigue life distribution under different groups; and (b) the fatigue life extension factor under different groups.

5.5.3 Fatigue Assessment of Welded Joints under UIT

Figure 5.32a shows the total fatigue life data of specimens tested against two widely used S-N curves based on nominal stress range, that is, the FAT80 curve for the as-welded specimens [54] and the FAT160 curve for the specimens treated by UIT without any prefatigued loading [55]. As can be seen in Figure 5.32a, the fatigue life data points of the as-welded specimens are all above the FAT80 curve, demonstrating that the FAT80 curve provides a conservative engineering design. Similarly, fatigue life data points of the UIT-treated specimens without any prefatigued loading are all above the FAT160 curve. However, for the total fatigue life of UIT joints with prefatigued loadings, some total fatigue life points fall below FAT 160, indicating that FAT 160 cannot provide safe engineering guidance for prefatigued specimens.

The two reduction factors are used to determine the residual fatigue life of prefatigued welded cruciform joints subjected to tension loading treated by UIT. One reduction factor N_{hs} is used to qualify the prefatigued loading cycles, that is, determine the percentage of the prefatigued loading cycles to the fatigue life of the as-welded joint. Another reduction factor N_R is used to qualify the residual fatigue life of the prefatigued cruciform joint treated by UIT, that is, determine the percentage of the residual fatigue life of the UIT-treated prefatigued cruciform joint to the fatigue life of the UIT-treated un-prefatigued one. The reduction factor N_{hs} is defined as follows:

$$N_{hs} = \frac{N_{AW\text{-}PF}}{N_{AW}} \tag{5.12}$$

where N_{AW-PF} is the prefatigued loading cycles of the as-welded cruciform joint ($N_{fi}^P, i = 2,3,4$ in this study), and N_{AW} is the fatigue life of the as-welded joint (N_{f1} in this chapter). The reduction factor N_R is expressed as follows:

$$N_R = \frac{N_{UIT\text{-}RS}}{N_{UIT}} \tag{5.13}$$

where N_{UIT-RS} is the residual fatigue life of the prefatigued cruciform joint treated by UIT ($N_{fi} - N_{fi}^P, i = 2,3,4$ in this study), and N_{UIT} is the fatigue life of the un-prefatigued as-welded joint treated by UIT (N_{f2} in this study).

Figure 5.32b compounds the two reduction factors based on the experimental results in this study, in the previous study [56] and reported by other researchers [57]. In the previous study, Li *et al.* [56] conducted a similar investigation on the effect of prefatigued loadings on the fatigue life of UIT-treated butt-joint specimens made of Q235 steel and welded using automatic tungsten argon-shielded welding. Welded butt joints were treated by UIT after pre-fatiguing and then tested till failure under tension fatigue loading. Prefatigued cycles are 30%, 50%, and 70% of the fatigue life of the as-welded specimens. For all the butt joints tested, the stress ratio is 0.1 and the stress range is 227 MPa. Zhang *et al.* [57] also conducted a similar investigation on the effect of prefatigued loadings on the fatigue life of UIT-treated butt-joint specimens made of S690QL high-strength steel and welded using flux-cored arc

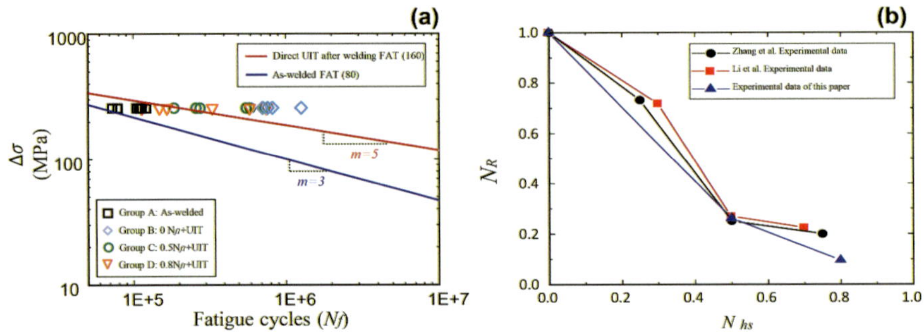

FIGURE 5.32 (a) The experimental fatigue life versus recommended FAT curves for all test groups; and (b) the experimental variation of the two reduction factors.

welding. Prefatigued loadings are 25%, 50%, 75%, and 100% of the statistical fatigue life of the as-welded specimens. The stress ratio is 0.1 and the stress range is 250 MPa for all the specimens tested by Zhang *et al.* [57]. As can be seen in Figure 5.32b, prefatigued loading has a remarkable adverse effect on the fatigue life enhancement of welded joints treated by UIT and the fatigue life enhancement brought by UIT decreases with the prefatigued loading increases, however, the decreasing rate becomes slow with the prefatigued loading increase. The experimental data from the existing studies are close to each other and follow a similar trend, which verifies that the two-reduction-factor curve proposed in this study can capture the effect of the UIT on prefatigued welded joints.

The fatigue performance of the welded plate joints consists of fatigue crack initiation and fatigue crack growth [Eq. (5.14)]. The fatigue crack initiation does not include any crack propagation; therefore, it is suitable to be described by the virtual damage accumulation indicators [Eq. (5.15)]. The fatigue crack propagation phase can be approached using Paris' law once the micro-slips merge into the micro-crack (a_0).

$$N_f = N_i + N_p \tag{5.14}$$

$$N_i = \int_0^1 f(D, N)\,dD \tag{5.15}$$

$$N_p = \int_{a_0}^{a_f} f(a, N)\,da \tag{5.16}$$

This chapter aimed to deal with the two-reduction factor curve following the two schemes in Figure 5.33. For the prefatigued cycles (N_{AW-PF}) less than the fatigue initiation life of as-welded plate joints N_i^{AW} , the remaining fatigue cycles to initiate the initial crack (a_0) is denoted as N_2^{UIT} . The fatigue crack propagation life equals to N_p^{UIT} . Therefore,

$$N_p^{UIT} = \int_{a_0}^{a_f} f\left(C^{UIT}, m^{UIT}, \Delta K, K_{th}^{UIT}\right)da \tag{5.17}$$

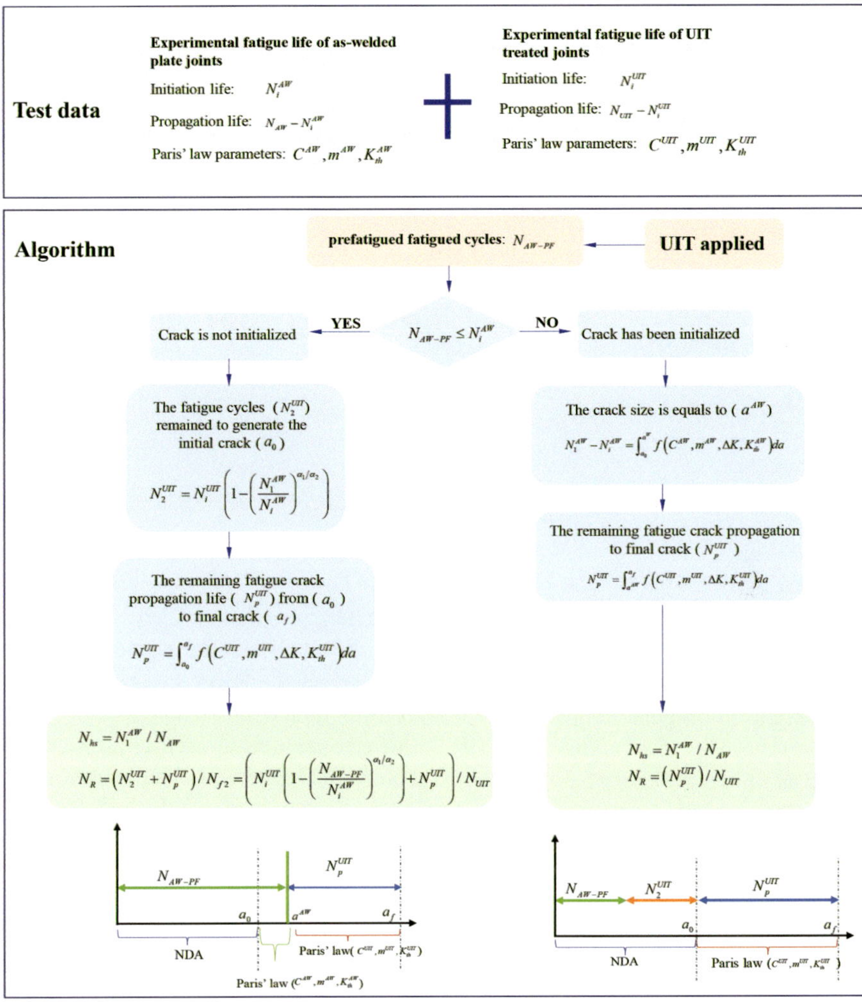

FIGURE 5.33 Block diagram of the computation algorithm for the two-reduction factor curves.

$$N_{hs} = N_{AE-PF} \,/\, N_{AW} \tag{5.18a}$$

$$N_R = \left(N_2^{UIT} + N_p^{UIT}\right) / \, N_{f2} = \left(N_i^{UIT}\left(1 - \left(\frac{N_{AW-PF}}{N_i^{AW}}\right)^{\alpha_1/\alpha_2}\right) + N_p^{UIT}\right) / \, N_{UIT} \tag{5.18b}$$

where N_i^{UIT} and N_i^{AW} are the fatigue crack initiation life of UIT-treated specimens and as-welded specimens, N_{AW} represents the total fatigue life of as-welded plate joints, equal to N_{f1} in this study, N_{UIT} represents the total fatigue life of UIT-treated plate joints, equal to N_{f2} in this study.

For the prefatigued cycles (N_{AW-PF}) larger than the fatigue initiation life of as-welded plate joints N_i^{AW}, the microcrack has been generated and propagated to a certain depth (a^{AW}), the remaining fatigue crack propagation life can be defined

$$N_p^{UIT} = \int_{a^{AW}}^{a_f} f\left(C^{UIT}, m^{UIT}, \Delta K, K_{th}^{UIT}\right) da \tag{5.19}$$

where a^{AW} can be derived from Eq (5.20)

$$N_{AW-PF} - N_i^{AW} = \int_{a_0}^{a^{AW}} f\left(C^{AW}, m^{AW}, \Delta K, K_{th}^{AW}\right) da \tag{5.20}$$

Therefore,

$$N_{hs} = N_1^{AW} / N_{AW} \tag{5.21a}$$

$$N_R = \left(N_p^{UIT}\right) / N_{UIT} \tag{5.21b}$$

This study assumes the fatigue initiation period to be 30% of the total fatigue life [53]. Therefore, $N_i^{AW} = 0.3N_{AW} = 0.3N_{f1}, N_i^{UIT} = 0.3N_{UIT} = 0.3N_{f2}$. With the assumption of a_0 with a value of 0.001 mm, as well as the nonlinear damage accumulation law in the fatigue initiation life [53]. This study derives the Pairs' law-related parameters [53] based on genetic optimization algorithm, as listed in Table 5.12.

Following Eq. (5.17) and Eq. (5.18), the two-reduction-factor curve can be derived as shown in Figure 5.34. As can be seen from Figure 5.34. The two-reduction-factor curve shows two nonlinear phases, which can capture the effects of the UIT on the fatigue life extension factor for as-welded and in-service joints. The coefficient of correlation (R^2) equals to 0.98. Based on the theoretical derivation [Eq. (5.17) and Eq. (5.18)], the UIT effect is significant when it is conducted during the fatigue initiation phase. The beneficial effect of UIT conducted after the fatigue crack initiation is still there, but quickly decreases with the initial crack size subjected to the UIT.

TABLE 5.12
The Fitted Parameters for Fatigue Crack Propagation Life.

Paris' law	Specimens	C (10^{-13})	m	K_{th} ($N \cdot mm^{-2/3}$)	Fitted fatigue crack growth life (N_p)
Paris' law 2 $\frac{da}{dN} = C\left(\Delta K - K_{th}\right)^m$	As-welded (Group A)	0.1728	3.7387	74.1061	61000
	UIT (Group B)	2.2656	3.0260	100.7564	505000

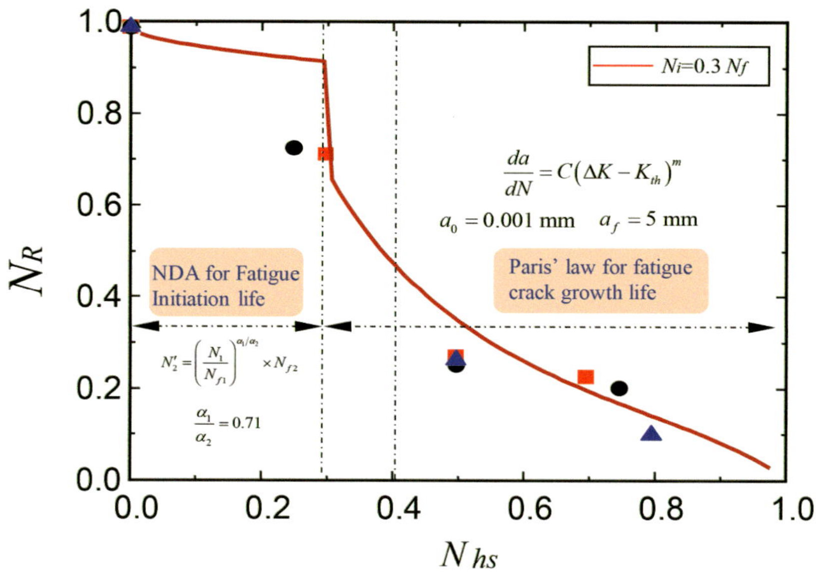

FIGURE 5.34 The two-reduction factors curve based on the proposed damage accumulation and Paris' law theory.

REFERENCES

[1] A.P.I. (API), Recommended practice for planning, designing and constructing fixed offshore platform, in, API RP 2A WSD, 2000.

[2] I.O.f.S. (ISO), ISO 19902: 2007 Petroleum and natural gas industries – fixed steel offshore structures, in, International Organization for Standardization, 2007.

[3] D.N.V. Fatigue design of offshore steel structures, in, DNV-RP-C203, 2011.

[4] A. Hobbacher, Recommendations for fatigue design of welded joints and components, Springer, 2016.

[5] S.-P. Chiew, S.-T. Lie, C.-K. Lee, Z.-W. Huang, Fatigue performance of cracked tubular T joints under combined loads. I: Experimental, Journal of Structural Engineering, 130 (2004) 562–571.

[6] W. Fricke, Fatigue analysis of welded joints: State of development, Marine Structures, 16 (2003) 185–200.

[7] X. Qian, Stress-intensity factors for circular hollow section V-joints with a rack-plate chord, Fatigue & Fracture of Engineering Materials & Structures, 32 (2009) 61–79.

[8] X. Qian, C.T. Nguyen, Y. Petchdemaneengam, Z. Ou, S. Swaddiwudhipong, P. Marshall, Fatigue performance of tubular X-joints with PJP+ welds: II – Numerical investigation, Journal of Constructional Steel Research, 89 (2013) 252–261.

[9] K. Wang, L.-W. Tong, J. Zhu, X.-L. Zhao, F.R. Mashiri, Fatigue behavior of welded T-joints with a CHS brace and CFCHS chord under axial loading in the brace, Journal of Bridge Engineering, 18 (2013) 142–152.

[10] D. Radaj, C.M. Sonsino, W. Fricke, Fatigue assessment of welded joints by local approaches, Woodhead publishing, 2006.

[11] A.S.W. Committee, Structural welding code-steel: ANSI/AWS D1. 1–90, an American national standard approved by American national standards institute, American Welding Society, January 5, 1990.
[12] X. Qian, P. Marshall, W. Cheong, Y. Petchdeman, Z. Chen, Partial joint penetration plus welds for tubular joints: Fabrication and SCFs, Welding World, 53 (2009) 143–148.
[13] A.P.I. (API), Recommended practice for ultrasonic and magnetic examination of offshore structural fabrication and guidelines for qualification of technicians, in, API RP 2X, 2004.
[14] M. Pradana, X. Qian, S. Swaddiwudhipong, An extrapolation method to determine the effective notch stress in welded joints, Fatigue & Fracture of Engineering Materials & Structures, 38 (2015) 1118–1135.
[15] A. Wordsworth, Aspects of the stress concentration factors at tubular joints, Elsevier Science Publishers, 1987, pp. 349–361.
[16] J. Wylde, Fatigue tests on welded tubular T-joints with equal brace and chord diameters, in: Offshore technology conference, OTC, 1983, pp. OTC-4527-MS.
[17] A.S.U.s. Manual, Abaqus 6.11, http://130.149, 89 (2012) v6.
[18] O. Dijkstra, J. De Back, Fatigue strength of tubular T-and X-joints, in: Offshore technology conference, OTC, 1980, pp. OTC-3696-MS.
[19] T. Maeda, K. Uchino, H. Sakurai, Experimental study of the fatigue strength of welded tubular X-joints with and without ring stiffener rings, Supplement to IIW Doc. XV-270-69, 1970.
[20] S. Senz, Essais de fatigue sur dix gros noeuds tubulaires, in: Conference international "L'Acier Dans les structure marines", INSTITUT DE RECHERCHES DE LA SIDÉRURGIE FRANÇAISE, Paris, 1981.
[21] R. Olivier, V. Köttgen, T. Seeger, Schweißverbindung II, Schwingfestigkeitsnachweise [Welded joints II, fatigue assessments], Frankfurt: FKM-Forschungsheft, 1994.
[22] A. Esin, A method for correlating different types of fatigue curve, International Journal of Fatigue, 2 (1980) 153–158.
[23] P.l.G. Bergan, I. Lotsberg, W. Fricke, M. Franc¸ ois, H. Pisarski, Overview of the FPSO: fatigue capacity JIP, in: International conference on offshore mechanics and arctic engineering, 2002, pp. 411–418.
[24] E. Chang, W. Dover, Prediction of degree of bending in tubular X and DT joints, International Journal of Fatigue, 21 (1999) 147–161.
[25] W. Shen, Y.S. Choo, Stress intensity factor for a tubular T-joint with grouted chord, Engineering Structures, 35 (2012) 37–47.
[26] G.R. Halford, Fatigue and durability of structural materials, ASM International, 2006.
[27] R. Kuguel, A relation between theoretical stress concentration factor and fatigue notch factor deduced from the concept of highly stressed volume, in: Proceedings of the ASTM, 1961, pp. 732–748.
[28] R. Kuguel, The highly stressed volume of material as a fundamental parameter in the strength of metal members notch predictions based on weakest-link failure models, International Journal of Fatigue, 25 (2003) 835–841.
[29] P. Dong, A structural stress definition and numerical implementation for fatigue analysis of welded joints, International Journal of Fatigue, 23 (2001) 865–876.
[30] C. Miki, T. Mori, K. Sakamoto, H. Kashiwagi, Size effect on the fatigue strength of transverse fillet welded joints, Journal of Structural Engineering, 33 (1987) 393–402.
[31] B. Standard, Guide on methods for assessing the acceptability of flaws in fusion welded structures, BS7910, September 2007.
[32] E. Mikkola, Y. Murakami, G. Marquis, Equivalent crack approach for fatigue life assessment of welded joints, Engineering Fracture Mechanics, 149 (2015) 144–155.

[33] T. Omura, M. Koyama, Y. Hamano, K. Tsuzaki, H. Noguchi, Generalized evaluation method for determining transition crack length for microstructurally small to microstructurally large fatigue crack growth: Experimental definition, facilitation, and validation, International Journal of Fatigue, 95 (2017) 38–44.
[34] Z.-G. Xiao, K. Yamada, A method of determining geometric stress for fatigue strength evaluation of steel welded joints, International Journal of Fatigue, 26 (2004) 1277–1293.
[35] ABAQUS, ABAQUS/Standard: User's manual, Hibbitt, Karlsson & Sorensen, 2012.
[36] T.R. Gurney, The fatigue strength of transverse fillet welded joints: A study of the influence of joint geometry, Elsevier, 1991.
[37] D.P. Kihl, S. Sarkani, Mean stress effects in fatigue of welded steel joints, Probabilistic Engineering Mechanics, 14 (1999) 97–104.
[38] J.W. Knight, Some basic fatigue data for various types of fillet welded joints in structural steel, in, The Welding Institute, Abington Hall, Abington, Cambridge, 1976.
[39] Z.P. Chen, Fatigue initiation life assessment for offshore structures, in, National University of Singapore, 2015.
[40] E. Haibach, Die Schwingfestigkeit von Schweißverbindungen aus der Sicht einer örtlichen Beanspruchungsmessung, Fraunhofer-Ges. z. Förderung d. Angewandten Forschung eV, 1968.
[41] I. Lotsberg, Overview of the FPSO fatigue capacity JIP, in: 20th international conference on offshore mechanics and arctic engineering, 2001, pp. 3–8.
[42] Y. Kim, J.-S. Oh, S.-H. Jeon, Novel hot spot stress calculations for welded joints using 3D solid finite elements, Marine Structures, 44 (2015) 1–18.
[43] N. Veritas, Fatigue strength analysis of offshore steel structures, Det Norske Veritas, 2001.
[44] A. Hobbacher, IIW recommendations for fatigue design of welded joints and components WRC, Welding Research Council Bulletin, WRC, 520 (2009).
[45] P. Marshall, X. Qian, C. Nguyen, Y. Petchdemaneengam, Welder-optimized CJP-equivalency welds for tubular connections, Welding in the World, 57 (2013) 569–579.
[46] X. Qian, Y. Petchdemaneengam, S. Swaddiwudhipong, P. Marshall, Z. Ou, C.T. Nguyen, Fatigue performance of tubular X-joints with PJP+ welds: I – Experimental study, Journal of Constructional Steel Research, 90 (2013) 49–59.
[47] X. Qian, Z. Ou, S. Swaddiwudhipong, P.W. Marshall, Brittle failure caused by lamellar splitting in a large-scale tubular joint with fatigue cracks, Marine Structures, 34 (2013) 185–204.
[48] G. Kim, K. Koo, Effects of hydration heat velocity on autogenous shrinkage of high-strength concrete, Magazine of Concrete Research, 65 (2013) 1385–1395.
[49] H.C. Yildirim, G.B. Marquis, Z. Barsoum, Fatigue assessment of high frequency mechanical impact (HFMI)-improved fillet welds by local approaches, International Journal of Fatigue, 52 (2013) 57–67.
[50] H. Al-Karawi, M. Al-Emrani, Fatigue life extension of existing welded structures via high frequency mechanical impact (HFMI) treatment, Engineering Structures, 239 (2021) 112234.
[51] H. Liang, Y. Kan, H. Chen, R. Zhan, X. Liu, D. Wang, Effect of cutting process on the residual stress and fatigue life of the welded joint treated by ultrasonic impact treatment, International Journal of Fatigue, 143 (2021) 105998.
[52] E.S. Statnikov, O.V. Korolkov, V.N. Vityazev, Physics and mechanism of ultrasonic impact, Ultrasonics, 44 (2006) e533–e538.
[53] T. Li, Y. Wang, L. Feng, J. Gan, Y. Shao, Fatigue life extension of the prefatigued welded joints by ultrasonic impact treatment, Journal of Constructional Steel Research, 229 (2025) 109508.

[54] X.W. Liao, Y.L. Zong, Fatigue analysis of steel bridge welding details based on effective notch stress method, Journal of Zhejiang University, 51 (2017) 8.
[55] H.C. Yildirim, G.B. Marquis, Overview of fatigue data for high frequency mechanical impact treated welded joints, Welding in the World, 56 (2012) 82–96.
[56] T. Li, C. Zhang, G. Huang, L. Feng, Fatigue life prediction of prefatigued welded Q235 steel butt-joints treated by high frequency mechanical impact, International Journal of Steel Structures, 24 (2024) 839–848.
[57] H. Zhang, D. Wang, L. Xia, Z. Lei, Y. Li, Effects of ultrasonic impact treatment on pre-fatigue loaded high-strength steel welded joints, International Journal of Fatigue, 80 (2015) 278–287.

6 Low-Cycle Fatigue of Welded Joints

6.1 INTRODUCTION

In Arctic areas, rich oil and gas resources stimulate the demand of ice-classed offshore structures and facilities for storing and transporting liquefied natural gases. The temperature in most Arctic areas varies between $-30^{\circ}C$ and $-40^{\circ}C$ in the winter seasons [1]. The lowest temperature can drop to $-60^{\circ}C$ over large parts of the Arctic. The harsh and cold environments there impose additional challenges on the failure assessment of these structures and components. The high strength steel, as the popular choice for infrastructures, offshore structures, and facilities for storing and transporting liquefied natural gases [2], expects more research attention at low temperatures to understand its mechanical properties.

As important structural components, welded plate or tubular joints in offshore structures, bridges, and large span structures often experience cyclic actions caused by the current, wind, and traffic loads. These welded joints tend to fail due to stress concentration under cyclic environmental loading. Cyclic actions with fairly large magnitudes usually initiate low-cycle fatigue cracks at the geometrically discontinuous locations of welded joints, for example, near the weld toes. These fatigue cracks, with relatively short fatigue lives, impose potential threats to the safety of structures, which can cause unstable failure in the entire structure. Geometric parameters, loading conditions, and material impose strong influences on the fatigue life of structural components. In comparison to the thorough research on the high-cycle fatigue assessment of welded plate connections, the thickness effect coupled with the elastic-plastic material property in the low-cycle fatigue requires more experimental and numerical attention [3]. To tackle the low-cycle fatigue issues in welded connections, the local fatigue driving forces [4–8] based on the stress, strain, or strain energy density at the location of the geometric discontinuity have emerged.

To zoom into the detailed fatigue crack initiation and propagation, existing theories (fracture mechanics and continuum damage mechanics) tend to treat low- and high-cycle fatigue assessments separately. Linear fracture mechanics [9] quantifies the high-cycle fatigue crack propagation utilizing the stress intensity factor, while elastic-plastic fracture mechanics [10] quantifies the low-cycle fatigue crack propagation utilizing the *J*-integral. Damage-based fatigue driving forces tend to be different for high- and low-cycle fatigue. Lemaitre and Chaboche [11] incorporate the stress-related parameters into the damage evolution for high-cycle fatigue assessment. Remes *et al.* [12] utilize the damage parameters by Socie and Marquis [13] to quantify every crack growth in a characteristic material length under high-cycle fatigue actions. Strain-based damage indicators are widely used for low-cycle fatigue evaluation. Lemaitre and Plumtree [14] suggest that the plastic strain is the leading

DOI: 10.1201/9781003509806-6

indicator of low-cycle fatigue life. Feng and Qian [15] demonstrate the potential of the strain energy density-based damage indicators in the rapid determination of the *S-N* curve for low-cycle fatigue. These non-uniform damage indicators for high and low-cycle fatigue impose challenges in the quick assessment of welded plate joints under various environmental loads. Feng *et al.* [16] recently present a uniform fatigue driving force for the high and low-cycle fatigue assessment, which is validated through the smooth and notched specimens. However, the tedious procedure in building multiple FE models with different crack profiles leads to the proposed method's limited application in complicated structures. Therefore, the fast and simple uniform high- and low-cycle fatigue assessment approach suitable for FE analysis for complicated welded joints requires further research efforts.

Welding residual stress exists at the welded joints. Understanding and predicting the effect of welding residual stresses on the fatigue life of welded connections is paramount for ensuring the reliability and safety of engineering structures [17–19]. Acevedo and Nussbaumer [20] conduct experimental and numerical studies on welding residual stresses in CHS K-joints used in bridges. Do *et al.* [21] conduct a numerical simulation study on butt-welded plates made of SM490 steel to predict the high-cycle fatigue life considering welding residual stresses. Despite notable progress in understanding the effects of welding residual stresses on fatigue life prediction, challenges and uncertainties remain. Factors such as the types of welded joints, maximum cyclic load, and the presence of fatigue cracks can influence the magnitude and distribution of residual stresses during the entire fatigue life. However, a cyclic loading has not yet been developed.

6.2 LOW-CYCLE FATIGUE OF WELDED JOINTS UNDER LOW TEMPERATURE

6.2.1 Experimental Setup

The material investigated in this chapter is the high-strength steel S550, with the chemical composition listed in Table 6.1. Figure 6.1 presents the geometrical details of the non-loading cruciform welded plate joints with the geometrical values listed in Table 6.2. The experimental procedure refers to ASTM E606 [22] using the MTS 1000 kN closed-loop servo-hydraulic test machine. A calibrated extensometer with a 100 mm gauge length monitors the nominal strain during the low-cycle fatigue test. To prevent slippage between the extensometer edges and the specimen caused by the cyclic action and crack initiation, we adopted springs and rubber bands to tighten the extensometer edges around the specimens. The data acquisition and storage utilize the dynamic oscilloscope DL850 with about 200 ~ 500 data acquisition per second to monitor the failure of these joints. Figure 6.2 shows the experimental set-up under the room temperature. At the room temperature, the crack depth in the welded plate connections is monitored by the Alternating Current Potential Drop (ACPD) and by visual inspection. Among the four weld toes in each plate connection, three of them have fixed ACPD probes near the weld toe, and the other has an attached strain gauge. Figure 6.3 presents the experimental set-up under low temperature. The liquified nitrogen provides the media to control the ambient temperature around

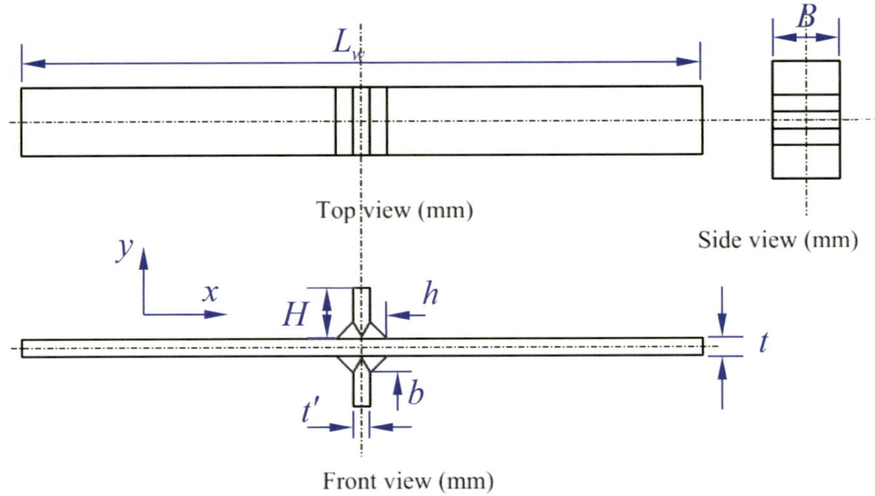

FIGURE 6.1 The geometrical details of the welded plate connections.

TABLE 6.1
Weight Percentage for Chemical Compositions of S550 High-Strength Steel.

S550	Fe	C	Si	Mn	P	S	Cu	Cr	Ni	Mo
Weight(%)	96.19	0.106	0.327	1.41	0.010	0.0015	0.121	0.454	0.918	0.462

TABLE 6.2
Geometrical Details of the Cruciform Welded Plate Connections.

	t (mm)	t' (mm)	B (mm)	H (mm)	b (mm)	h (mm)	L_W (mm)
PT 10	10	10	40	40	10	10	400
PT 22	22	20	40	40	10	10	400
PT 40	40	20	22	40	10	10	400

the specimen. The designed low temperature chamber provides an insulated atmosphere. The Proportional-Integral-Derivative (PID) temperature controller monitors the temperature in the chamber through the J thermocouple and controls the Asco Solenoid Valve to maintain a stable low temperature -60°C in the chamber by adjusting the flow rate of liquid nitrogen into the chamber. Figure 6.3b shows the geometric details of the low temperature chamber. The insulation layer with a thickness of 50 mm in the low-temperature chamber is fabricated from the spray

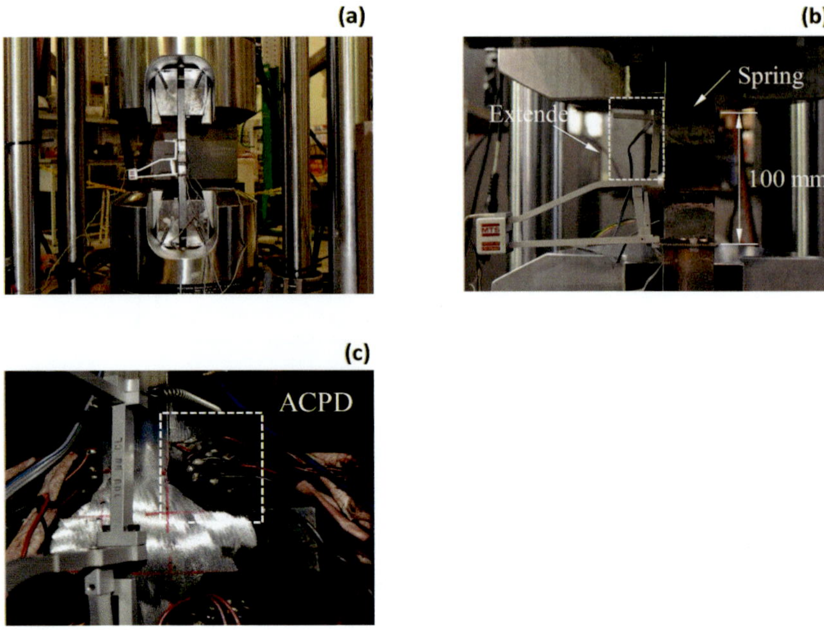

FIGURE 6.2 (a)–(c) Experimental set-up of the low-cycle fatigue under room temperature.

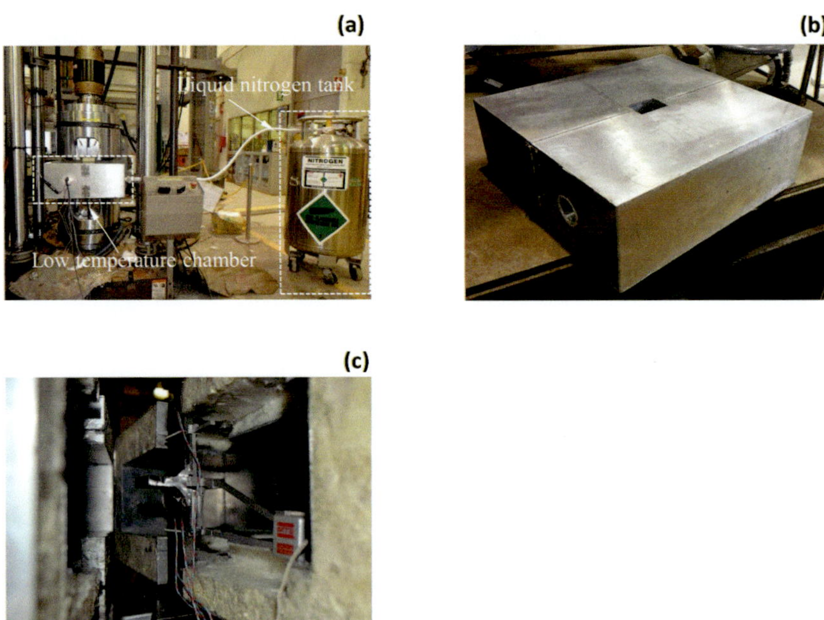

FIGURE 6.3 (a)–(c) Experimental set-up of the low-cycle fatigue under the low temperature (-60°C).

TABLE 6.3
Details of the Experimental Programs.

		Coupon specimen	PT10			PT22			PT40		
28°C	ε_a	0.5%	0.2%	0.3%	0.45%	0.2%	0.3%	0.45%	0.2%	0.3%	0.45%
	No.	3	3	3	3	3	3	3	3	3	3
-60°C	ε_a	0.5%	0.2%	0.25%	0.3%	0.2%	0.25%	0.3%			
	No.	3	3	3	3	3	3	3			

polyurethane foam. The holes used for wireline input of liquid nitrogen are filled with fiber cottons during the experiment to reduce the leakage of the nitrogen gas.

The main experimental program includes three parts: low-cycle fatigue tests of 1) coupon specimens under the room ($28^{\circ}\mathrm{C}$) and low temperature ($-60^{\circ}\mathrm{C}$) for calibrating the cyclic plastic material properties; 2) the welded plate connections with thickness of 10 mm, 22 mm, and 40 mm at the room temperature of $28^{\circ}\mathrm{C}$ and 3) the welded plate connections with thickness of 10 mm and 22 mm at the low temperature of $-60^{\circ}\mathrm{C}$; The nominal strain amplitudes of the welded plate joints range from 0.2% to 0.45% with the strain ratio of –1. Each strain amplitude and thickness entails three duplicated specimens for repeatability. The frequency of low-cycle fatigue ranges from 0.2 Hz to 0.5 Hz to maintain a constant slow strain change rate. Table 6.3 lists the experimental program details.

6.2.2 Calibrated Cyclic Material Properties under Low Temperature

To calibrate the cyclic plastic material properties, this chapter adopts standard coupon specimens shown in Figures 6.4a and b to conduct the low-cycle fatigue test under the room ($28^{\circ}\mathrm{C}$) and low temperature ($-60^{\circ}\mathrm{C}$), respectively. The tested strain amplitude is 0.5% with the strain ratio of –1. The length of the extensometer in the coupon specimen test is 50 mm in Figure 6.4c. Figure 6.4d illustrates the experimental set-up of the coupon test under low temperatures.

The present study considers both isotropic and kinematic hardening in the material constitutive formulation originally proposed by Armstrong and Frederick [23] and subsequently modified by Chaboche [24, 25]. The yield surface implemented in ABAQUS [26] has the following form,

$$f(\boldsymbol{\sigma}-\boldsymbol{\alpha})-\sigma_y=0 \tag{6.1a}$$

$$f(\boldsymbol{\sigma}-\boldsymbol{\alpha})=\left|\frac{3}{2}(\boldsymbol{\sigma}'-\boldsymbol{\alpha}'):(\boldsymbol{\sigma}'-\boldsymbol{\alpha}')\right| \tag{6.1b}$$

$$\boldsymbol{\sigma}'=\boldsymbol{\sigma}-\frac{1}{3}Tr(\boldsymbol{\sigma})I \tag{6.1c}$$

$$\boldsymbol{\alpha}'=\boldsymbol{\alpha}-\frac{1}{3}Tr(\boldsymbol{\alpha})I \tag{6.1d}$$

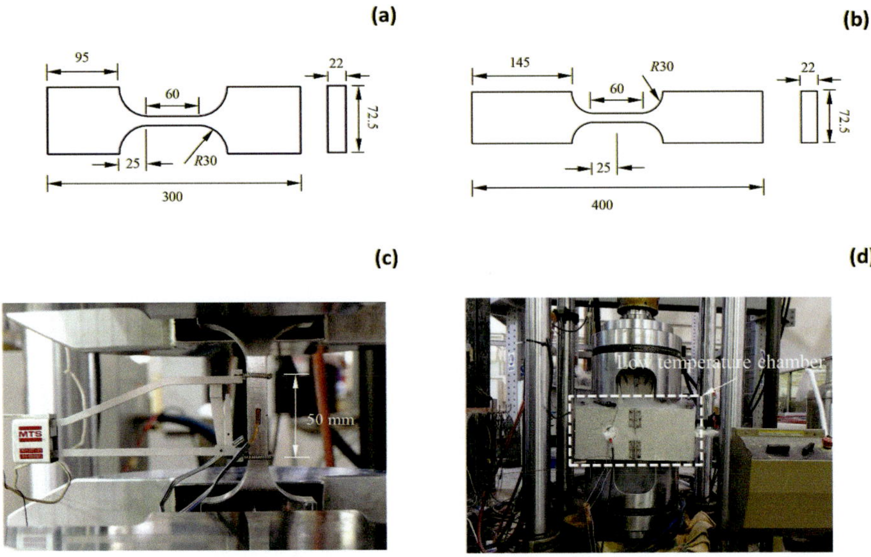

FIGURE 6.4 (a, b) Geometrical details of the standard coupon specimens under the room and low temperature, respectively; (c, d) experimental details of the standard coupon specimen test.

where $\boldsymbol{\sigma}$ and $\boldsymbol{\sigma}'$ denote the stress tensor and deviatoric stress tensor; $\boldsymbol{\alpha}$ and $\boldsymbol{\alpha}'$ represent the backstress tensor and the deviatoric backstress tensor, while σ_y refers to the current yield strength.

The isotropic hardening of the model defines the size evolution of the yield surface, as a function of the effective cumulative plastic strain ε^{pl},

$$\sigma_{\mathrm{y}} = \sigma_{\mathrm{y}}^{\mathrm{o}} + Q_{\infty}\left(1 - \exp\left(-g\varepsilon^{\mathrm{pl}}\right)\right) \tag{6.2}$$

where $\sigma_{\mathrm{y}}^{\mathrm{o}}$ is the yield stress at zero plastic strain, Q_{∞} refers to the maximum expansion of the yield surface and g denotes the rate at which the size of the yield surface changes as the plastic strain develops.

The nonlinear kinematic hardening of the model represents the translation of the center of the yield surface without alternating the size of the surface, which illustrates the Bauschinger [27] effect using the backstress tensor α under the cyclic loading. The incremental backstress tensor depends on the backstress tensor and the incremental plastic strain,

$$d\boldsymbol{\alpha} = \frac{2}{3} C d\varepsilon_{\mathrm{ij}}^{\mathrm{pl}} - \gamma \boldsymbol{\alpha} d\varepsilon^{\mathrm{pl}} \tag{6.3a}$$

$$d\varepsilon^{\mathrm{pl}} = \left(\frac{2}{3} d\varepsilon_{\mathrm{ij}}^{\mathrm{pl}} : d\varepsilon_{\mathrm{ij}}^{\mathrm{pl}}\right)^{1/2} \tag{6.3b}$$

where C denotes the initial kinematic hardening modulus and γ defines the rate at which the kinematic hardening modulus decreases with increasing plastic deformation.

This chapter calibrates the isotropic hardening parameters based on the stress-strain curves from the first cycle to the $0.5N_f$ cycle, and kinematic hardening parameters based on the stable hysteresis loop at $0.5N_f$ cycle. Table 6.4 lists the obtained plasticity parameters based on the least square method under the room and low temperature. Figures 6.5a and b illustrate good agreements between the stress-strain curve in the first few cycles under the strain amplitude of 0.5% in the first few cycles by the experiments and numerical results based on the calibrated plasticity parameters under the room temperature and the low temperature, respectively.

Based on the material properties calibrated in the above session, the elastic-plastic analysis for the cruciform welded joints in ABAQUS with the finite element model in Figure 6.6a. The 20-node brick elements are adopted with local refinement of mesh following the recommendation in Parada *et al.* [28]. Figures 6.6b–d compare the load-displacement ($P-\Delta$) by numerical simulation and experimental

TABLE 6.4

Parameters of the Nonlinear Combined Isotropic/Kinematic Hardening Model of Cyclic Plasticity.

	E (MPa)	σ_y^o (MPa)	Q_∞ (MPa)	g	C_1 (MPa)	γ_1	C_2 (MPa)	γ_2
28°C	190877	479.46	−57.36	0.430	39974	116.00	26905	1131.9
−60°C	198503	421.57	−55.35	0.335	29140	65.55	83613	991.06

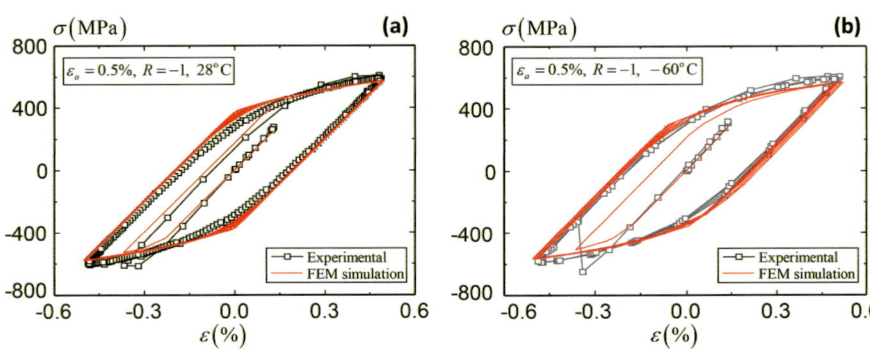

FIGURE 6.5 (a, b) Comparison of the stress-strain variation by experimental data and numerical simulation under the room and low temperature, respectively.

measurements at the first few cycles under room and low temperatures. A reasonable agreement validates the calibrated elastic-plastic constitutive material properties.

6.2.3 Size Effects on Low-Cycle Fatigue Life of Welded Joints

The different geometric sizes lead to different stress, strain, and strain energy density concentration at the weld notch. In the high-cycle fatigue of welded plate joints, a larger main plate thickness generally reduces the fatigue strength. Different approaches, including the thickness reduction factor [29], hot-spot indicator [4, 30–33], and effective notch indicators [34, 35] have emerged to consider the size effect in the high-cycle fatigue of the welded plate joints.

This chapter defines the low-cycle fatigue life as the 80% drop of the external load in the experimental test, as shown in Figure 6.6. For cruciform welded plate joints under low-cycle actions, the near-toe material experiences a multiaxial stress condition, and the orientation of the principal stress and strain evolves as

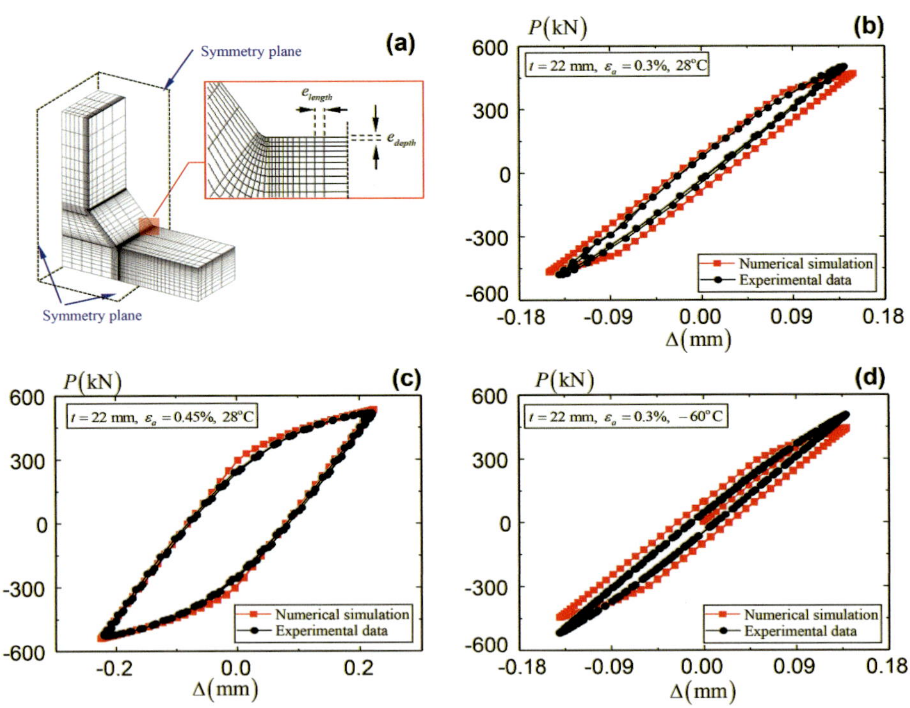

FIGURE 6.6 (a) FE model of the cruciform welded plate joints; (b)–(d) comparison in the load-displacement by the elastic-plastic numerical analysis and experimental test data for room and low temperatures.

the plastic deformation increases. The energy fatigue driving force at the weld toe [36] is

$$\Delta\sigma_{ij}\Delta\varepsilon_{ij} = \Delta\sigma_{ij}\Delta\varepsilon_{ij}^{e} + \Delta\sigma_{ij}\Delta\varepsilon_{ij}^{p} = 2\Delta W^{e} + \frac{1}{1+n}\Delta W^{p} \tag{6.4}$$

Figures 6.7b–d assess the total fatigue life based on the nominal strain amplitude (ε_a), the nominal energy fatigue driving force ($\Delta\sigma_{nom}\Delta\varepsilon_{nom}$) and the local energy fatigue driving force ($\Delta\sigma_{ij}\Delta\varepsilon_{ij}$) at the weld toe, respectively. Different from the size effect in the high-cycle fatigue life, the influence of the main plate thickness in low-cycle fatigue life appears negligible based on the nominal strain amplitude (ε_a) and nominal energy fatigue driving force ($\Delta\sigma_{nom}\Delta\varepsilon_{nom}$). In contrast, based on the local energy fatigue driving force ($\Delta\sigma_{ij}\Delta\varepsilon_{ij}$) at the weld toe, the size effect on the low-cycle fatigue is notable. In low-cycle fatigue, the larger main plate thickness seems beneficial to the fatigue capacity.

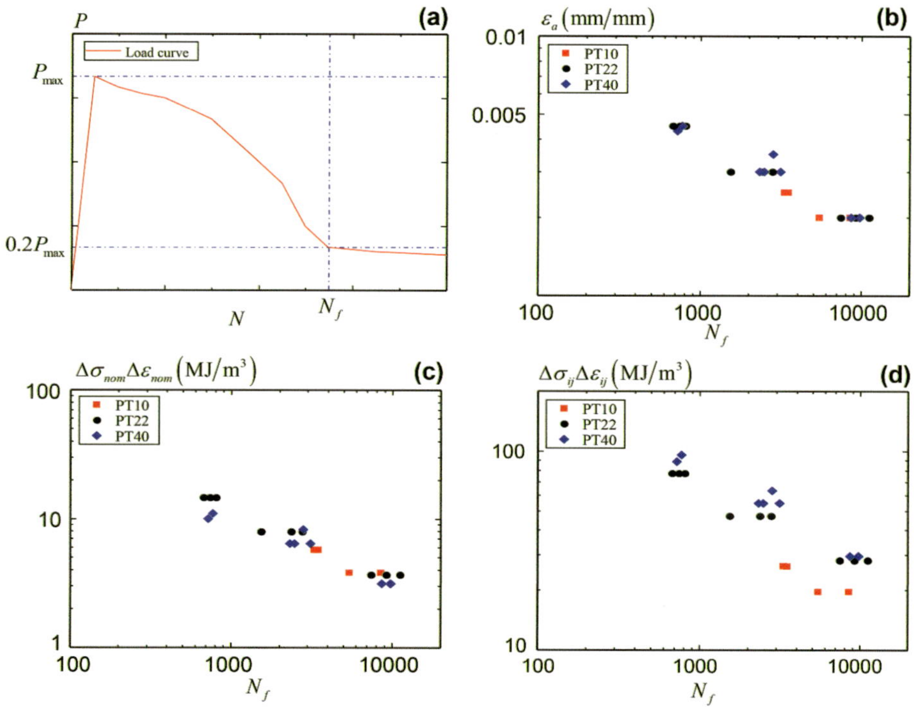

FIGURE 6.7 (a) Definition of the total low-cycle fatigue life; (b) fatigue life assessment based on the nominal strain amplitude (ε_a), the nominal energy fatigue driving force ($\Delta\sigma_{nom}\Delta\varepsilon_{nom}$) and the local energy fatigue driving force ($\Delta\sigma_{ij}\Delta\varepsilon_{ij}$) at the weld toe, respectively.

The total fatigue life consists of fatigue crack initiation life and fatigue crack propagation life. Therefore, this study elaborates the size effect separately for the fatigue crack initiation and crack propagation. For welded plate joints with different geometries under the same nominal loadings, the size effect in the fatigue crack initiation life is due to different local stress and strain fields at the weld toe, while the size effect in the fatigue crack propagation relies on different crack driving forces with growing cracks. For high-cycle fatigue issues, the initiation life contributes to a significant portion of the total fatigue life [33, 37–40]. In addition, the high-cycle fatigue test is usually under stress-controlled loading. Once the fatigue crack initiates at the weld notch, the fatigue crack growth accelerates, driven by the increasing local fatigue driving force with propagating cracks (e.g., stress intensity factor) with growing cracks. This speeds up the crack propagation process and leads to a relatively quicker propagation procedure than the initiation life. Therefore, the size effect in the total high-cycle fatigue life can be represented by the size effect in the fatigue initiation life to a large extent.

In contrast, the strain-controlled low-cycle fatigue test introduces the large-scale elastic-plastic deformation. The local stress and strain at the weld toe greatly exceed the yield strength and strain. Therefore, the crack at the weld toe can be quickly initiated. After the crack initiates, the nominal load decreases under constant controlled strain due to the presence of a crack. The crack propagation thus does not continue to accelerate due to the decreasing nominal loads. Consequently, the crack propagation life becomes a critical phase in the total low-cycle fatigue life and requires detailed investigations. Overall, based on the nominal fatigue driving forces, the size effect in low-cycle fatigue is complicated and depends on the interaction of the size effect in fatigue initiation life and propagation life.

Figures 6.8a–c present the fatigue test initiation life at the weld toe based on the nominal strain amplitude (ε_a), the nominal energy fatigue driving force ($\Delta\sigma_{nom}\Delta\varepsilon_{nom}$), and the local energy fatigue driving force ($\Delta\sigma_{ij}\Delta\varepsilon_{ij}$) at the weld toe, respectively. Similar to the size effect in the high-cycle fatigue assessment, the size effect exists in the low-cycle fatigue initiation life assessment based on nominal fatigue driving force (ε_a and $\Delta\sigma_{nom}\Delta\varepsilon_{nom}$) as shown in Figures 6.8a and b. The larger the main plate thickness is, the shorter is the fatigue initiation life. In contrast, the local energy fatigue driving force ($\Delta\sigma_{ij}\Delta\varepsilon_{ij}$) at the weld toe can integrate the fatigue initiation life regardless of different sizes. Figure 6.8d illustrates the percentage of the fatigue initiation life in the total low-cycle fatigue life. The fatigue initiation life does not dominate the total low-cycle fatigue life in PT22 and PT40 except for PT10. The difference in the percentage (N_i/N_f) between PT10, PT22, and PT40 is probably attributed to the small concentration factor at the weld toe and the small cross-sectional area of PT10. The nominal fatigue driving forces (ε_a and $\Delta\sigma_{nom}\Delta\varepsilon_{nom}$) cannot reveal the concentration factor at the weld toe, and also cannot reflect the crack propagation behavior accurately. Therefore, based on the nominal fatigue driving forces, the size effect in the low-cycle fatigue of cruciform welded plate joints is case-sensitive and should be considered separately for fatigue crack initiation and crack propagation life. The local fatigue driving force at the weld toe ($\Delta\sigma_{ij}\Delta\varepsilon_{ij}$) captures the concentration effect, which is a decisive factor in the fatigue initiation life.

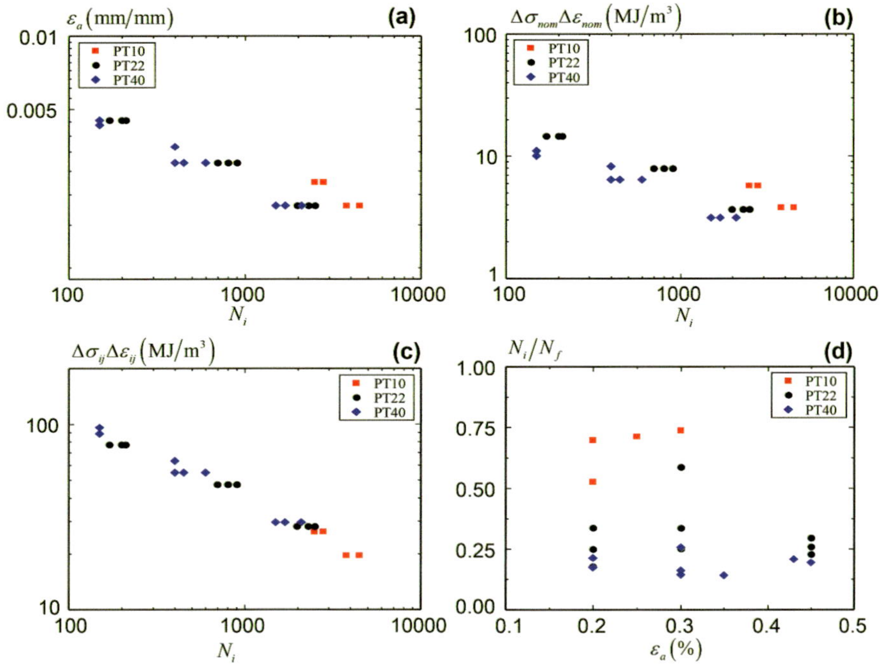

FIGURE 6.8 (a)–(c) Fatigue crack initiation life assessment based on the nominal strain amplitude, nominal energy indicator and the local energy indicator at the weld toe, respectively; (d) percentage of the fatigue initiation life in the total low-cycle fatigue life under different welded plate connections and different nominal strain amplitudes.

6.2.4 Effects of Low Temperature on Low-Cycle Fatigue Life of Welded Joints

Relying on the energy fatigue driving force in Eq. (6.4), this study further investigates the effect of low temperature on the low-cycle fatigue life of welded plate joints. The thicknesses of the studied welded plate joints at the low temperature are 10 mm and 22 mm. In addition, the amplitude of the applied nominal strain has a lower value than that at room temperature. Due to the non-applicability of the ACPD under the low temperature, the only result presented in this chapter is the total low-cycle fatigue life. Figures 6.9a and b compare the fatigue life of PT10 and PT22, respectively, under the room and low temperature based on the nominal strain amplitude. Figures 6.9c and d compare the fatigue life of PT10 and PT22, respectively, under the room and low temperature based on the local energy fatigue driving force at the weld toe. As illustrated in Figure 6.9, the difference in the total low-cycle fatigue life of welded plate joints due to the low temperature's effect is negligible, which is different from the previous work [41] about the low temperature effect on the high-cycle fatigue of butt welded joints.

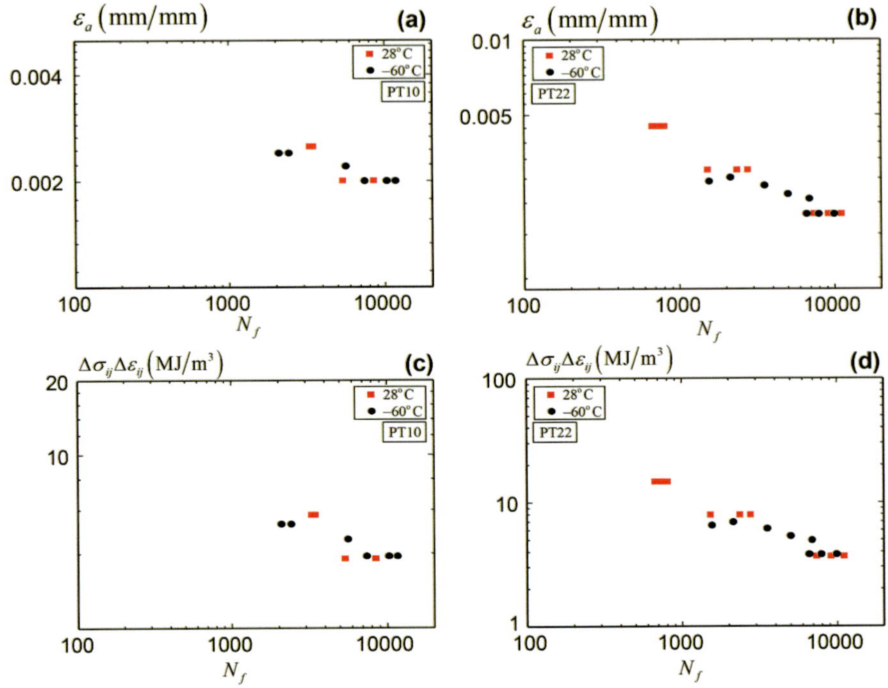

FIGURE 6.9 (a, b) Comparison of the total fatigue life of welded plate joints with the thickness of 22 mm and 40 mm, respectively, based on the nominal strain amplitude; (c, d) comparison of the total fatigue life of welded plate joints with the thickness of 22 mm and 40 mm, respectively, based on the local energy fatigue driving force at the weld toe.

6.3 ENERGY-BASED APPROACH TO INTEGRATE LOW- AND HIGH-CYCLE FATIGUE

6.3.1 Continuum Damage Mechanics Methodology

Compared with conventional fracture mechanics, which requires a prior defined crack shape or growth, continuum damage mechanics is a tractable numerical approach to simulate the crack initiation and propagation (damage indicator reaching threshold value [42]) under fatigue cyclic loadings. Different damage-related theories have emerged in assessing fatigue crack initiation and propagation separately. The common challenges and controversies in different continuum damage mechanics theories exist due to different levels of material plasticity caused by the fatigue loading [10, 11] (e.g., high-cycle fatigue versus low-cycle fatigue), fatigue driving forces [30, 43, 44] (stress, strain, and strain energy density), the distinction between fatigue crack initiation and propagation [45, 46]. To alleviate the above challenges and controversies, our previous study [47] proposed a fatigue life estimation strategy based on a uniform fatigue driving force for the crack initiation and propagation in notched specimens under high- and low-cycle fatigue loadings. Inspired by the weakest link theory [48], as shown in Figure 6.10a, the fracture process zone comprises a large number of random cleavage initiation sites (microcracks), the fatigue crack propagation in the homogeneous material is a consequence of a series of microscopic fatigue crack growth, namely, the fatigue crack "initiation" ahead of a growing tip [12].

The energy-based fatigue driving force provides a uniform treatment of the high- and low-cycle fatigue. Based on the previous study [16], a new fatigue indicator involving the plastic strain energy density, positive elastic strain energy density, and mean stress effect follows the expression in Eq. (6.5), as also shown in Figure 6.10b

$$\Delta W^t = \Delta W^p + \left(\frac{\sigma_{\max}}{\sigma_{ref}}\right)^{\alpha} \times \Delta W^{e+} \tag{6.5a}$$

$$A \cdot \left(\Delta W^t\right)^B = N \tag{6.5b}$$

where ΔW^p represents the range of the plastic hysteresis energy; ΔW^{e+} denotes the positive elastic strain energy range; $\sigma_{\max}$ is the maximum stress; and σ_{ref} is the reference stress to normalize the effect of the maximum stress. The reasonability of Eq. (6.5) is validated in detail in the previous paper [16]. Therefore, this study utilizes the total energy ΔW^t to quantify the fatigue driving force and damage indicator. However, integrating the new uniform fatigue indicator into existing commercial software lacks research efforts, which leads to tedious and repeated work in building different FE models.

Therefore, this chapter proposed a numerical framework (Figure 6.11) to conduct the damage evaluation and updating scheme in the numerical simulation by incorporating the uniform fatigue assessment indicator (Eq. (6.5)). At the end of each fatigue cycle (each cycle with a period of T), the numerical procedure calculates the fatigue

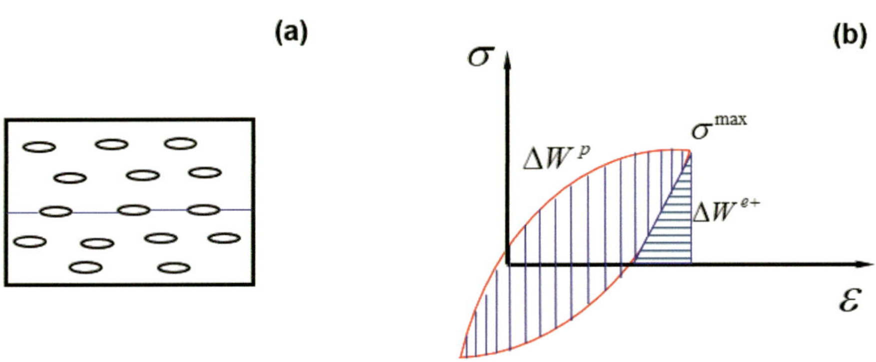

FIGURE 6.10 (a) Concept of the weakest link theory; (b) definition of total strain energy density.

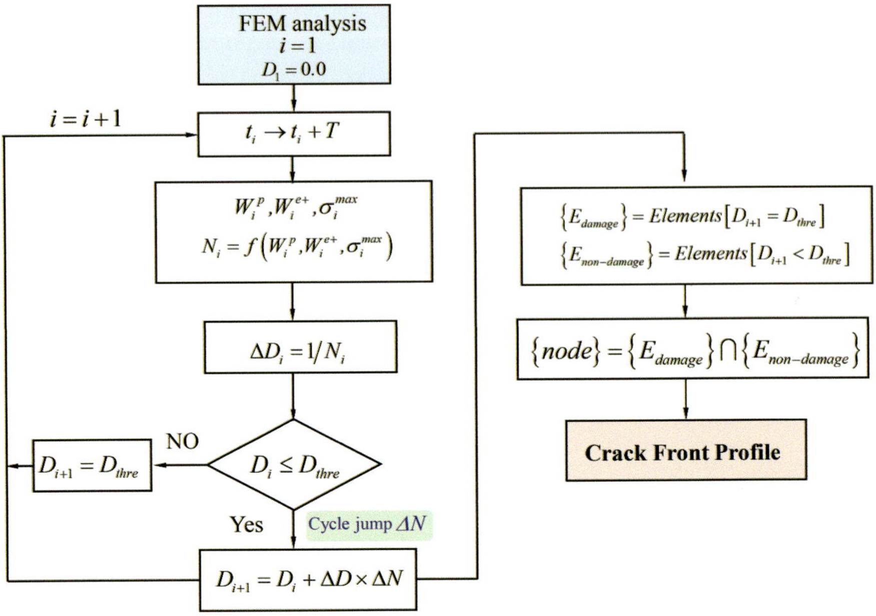

FIGURE 6.11 The numerical framework of the damage evaluation and updating in the developed algorithm.

damage, the plastic strain energy density, positive strain energy density, and positive maximum principal stress. Following Eq. (6.5), the corresponding fatigue life N_i derives from

$$N_i = \left(\frac{\Delta W_i^t}{A}\right)^{1/B} \tag{6.6}$$

and the damage accumulated in one cycle is

$$\Delta D_i = 1/N_i \tag{6.7}$$

This study adopts the threshold value of the damage indicator (D_{thre}) as 0.98 to examine the condition of the damage indicator D_i after each calculation. If the damage indicator is less than 0.98, the damage will continue to accumulate. This study adopts the threshold value of 0.98 instead of 1.0 or 0.99 for achieving a convergent solution in the elastic-plastic low-cycle fatigue analysis. To reduce the computational cost, the numerical simulation employs the cycle jump strategy [49, 50]. Therefore, the updated value of the damage indicator follows

$$D_{i+1} = D_i + \Delta D \times \Delta N \tag{6.8}$$

where ΔN denotes the jumped cycles in the FE analysis. After each step, the elements with a damage value equal to 0.98: $\left\{E_{damage}\right\}$ or less than 0.98: $\left\{E_{non-damage}\right\}$ can be determined. Finally, the nodes along the boundary between the damaged elements and undamaged elements define the crack front profile.

Through incorporating the damage indicator D in the cyclic constitutive model of the S550, the material stiffness and strain evolve

$$\tilde{\sigma} = \frac{\sigma_0}{1-D} \tag{6.9}$$

$$\varepsilon_{ij}^{e} = \frac{1+\nu}{E}\frac{\sigma_{ij}}{1-D} - \frac{\nu}{E}\frac{\sigma_{kk}}{1-D}\delta_{ij} \tag{6.10}$$

where $\tilde{\sigma}$ denotes the effective stress in the damaged material; σ_0 is the original stress; ν is the Poisson's ratio; ε_{ij}^{e} and δ_{ij} are the elastic strain tensor and Kronecker delta. Correspondingly, the plastic criterion function follows

$$f = \sqrt{\frac{3}{2}\left(\frac{\sigma_{ij}'}{1-D} - X_{ij}'\right)\left(\frac{\sigma_{ij}'}{1-D} - X_{ij}'\right)} - R - \sigma_y \tag{6.11a}$$

$$R = R_{\infty}\left(1 - e^{-h \bullet p}\right) \tag{6.11b}$$

$$dX_{ij} = \frac{2}{3}C\dot{\varepsilon}_{ij}^{pl} - \gamma \times X_{ij} \bullet \dot{p} \tag{6.11c}$$

$$\dot{p} = \left(\frac{2}{3}\dot{\varepsilon}_{ij}^{pl}\dot{\varepsilon}_{ij}^{pl}\right)^{0.5} \tag{6.11d}$$

where σ_{ij}' and X_{ij}' are deviatoric stress and backstress tensor; σ_y is the yield strength; R and X refer to the isotropic hardening and kinematic hardening stress; $\dot{\varepsilon}_{ij}^{pl}$ denotes the plastic strain tensor; h is the rate of the yield surface changes with increasing plastic strains; p and $\dot{p}$ are the accumulative plastic strain and corresponding rate, respectively; C and γ are the initial kinematic hardening modulus and corresponding changing rate, respectively.

6.3.2 Calibration of Damage-Related Material Properties

The damage simulation of welded plate joints requires the calibration of cyclic material properties and damage-related material parameters from the smooth coupon specimen. This study adopts the cyclic elastic-plastic material parameters of the S550 from [51]. According to the experiment results in [51], the plastic strain energy density, positive strain energy density, and maximum positive stress correspond to the values at half of the fatigue life. Based on Eq. (6.5), this study calibrates the parameters between the total energy density and the total fatigue life, as shown in Table 6.5. Figure 6.12 compares the predicted fatigue life with the experimental fatigue life, which indicates a good agreement based on the total strain density indicators.

TABLE 6.5
Damage-Related Material Parameters in Eq. (6.5) of S550 Steel.

	σ_{ref} (MPa)	α	A	B
S550	250	2.5	104363.489	−1.9484

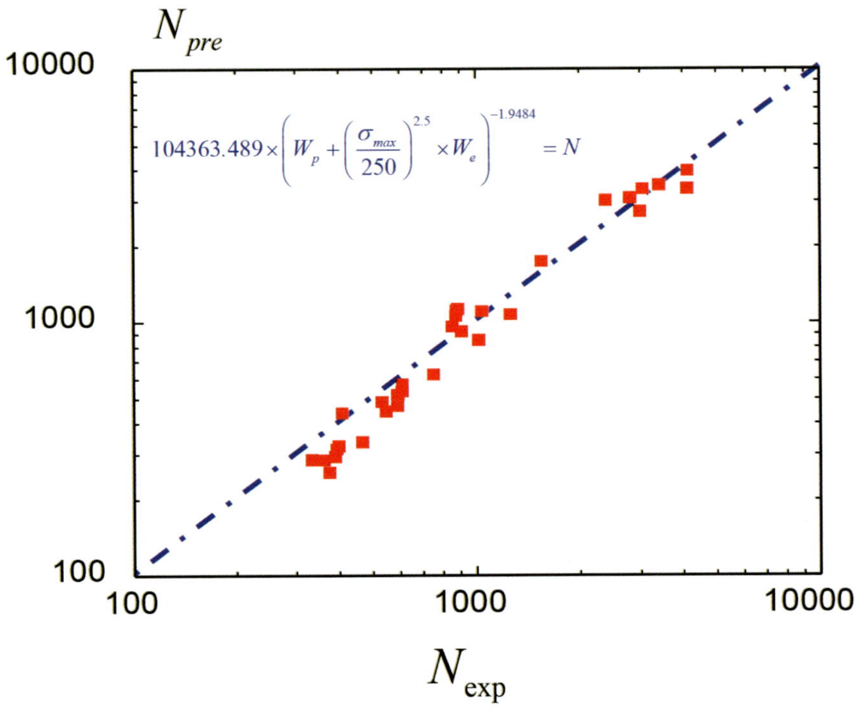

FIGURE 6.12 Comparison between the predicted fatigue life and experimental results based on Eq. (6.5).

This study developed the UMAT subroutine in ABAQUS [52] to verify the calibrated damage-related material properties, incorporating the damage assessment and updating scheme in Figure 6.11. The low-cycle fatigue simulation of the smooth coupon specimen is under strain control. With different nominal strain amplitudes of 0.5%, 0.75%, 1.0%, and 1.25%, Figure 6.13a presents the evolution of the damage indicator D of one element in the coupon specimen. The jumped cycle ΔN is ten cycles. Figure 6.13b describes the variation of the total reaction force with increasing fatigue cycles under the strain amplitude of 1.0%. This study identifies the numerical simulation of fatigue life N_f^{FEM} as indicated by the red dot in Figure 6.13b. The numerical simulation of the fatigue life based on the total strain energy density is listed in Table 6.6. Figure 6.13c compares the experimental fatigue life with the numerically calculated fatigue life for smooth coupon specimens, which illustrates the accuracy of the calibrated damage-related material parameters. To examine the sensitivity of different threshold values of damage, D_{thre}. Figure 6.13d compares the numerical fatigue life of standard coupon specimens N_f^{FEM}. The numerically calculated fatigue life reaches a convergent value once the threshold value of damage D_{thre} exceeds 0.9. The effect of the threshold damage on the longer fatigue life is larger than that on the shorter fatigue life. In addition, when the threshold value is too close

TABLE 6.6

Numerical Fatigue Life Estimation of Standard Smooth Coupon Specimens.

	$R_\varepsilon = -1$			$R_\varepsilon = 0.1$			$R_\varepsilon = 0.5$			
ε_α		N_f^{Exp}			N_f^{Exp}			N_f^{Exp}		N_f^{FEM}
0.5%	1552	2384	3701	2818	3009	4127	3063	3421	4135	4000
0.75%	1020	911	-	891	877	881	856	1261	1042	1150
1.0%	549	409	535	593	610	-	753	611	594	640
1.25%	376	334	388	400	390	434	394	468	400	330

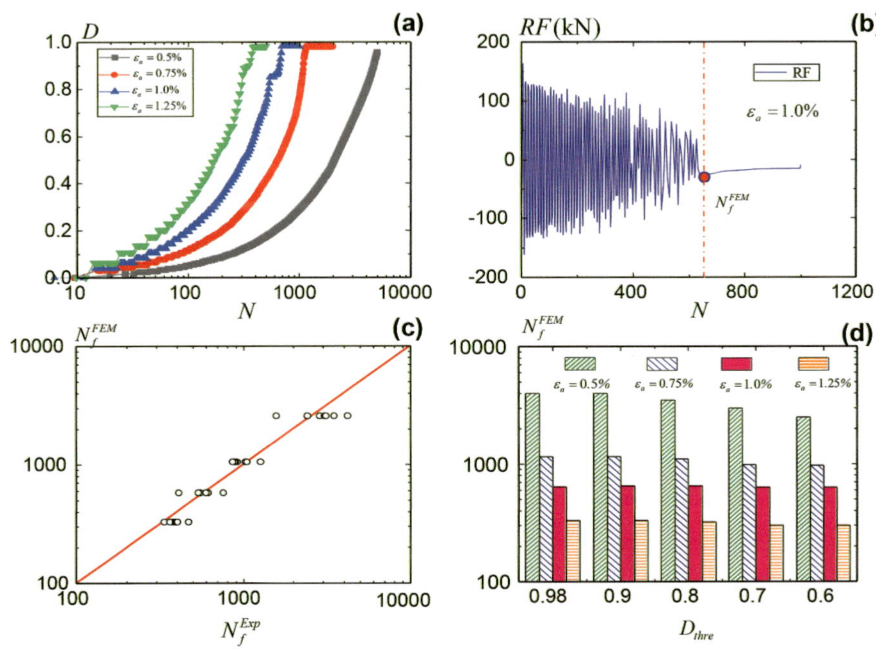

FIGURE 6.13 (a) Damage evolution in the smooth coupon specimen under different strain amplitudes; (b) the variation of the reaction force and definition of numerical fatigue life; (c) comparison between the numerical fatigue life and experimental results; (d) sensitivity of the numerical fatigue life under different threshold values of damage indicator.

to 1.0, the numerical solution of elastic-plastic low-cycle fatigue cannot be achieved due to significant stiffness change. Therefore, the subsequent numerical simulation of welded plate joints adopts the threshold of damage value of 0.98.

6.3.3 Uniform Assessment of Fatigue Life of Welded Plate Joints

The calibrated damage-related material parameters based on the smooth specimen lay the foundation to evaluate more complex structural joints, such as welded plate joints under different levels of fatigue loadings. Figures 6.14a and b present the geometry details and experimental setup of the welded plate joints tested in this study. The welded plate joint is under three-point bending, with the main plate simply supported at the ends, as shown in Figure 6.14. The supported span of the welded plate joint is 300 mm. The geometry details of the welded plate joints include two main plate thicknesses (20 mm and 40 mm) with the same width of 40 mm. The full penetration welds for all the welded plate joints entail a weld leg size of 10 mm. The total height of the specimen is 400 mm, with the thickness of the vertical attachment equal to 20 mm. The downward loading P creates tensile bending stress at the bottom of the welded plate joints. Therefore, fatigue-induced cracks tend to occur in these two weld toes (yellow block in Figure 6.14a). The experimental tests of the welded plate joints are separated into two stages: load-controlled cyclic loading and final fracture tests. The load-controlled cyclic loadings stop when the generated crack reaches 50% of the main plate thickness or the welded plate joints lose the stiffness to withstand the cyclic loads. Table 6.7 lists the fatigue loading information for all the welded plate joints.

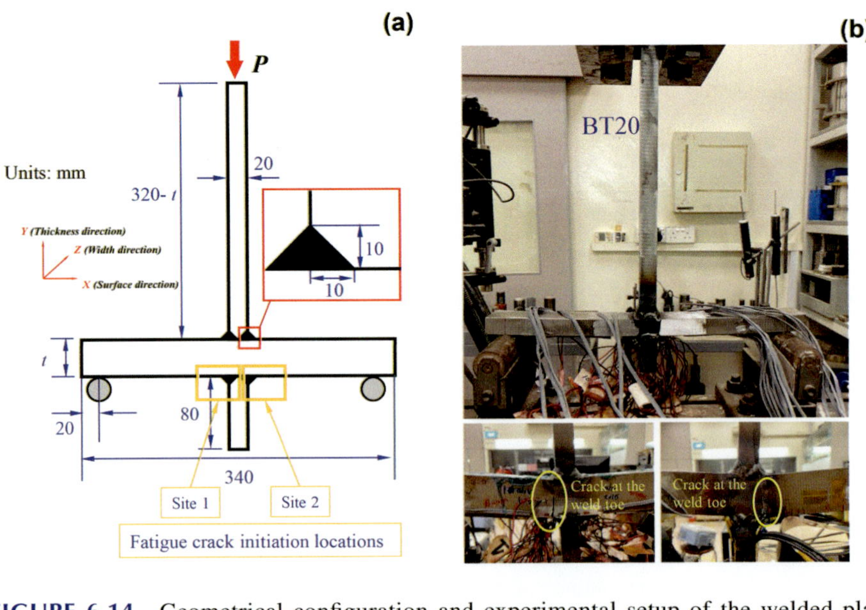

FIGURE 6.14 Geometrical configuration and experimental setup of the welded plate connections.

TABLE 6.7
Experimental and Numerical Simulation of Different Welded Plate Joints.

Specimens	Loading type	Load ratio	Main plate thickness *t* (mm)	Nominal loading	Test fatigue life N_f^{EXP}	FE fatigue life N_f^{FEM}
BT20	Three point bending [53]	0.1	20	270 MPa	268902	108000
				270 MPa	146997	108000
				225 MPa	416613	700000
				225 MPa	554572	700000
BT40		0.1	40	270 MPa	149093	170000
				270 MPa	145035	170000
				225 MPa	266875	303000
				225 MPa	244325	303000
BT60		0.1	40	270 MPa	108407	160000
				270 MPa	88629	160000
				225 MPa	252407	230000
				225 MPa	233940	230000
				300 MPa	99423	80000
BT20	Three point bending	0.1	20	548 MPa	15000	14200
				658 MPa	8618	6000
				658 MPa	5600	6000
BT40		0.1	40	250 MPa	117500	156000
				250 MPa	181000	156000
				250 MPa	108700	156000
				300 MPa	60111	60000
				400 MPa	29700	36000
				500 MPa	17776	23000
PT10	Axial loadings [46]	−1	10	0.2%	5504	6000
				0.2%	8610	6000
				0.25%	3442	4000
				0.25%	3457	4000
PT22		−1	22	0.2%	7769	9000
				0.2%	11199	9000
				0.2%	9613	9000
				0.3%	2745	4000
				0.3%	3665	4000
				0.3%	2681	4000
				0.45%	840	700
				0.45%	706	700
				0.45%	1434	700
PT40		−1	40	0.2%	9735	12000
				0.2%	9783	12000
				0.2%	10482	12000
				0.3%	2707	2400
				0.3%	3290	2400
				0.3%	2783	2400
				0.3%	2918	2400
				0.45%	780	500
				0.45%	805	500

This study constructs the full FE model of welded plate joints, using the continuous elements C3D8 in ABAQUS [52], with a total of 47292 nodes and 38190 elements, as shown in Figure 6.15a. The mesh near the weld toe has the smallest element size of 0.4 mm.

To validate the proposed damage methodology and assessment scheme, this study combines the experiments presented in this study and other fatigue tests of cruciform welded plate joints made of S550 steel from literatures [46, 53]. Figure 6.15 presents the FE model of BT specimens under three-point bending, and PT specimens under axial fatigue loadings. The mesh near the weld toe has the smallest element size of 0.4 mm. For the BT specimens under three-point bending loads, the damage material models are located only on one side of the joints, which replicates the crack propagation only on one side of the welded joint observed in the experiment. For the PT specimens under axial fatigue loadings, only half of the FE model is adopted to save computational cost. The detailed experimental information and numerical simulation results are listed in Table 6.7. Figure 6.16a compares the experimental test results and numerical simulation results for all the BT and PT specimens. The numerical simulation based on the proposed damage indicator and evaluation scheme can achieve a good agreement with the experimental fatigue life of welded plate joints under high- and low-cycle fatigue loadings, which verifies the accuracy and efficiency of the proposed methodology.

One of the popular damage-based methods to evaluate the low-cycle fatigue is the direct low-cycle analysis, proposed by Darveaux [54] and Lau *et al.* [55], which has been implemented in ABAQUS [52].

$$N_0 = c_1 \Delta w_0^{c_2} \tag{6.12}$$

$$\frac{dD}{dN} = \frac{c_3 \Delta w^{c_4}}{L} \tag{6.13}$$

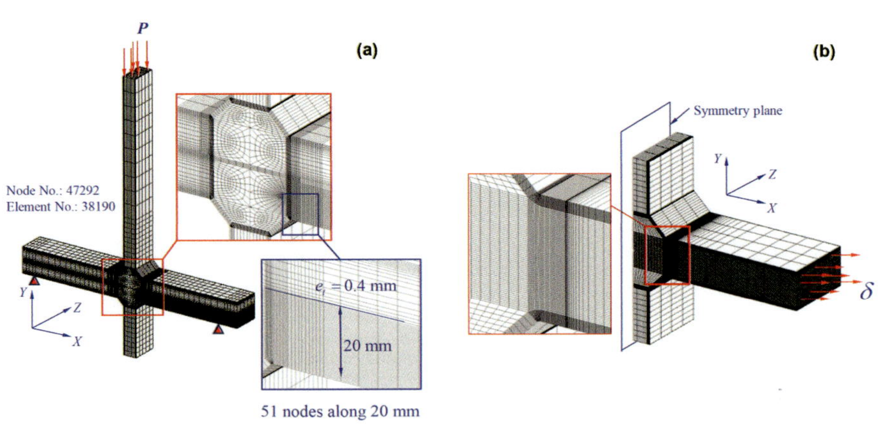

FIGURE 6.15 (a) The full FE model of welded plate joints under three-point bending; (b) the half FE model of PT specimen under axial loadings.

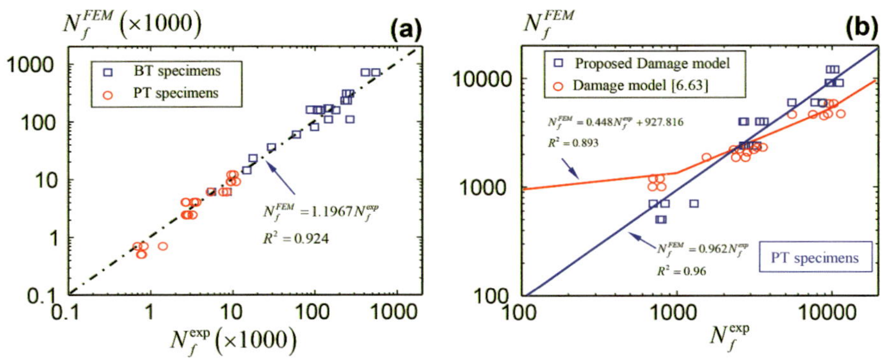

FIGURE 6.16 (a) Comparison between the numerical fatigue life and the experimental fatigue life of welded plate joints under high- and low-cycle fatigue actions; (b) comparison in low-cycle fatigue assessment of PT specimens based on the proposed damage approach and damage model in [52].

where N_0 represents the number of cycles to initiate the damage, Δw_0 is the hysteresis energy in the first few cycles, c_1, c_2, c_3, and c_4 refer to the material damage constants, and L is the characteristic length associated with the distance between adjacent integration points. The damage-related material parameters of S550 based on this model are the same as those of [46]. Therefore, this study compares the predicted fatigue life based on different damage methodologies with the low-cycle fatigue data of PT specimens. As shown in Figure 6.16b, the damage methodology proposed by this study can better predict the total fatigue life in comparison with the damage model adopted in [46]. Under smaller nominal strain amplitudes, the numerical simulation based on the ABAQUS damage model in Eqs. (6.12) and (6.13) overestimates the total low-cycle fatigue life under a smaller strain amplitude while underestimating the total low-cycle fatigue under a lager nominal strain amplitude. These deviations arise since the ABAQUS damage model [Eqs. (6.12) and (6.13)] only considers the plastic strain energy density and neglects the damage contributed by the elastic energy density. This study fits the low-cycle fatigue data for these two damage models (blue curve and red curve). The fitted blue curve based on the proposed damage methodology has a high coefficient of determination ($R^2 = 0.96$) in comparison with the fitted red curve based on the damage model from [52] ($R^2 = 0.89$). The slope of the fitted blue curve is much closer to 1.0, which indicates the good predictability based on the proposed damage approach.

6.4 WELDING RESIDUAL STRESS OF WELDED JOINTS

6.4.1 Experimental Examination of Welding Residual Stress

Twelve specimens of butt-welded plates made of S355 and two CHS X-shaped joints made of S550 are investigated to investigate the variation of welding residual stress throughout the entire process of fatigue loading. This study obtains the fatigue life of specimens under different maximum cyclic loads and load ratios, thereby laying the experimental data foundation for assessing the influence of welding residual stress on fatigue life. As shown in Figure 6.17, the specimen of the as-welded plate with a V-shaped groove has dimensions of 340 mm in length, 40 mm in width, and 20 mm in thickness after welding completion. The welded plates were produced directly through multipass welding by a highly skilled welder. In this case, the two end planes of the samples were carefully finished to ensure consistency in geometry. Figure 6.18a illustrates the geometric dimensions of the CHS X-joint, where the outer diameter of the chord is 168 mm with a wall thickness of 10 mm. The braces, each having a length of 616 mm, are connected to the chord via two circumferential weld seams. The outer diameter ratio of the brace to chord is 0.83, the wall thickness ratio of the brace to chord is 0.6, and the ratio of twice the chord length to its outer diameter is 16.7.

This study utilizes the Gas Metal Arc Welding (GMAW) process for the two types of specimens, that is, welded plates and CHS X-joints. For each weld bead, the arc was initiated and terminated directly on the specimens of welded plates and CHS X-joints. The starts and stops were managed carefully to minimize defects or discontinuities. For the welded plates, each weld seam runs in a single direction

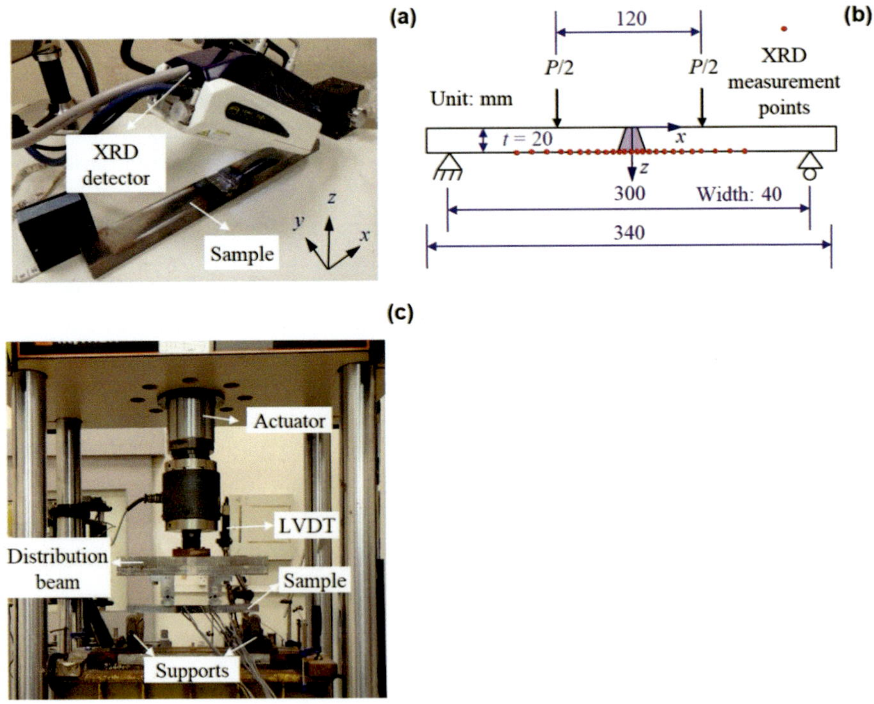

FIGURE 6.17 (a) Specimen of the welded plate and X-ray diffraction measurement; (b) schematic diagram of the welded plates (plate width of 40 mm); and (c) test setup of the welded plates under the cyclic four-point bending test.

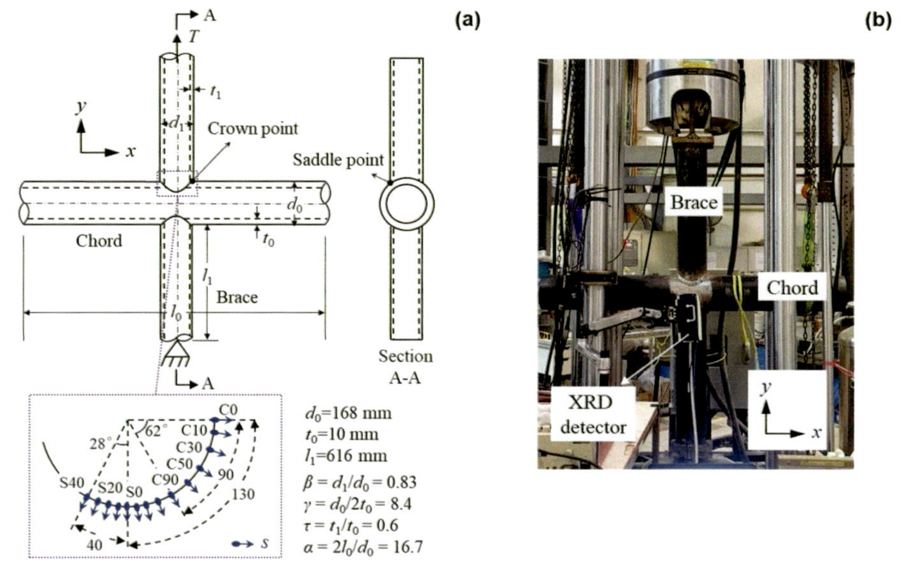

FIGURE 6.18 (a) Geometric details and XRD measurement points of the CHS X-joints; (b) test setup of the CHS X-joints under cyclic axial tension in the brace.

along the plate width, beginning at the same start point at one plate side surface and ending at the same end point at the other plate side surface. For the CHS X-joints, each weld seam starts at the saddle point, proceeds around the annular weld seam, and finishes at the same saddle point. Any inconsistencies or imperfections at the start and stop points on the side surfaces of the welded plates were addressed by additional grinding and polishing to achieve a smooth and uniform weld surface. Table 6.8 presents the actual recorded welding parameters including voltage, current, and welding speed for each welding pass for use in thermal-mechanical simulations. Figure 6.19 illustrates the base metal zone (BMZ), the heat-affected zone (HAZ), the weld zone (WZ), and the profiles of individual welding passes on the cross-sectional welds after the macro-etching treatment. The cross-sectional weld profile includes the slices of welds cut from the central location of the welded plates and the circumferential welds on the brace-chord interface of the CHS joints. The welded plate employs seven welding passes in the V-groove, and the CHS X-joint utilizes a four-pass configuration for the fillet weld.

Table 6.9 presents the loading schemes of the four-point bending fatigue tests conducted on the welded plate specimens in Figure 6.17. The weld toe of the specimens is located at the bending-tensile surface to investigate the initiation and propagation of fatigue cracks. The fatigue loads applied by the Instron actuator are evenly distributed to two loading points on the specimen using a loading beam, with a spacing of 120 mm between the two loading points and a 300 mm spacing between the end supports. Various loading conditions with different maximum cyclic loads (P_{max}) ranging from 20 kN to 45 kN under the load ratio of 0.1 or 0.5. The measurement

TABLE 6.8

Welding Parameters of the Welded Plates and the Circular Hollow Section (CHS) X-Joints.

Welded connections	Welding pass	Welding speed (m/s)	Welding voltage (V)	Welding current (A)
Welded plates	1	0.0021~0.0024	20	140
	2	0.0012~0.0014	20	125
	3	0.0014~0.0018	20	125
	4	0.0016~0.0017	20	130
	5	0.0031~0.0033	20	135
	6	0.0029~0.0031	20	125
	7	0.0021~0.0022	20	135
CHS X-joint	1	0.00175~0.0033	20~25	120~130
	2	0.0025~0.0041	20~25	120~130
	3	0.0025~0.0041	20~25	120~130
	4	0.0025~0.0041	20~25	120~130

points for welding residual stresses are located near the weld toe and at the central plate to avoid the influence of the start and stop of welding arcs. Adjacent measurement points are spaced 5 mm to 20 mm apart. The XRD measurements record the welding residual stresses before and during the fatigue loading process, after the following number of cycles $N =$ 1, 5, 10, 50, 100, 200, 1000, 10000, and so on.

Based on the experimental set-up of the CHS X-joint in Figure 6.18, the MTS electro-hydraulic servo testing machine applies a cyclic tensile loading to the top of the upper brace member in the CHS X-joint, with a maximum force of 200 kN with a load ratio of 0.1. The bottom end of the lower brace remains fixed. The XRD measurement follows the paths near the crown point depicted in Figure 6.18a, denoted as C0, C10, C30, C50, C70, and C90, at distances ranging from 10 mm to 90 mm from the crown, with C0 at the crown point. The XRD measurement paths near the saddle point, labeled as S0, S10, S20, S30, and S40, lie within 40 mm from the saddle point. Along each measurement path perpendicular to the weld line, the spacing between adjacent measurement points ranges from 5 mm to 10 mm, with a spacing of 5 mm in the weld region. The measurement direction of welding residual stresses is along the s-direction [as indicated in Figure 6.18a for each measurement path], which aligns along the chord surface. The distance along the measurement points from the weld refers to the distance recorded on the surface of the chord. The XRD measurements perform at the end of the cycles (N=0,1,100,1000,20000,30000, and 31000). To obtain statistically averaged residual stress measurements, this study takes at least three measurements at each measurement point. The following experimental results present the statistical averages and errors of welding residual stresses.

TABLE 6.9
Test Matrix, Experimentally Measured and Numerically Predicted Fatigue Lives of Welded Plates.

Specimen label	Length (mm)	Width (mm)	Thickness (mm)	Load ratio	Maximum load (kN)	Experimental fatigue life	Predicted fatigue life
WP-10	340	40	20	0.1	45	4687	3210
WP-17	340	40	20	0.1	45	3411	3210
WP-11	340	40	20	0.5	45	55910	48400
WP-18	340	40	20	0.5	45	56172	48400
WP-1	340	40	20	0.1	40	13098	8700
WP-15	340	40	20	0.1	40	9868	8700
WP-6	340	40	20	0.5	40	83855	81500
WP-16	340	40	20	0.5	40	132115	81500
WP-2	340	40	20	0.1	30	25424	26200
WP-14	340	40	20	0.1	30	32310	26200
WP-3	340	40	20	0.1	20	152352	120300
WP-19	340	40	20	0.1	20	168540	120300

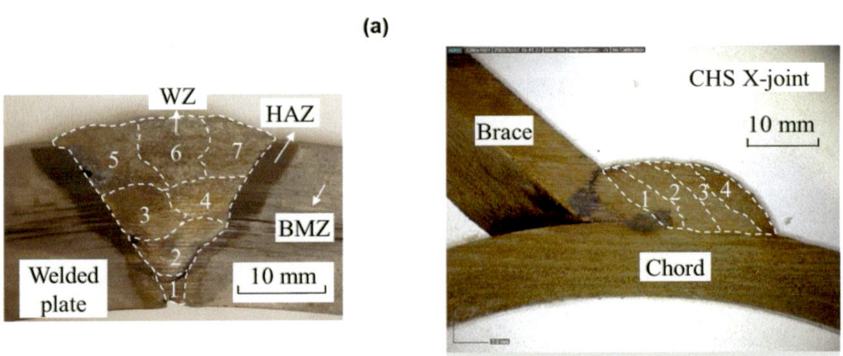

FIGURE 6.19 The macro-etched sections with the illustrations of welding passes in (a) the welded plate; and (b) the CHS X-joint.

Prior efforts [55, 56] described the distribution of welding residual stress on the welded plate surface in detail, highlighting that the peak residual stress occurs near the fusion boundary. This section focuses on the evolution of welding residual stresses throughout the entire fatigue life of the specimens. The following discussion focuses on the welded plate specimens used in the study. Figure 6.20a presents the evolution curves of peak residual stresses of the welded plate joints (Figure 6.17), measured by XRD for cyclic loads with the maximum loads of 40 kN and 30 kN. Before the

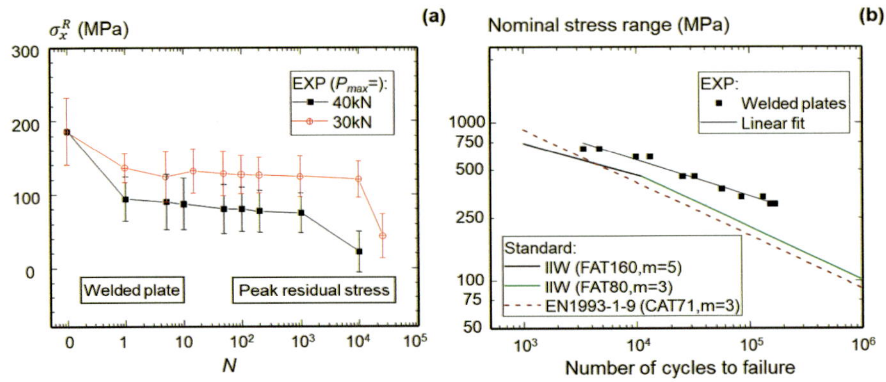

FIGURE 6.20 (a) XRD measured peak residual stresses during the fatigue life of the welded plates; and (b) comparison of the experimental fatigue life with the *S-N* curves in engineering standards.

fatigue crack appears, the most significant residual stress relaxation occurs after the first loading cycle because of the large plastic deformations, after which the residual stress remains relatively stable over subsequent cycles until the initiation of the fatigue crack. Thereafter, a significant decrease in residual stress magnitude due to the fatigue crack initiation on the specimen surface.

Figure 6.21a demonstrates the fatigue cracks at the end of the fatigue test in different specimens. Fatigue cracks emerge at the fusion boundary within the welds and occur at approximately 9900 cycles under a maximum load of 40 kN for specimen WP-15 and at approximately 30000 cycles under a maximum load of 30 kN for specimen WP-14. Those fatigue cracks propagate along the fusion boundary within the weld without penetrating into the base metal.

Figure 6.20b compares the experimental nominal stress range (S) and fatigue life (N_f) relationships of the welded plates with the IIW standard [57] and EN1993-1-9 standard [58]. The experimental nominal stress range is derived by calculating the nominal bending stress experienced by the center span of the plate using elastic theory, based on the four-point bending test illustrated in Figure 6.17. The IIW standard defines the fatigue class (FAT) as the fatigue strength at 2 million cycles. The FAT value of the unwelded component is 160 MPa and the butt-welded plate is 80 MPa. The slopes in the *S-N* curve are 5 and 3 for the unwelded plates and the butt-welded plates, respectively. Referring to the EN1993-1-9 standard, the butt-welded plates in this study belong to category 71 with a slope of 3. From the comparisons, all the experimental data points are above the *S-N* curves in the standards, which implies that the two standards provide conservative assessments. Equation (6.14) describes the linearly fitted *S-N* curve based on the experimental fatigue life of the 20 mm thick welded plates under the cyclic four-point bending tests.

$$\log N_f = 16.114 - 4.386 \log S \tag{6.14}$$

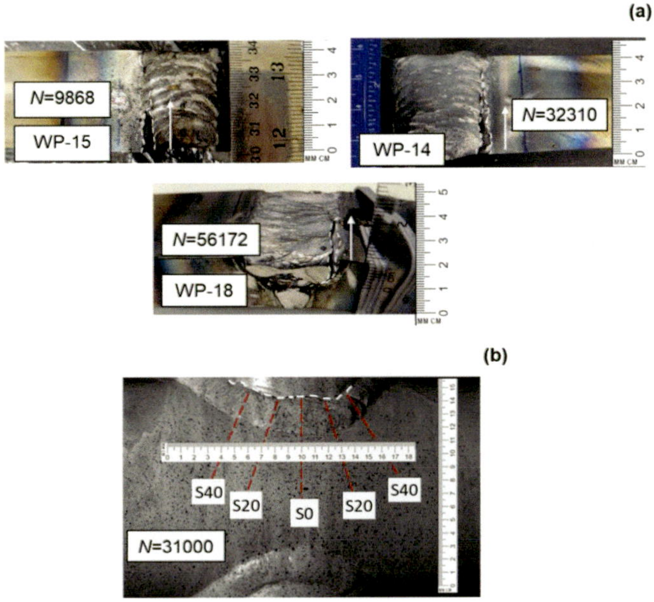

FIGURE 6.21 (a) Fatigue cracks at the number of cycles to failure in different welded plates; and (b) fatigue cracks at the number of cycles to failure in the CHS X-joint.

For the welded CHS X-joints, Figure 6.22a presents the measured profiles of welding residual stresses along the measurement paths in the CHS X-joints prior to the fatigue tests. The magnitudes of welding residual stresses are the highest near the fusion boundary in the weld seam area and decrease with increasing distance away from the weld. Welding residual stresses measured near the saddle point are higher than those measured near the crown, due to the larger weld volume and closer spacing of the two welds joining the brace and chord than the crown position (see Figure 6.21b), resulting in a higher input of welding heat. Figure 6.22b compares the welding residual stresses from the weld to the chord at different measurement locations in the CHS X-joint before and after the application of the first cycle of cyclic loading. After the first loading cycle, there is a noticeable decrease in residual stresses at the weld, while the decrease in residual stresses in the chord away from the weld is not as significant.

Figure 6.22c illustrates the evolution of peak residual stresses throughout the entire fatigue test of the CHS X-joint. The saddle position has a significant decrease in the peak residual stress after the first cycle, followed by a slower decrease in the subsequent cycles. Around 31000 loading cycles, there is a notable relaxation in the peak residual stress because of the initiation and propagation of the fatigue crack at the saddle point. Figure 6.21b presents the morphology of the fatigue crack at N = 31000, with the crack appearing at the fusion boundary between the weld and the brace.

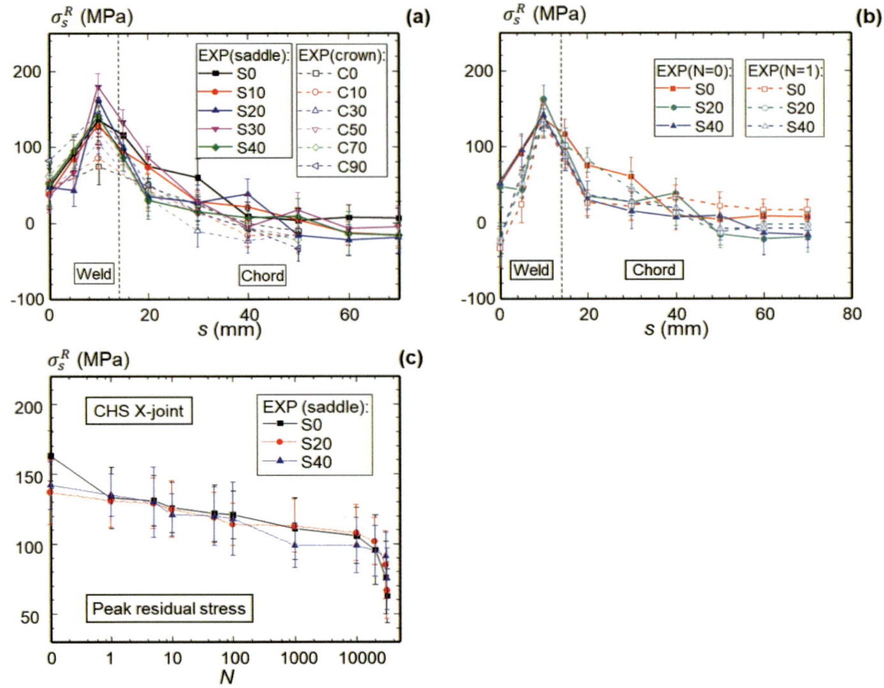

FIGURE 6.22 XRD measured residual stresses in the CHS X-joints: (a) as-welded residual stresses near both the saddle and crown; (b) residual stress profiles near the saddle before and after $N = 1$; and (c) peak residual stresses from the beginning to the end of the fatigue test.

6.4.2 Damage-Incorporated Thermal-Mechanical Simulation

Figure 6.23 demonstrates the numerical framework of the damage-incorporated thermal-mechanical simulation method. To replicate the welding heat inputs of the Gas Metal Arc Welding (GMAW), this study utilizes Goldak's double-ellipsoidal heat source model [59], which assumes that the heat distributions follow Eqs. (6.15) and (6.16) at the front end and the rear end of the heat source center along the welding pass, respectively. Prior work [60] utilized data-driven machine learning and optimization algorithms of Gaussian Process Regression (GPR) and Genetic algorithm (GA) to obtain the optimized shape parameters. The optimization algorithm leads to the following shape parameters to describe the heat source model for the GMAW process used in the current study. The simulation adopts the user subroutine DFLUX in ABAQUS [61] to define the heat source model and quantify the heat distributions during the welding process.

$$q(x,y,z) = \frac{6\sqrt{3}}{\pi^{3/2}} \cdot \frac{\eta UI}{a_1 bc} \cdot e^{-3(x-x_0)^2/a_1^2} e^{-3(x-x_0)^2/a_2^2} e^{-3(y-y_0)^2/b^2} e^{-3(z-z_0)^2/c^2} \quad (6.15)$$

$$q(x,y,z) = \frac{6\sqrt{3}}{\pi^{3/2}} \cdot \frac{\eta UI}{a_2 bc} \cdot e^{-3(x-x_0)^2/a_1^2} e^{-3(x-x_0)^2/a_2^2} e^{-3(y-y_0)^2/b^2} e^{-3(z-z_0)^2/c^2} \quad (6.16)$$

where η, U, I refer to the welding parameters (efficiency, voltage, and current); a_1, a_2, b, c demonstrate the heat model shape parameters. The coordinates (x_0, y_0, z_0) represent the position of the welding heat source center at the time t.

The thermal-mechanical simulation method [56] involves transient heat transfer analysis [Eq. (6.17)] and nonlinear mechanical analysis to predict the welding residual stresses. This approach involves thermal analysis to obtain the welding temperature field and mechanical analysis to derive the as-welded residual stress field by using the temperature field as an input. Based on the same meshed FE model, the sequential coupling of thermal and mechanical analysis utilizes the DC3D8 and C3D8 element types from the ABAQUS element library [61] for thermal and mechanical analysis, respectively. The non-linear transient thermal equilibrium equation [Eq. (6.17)] describes the time-dependent heat conduction process where thermal conductivity and heat sources vary with temperature and position. This equation captures the complex interactions and non-linear characteristics of heat transfer.

$$\rho c_p \frac{\partial T}{\partial t} - \dot{Q}_v = \frac{\partial}{\partial x}(k\frac{\partial T}{\partial x}) + \frac{\partial}{\partial y}(k\frac{\partial T}{\partial y}) + \frac{\partial}{\partial z}(k\frac{\partial T}{\partial z}) \quad (6.17)$$

where ρ demonstrates the material density; c_p refers to the thermal capacity; T denotes the welding temperature; t is the analyzing time; $\dot{Q}_v$ represents the welding heat input; k defines the heat conductivity.

As shown in Figure 6.23, the fatigue progressive damage parameter (D) will be introduced after determining the as-welded residual stress field. The evolution of the damage follows the uniform approach in previous Section 6.2. The user subroutine

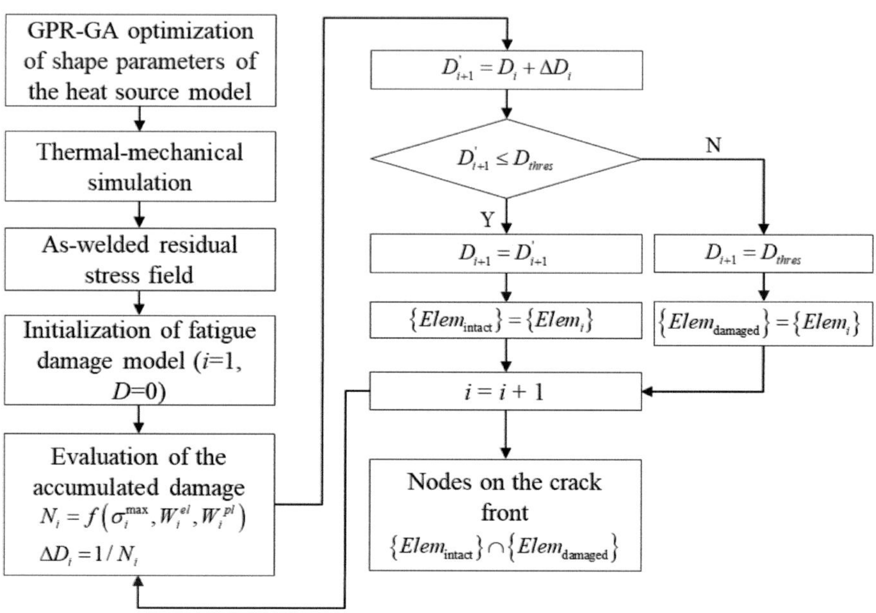

FIGURE 6.23 Numerical framework of damage incorporated thermal-mechanical analysis.

UMAT in ABAQUS defines the new fatigue damage model based on the uniform energy fatigue indicators.

The replication of welding residual stresses is inherently linked to the material property variations due to temperature changes at different stages of the welding process. Based on literature results [46, 47] and engineering standard EN 1993-1-2 [58], Figure 6.24 illustrates the temperature-dependent thermal and mechanical properties of S355 and S550 steels for the damage-incorporated thermal-mechanical simulations. Table 6.10 lists the isotropic hardening and kinematic hardening parameters for S355 steel and S550 steel. The previous paper [47] calibrates the hardening parameters for S355 steel based on the one kinematic hardening law using fatigue coupon specimens and neglects the effect of isotropic hardening. Additionally, the previous study [46] calibrates the parameters for S550 steel based on the two kinematic hardening laws using S550 coupon specimens.

Figure 6.25a presents the FE model with damage-incorporated elements of the welded plate. Based on the macro-etched weld section in Figure 6.19a, the V-groove weld involves seven welding passes in the plate thickness direction. This FE model comprises 12,960 elements and 16,023 nodes, with a minimum element size of 2 mm near the weld in the x-y plane and a uniform element size of 1 mm in the thickness (z) direction. Since the fatigue crack initiates at the weld toe within the weld material and propagates along the plate thickness direction, the fatigue damage-incorporated elements extend from the weld toe to half of the plate thickness to mitigate the computational costs. Since the observed fatigue cracks in the experiments, shown in Figure 6.21a, propagate along

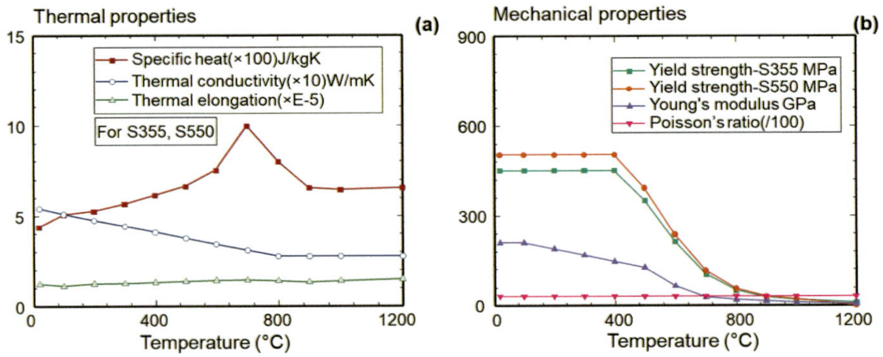

FIGURE 6.24 (a) Temperature-dependent thermal properties; and (b) temperature-dependent mechanical properties of S355 and S550 steels.

TABLE 6.10
The Isotopic and Kinematic Hardening Parameters of S355 and S550 Steels.

Steel	Q_∞ (MPa)	h	C_1 (MPa)	λ_1	C_2 (MPa)	λ_2
S355	-	-	24684	221.17	-	-
S550	−67.44	1.048	16907	64.008	21607	461.79

the fusion boundary within the welds, without penetrating into the base metal, the FE simulation assigns the damage-incorporated elements to the weld materials only.

Figure 6.25b showcases the FE model of the CHS X-joint and the definition of fatigue damage elements. Based on the macro-etched section results depicted in Figure 6.19b, the FE model incorporates four independent passes for the fillet welds. The FE model comprises 162,654 nodes and 137,120 solid elements, with a 2 mm element size in the thickness direction and the smallest element size of 1.6 mm near the weld. The element size gradually increases with increasing distance from the weld. This study defines the damage-incorporated elements at the saddle point, where the fatigue crack occurs in the experiment. The damage-incorporated elements extend from the weld root to the weld toe across the full thickness. The damage-incorporated elements apply to the welded plate and CHS X-joint under fatigue actions, disregarding the damage from the welding process.

Figure 6.26a compares the experimentally measured residual stresses and the numerical results obtained using the damage-incorporated thermal-mechanical simulation method in the welded plate under fatigue actions with a maximum load of 40 kN. The numerical results at the beginning of the experiment ($N = 0$), after completing the first cycle ($N = 1$), and the fifth cycle ($N = 5$) show good agreement with the experimental measurements. The difference in welding residual stresses within the weld region is not significant after the first and fifth cycles.

Figure 6.26b displays the peak residual stresses obtained through damage-incorporated thermal-mechanical simulation at different stages of fatigue loading and under different maximum loads (40 kN and 30 kN). The predicted residual stresses from the proposed damaged-incorporated simulation method exhibit close agreement with the XRD measurements. The predicted residual stresses effectively capture the relaxation of residual stress before fatigue crack initiation and predict the rapid decrease in residual stresses towards the end of loading, indicating the onset of fatigue crack formation.

Figure 6.27a compares the simulated as-welded residual stresses with the XRD measurement for CHS X-joints prior to the fatigue loading. The predicted residual stress within the weld from the thermal-mechanical simulation closely matches the experimental

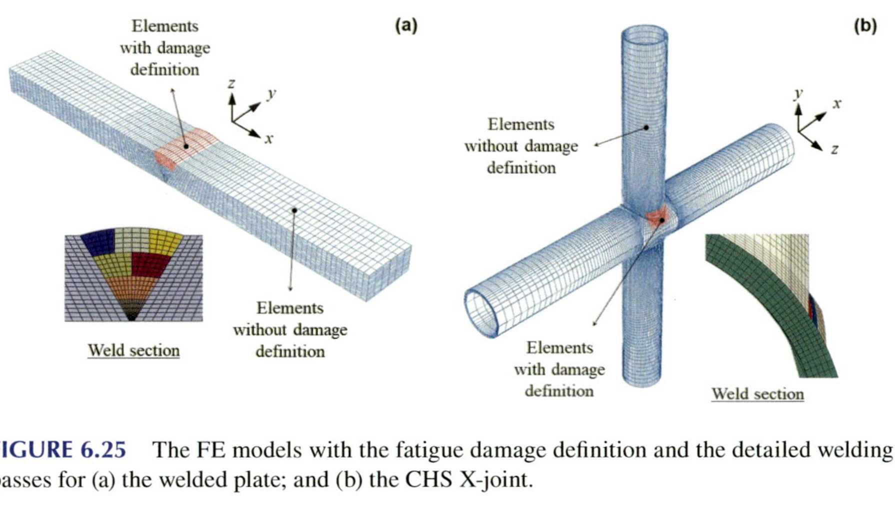

FIGURE 6.25 The FE models with the fatigue damage definition and the detailed welding passes for (a) the welded plate; and (b) the CHS X-joint.

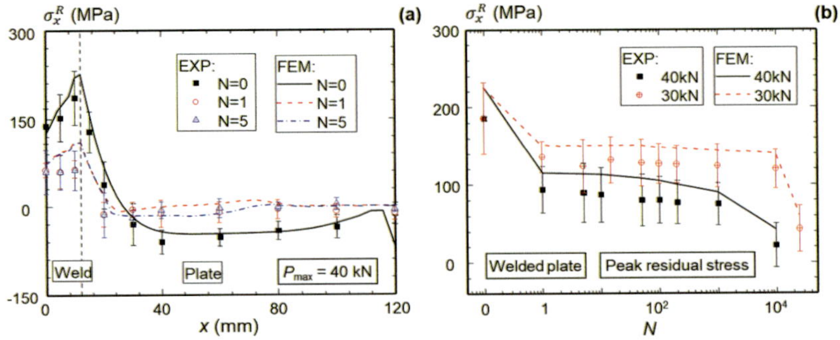

FIGURE 6.26 (a) Comparison of the measured and simulated welding residual stresses in welded plates before the test and after the first, and fifth loading cycles; and (b) comparison of peak residual stresses in welded plates from the experimental measurements and damage-incorporated simulations.

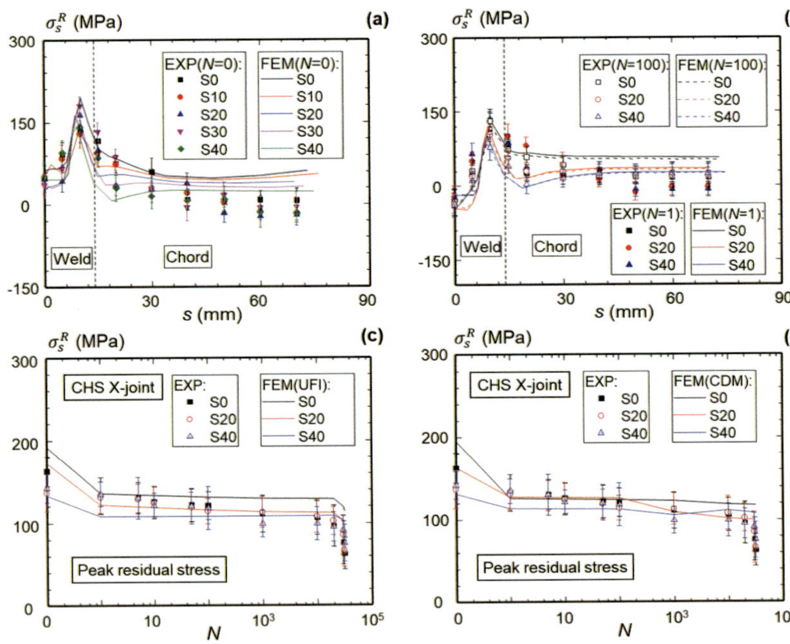

FIGURE 6.27 Comparisons of (a) the as-welded residual stresses in the CHS X-joints; (b) residual stress profiles after the 1st and 100th loading cycle; (c) peak residual stress evolutions by using the proposed uniform fatigue indicator; and (d) peak residual stress evolutions by using CDM model in ABAQUS.

measurements. However, at locations away from the weld on the chord side, the predicted residual stresses are higher than the experimentally measured values. This discrepancy arises because the simulation considers the unfavorable welding conditions, that is, simultaneous welding of two brace-chord welds, resulting in higher residual stresses compared to measurements obtained from the actual sequential welding conditions.

Figure 6.27b presents the profiles of residual stresses measured by XRD and simulated after the 1st cycle and the 100th loading cycle. Compared to the as-welded residual stresses, both the experimentally measured and simulated residual stresses exhibit a decrease after the first cycle and the 100th cycle. The decrease of tensile residual stresses is small from the 1st cycle to the 100th cycle because of the limited accumulation of plastic deformations during high-cycle fatigue. Figure 6.28c illustrates the evolutions of peak residual stresses throughout the entire fatigue life obtained using the damage-incorporated simulation based on the uniform fatigue indicator (UFI). The simulation results not only capture the magnitude of residual stresses before fatigue crack initiation but also predict the amount of relaxed residual stresses at $N = 31000$ when fatigue cracks occurs.

6.4.3 Effect of Welding Residual Stresses on Fatigue Life Estimations

Figure 6.28a illustrates the high-cycle and low-cycle fatigue life assessment of welded plates using the damage-incorporated thermal-mechanical analysis proposed in this chapter. For the analysis with the welding residual stress field, the predicted fatigue life of welded plates under different maximum loads closely approximates and slightly underestimates the experimentally measured fatigue life from both the high-cycle fatigue (HCF) and the low-cycle fatigue (LCF) tests. For the analysis without the welding residual stress field, the predicted fatigue life in LCF scenarios is similar to the experimentally measured value, while in HCF scenarios, the predicted fatigue life is conservatively higher than the actual fatigue life, with some estimations exhibiting a scatter factor of approximately two, which defines the range of errors between 50% to 200% of the experimentally measured fatigue life. Figure 6.28b compares the relationship between the nominal stress range and the number of cycles to failure from the experimental measurements and the damage-incorporated thermal-mechanical analysis. Compared to the experimental fatigue lives in LCF and HCF tests, the predicted fatigue lives with the welding residual stress field are closest to the experimental results. On the other hand, the predicted fatigue lives without considering welding residual stress fields are acceptable for scenarios with larger nominal stress ranges, that is, in low-cycle fatigue prediction. However, the prediction errors of fatigue life without

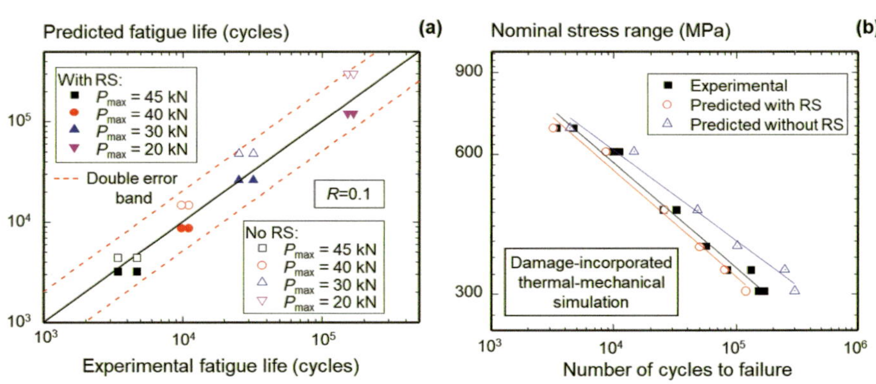

FIGURE 6.28 Comparisons of (a) fatigue lives under different maximum cyclic loads between the predictions and the experimental measurements; and (b) *S-N* curves predicted with and without welding residual stress field by the damage-incorporated thermal-mechanical simulation method.

considering welding residual stress fields are significant for HCF scenarios with small nominal stress ranges, indicating a non-conservative assessment. The LCF causes significant plastic deformations in the welded plates with a noticeable relaxation of welding residual stresses. In contrast, the smaller cyclic load in HCF leads to a less pronounced relaxation of welding residual stresses than that of LCF, maintaining a relatively high level of stabilized welding residual stresses after several loading cycles [56, 62].

Therefore, the damage-incorporated thermal-mechanical simulation considering welding residual stresses is suitable for fatigue life assessment in both low-cycle and high-cycle fatigue scenarios. The existence of welding residual stresses does not significantly affect the fatigue life prediction in low-cycle actions but results in substantial deviations in high-cycle fatigue prediction.

REFERENCES

[1] J.-B. Yan, J.R. Liew, M.-H. Zhang, J.-Y. Wang, Mechanical properties of normal strength mild steel and high strength steel S690 in low temperature relevant to Arctic environment, Materials & Design, 61 (2014) 150–159.

[2] Y.-B. Wang, G.-Q. Li, S.-W. Chen, F.-F. Sun, Experimental and numerical study on the behavior of axially compressed high strength steel box-columns, Engineering Structures, 58 (2014) 79–91.

[3] J. Raujol-Veillé, D. Thevenet, C. Doudard, S. Calloch, H. Minnebo, Rapid method for low cycle fatigue properties: Thickness effect on the fatigue crack initiation life of welded joints, Fatigue & Fracture of Engineering Materials & Structures, 38 (2015) 1492–1506.

[4] L. Feng, X. Qian, A hot-spot energy indicator for welded plate connections under cyclic axial loading and bending, Engineering Structures, 147 (2017) 598–612.

[5] C. LF., A study of the effect of cyclic thermal stresses on a ductile metal, Trans ASME, 76 (1954) 1–50.

[6] E. Macha, C. Sonsino, Energy criteria of multiaxial fatigue failure, Fatigue and Fracture of Engineering Materials and Structures, 22 (1999) 1053–1070.

[7] S. Manson, Fatigue: A complex subject – Some simple approximations, Experimental Mechanics, 5 (1965) 193–226.

[8] B. OH., The exponential law of endurance tests, American Society for Testing and Materials Proceedings, 10 (1910) 625–630.

[9] N. Pugno, M. Ciavarella, P. Cornetti, A. Carpinteri, A generalized Paris' law for fatigue crack growth, Journal of the Mechanics and Physics of Solids, 54 (2006) 1333–1349.

[10] T.L. Anderson, Fracture mechanics: Fundamentals and applications, CRC press, 2017.

[11] J. Lemaitre, R. Desmorat, Engineering damage mechanics: Ductile, creep, fatigue and brittle failures, Springer Science & Business Media, 2006.

[12] H. Remes, P. Varsta, J. Romanoff, Continuum approach to fatigue crack initiation and propagation in welded steel joints, International journal of Fatigue, 40 (2012) 16–26.

[13] D. Socie, G.B. Marquis, Multiaxial fatigue, Society of Automotive Engineers Warrendale, PA, 2000.

[14] J. Lemaitre, A. Plumtree, Application of damage concepts to predict creep-fatigue failures, Journal of Engineering Material Technology, 101 (1979) 284–292.

[15] L. Feng, Rapid SN type life estimation for low cycle fatigue of high-strength steels at a low ambient temperature, Steel and Composite Structures, an International Journal, 33 (2019) 777–792.

[16] L. Feng, X. Qian, R. Jiang, A uniform volume-based fatigue indicator for high-and low-cycle assessment of notched components, International Journal of Fatigue, 144 (2020) 106048.

[17] J. Hensel, T. Nitschke-Pagel, D.T. Ngoula, H.-T. Beier, D. Tchuindjang, U. Zerbst, Welding residual stresses as needed for the prediction of fatigue crack propagation and fatigue strength, Engineering Fracture Mechanics, 198 (2018) 123–141.
[18] J. Mei, A. Vasu, S. Jacob, S. Raghavan, J. Chung, Y. Mehta, P. Dong, A method for treating residual stress effects in traction structural stress based approach and validation on a welded component, International Journal of Fatigue, 167 (2023) 107347.
[19] G.S. Schajer, C.O. Ruud, Overview of residual stresses and their measurement, Practical Residual Stress Measurement Methods (2013) 1–27.
[20] C. Acevedo, A. Nussbaumer, Effect of tensile residual stresses on fatigue crack growth and S–N curves in tubular joints loaded in compression, International Journal of Fatigue, 36 (2012) 171–180.
[21] V.N. Van Do, C.-H. Lee, K.-H. Chang, High cycle fatigue analysis in presence of residual stresses by using a continuum damage mechanics model, International Journal of Fatigue, 70 (2015) 51–62.
[22] A. Standard, E606/E606M-12, Standard test method for strain-controlled fatigue testing, ASTM International, West Conshohocken, PA, 2012.
[23] C.O. Frederick, P. Armstrong, A mathematical representation of the multiaxial Bauschinger effect, Materials at High Temperatures (2014).
[24] J.-L. Chaboche, Time-independent constitutive theories for cyclic plasticity, International Journal of Plasticity, 2 (1986) 149–188.
[25] J.-L. Chaboche, Constitutive equations for cyclic plasticity and cyclic viscoplasticity, International Journal of Plasticity, 5 (1989) 247–302.
[26] A.U.s. Manual, Version 6.12, Karlsson & Sorensen, Hibbitt Inc, 2012.
[27] J. Bauschinger, Über die Veränderung der Elastizitätsgrenze und der Festigkeit des Eisens und Stahls durch Strecken und Quetschen, durch Erwärmen und Abkühlen und durch oftmals wiederholte Beanspruchung, Mitt. Mech.-Techn. Lab. K. Techn. Hochsch. München, 13 (1886) 108–112.
[28] M. Pradana, X. Qian, S. Swaddiwudhipong, n extrapolation method to determine the effective notch stress in welded joints, Fatigue & Fracture of Engineering Materials & Structures, 38 (2015) 1118–1135.
[29] A. Hobbacher, IIW recommendations for fatigue design of welded joints and components WRC, Welding Research Council Bulletin. WRC, 2009, p. 520.
[30] P. Dong, A structural stress definition and numerical implementation for fatigue analysis of welded joints, International Journal of Fatigue, 23 (2001) 865–876.
[31] E. Haibach, Die Schwingfestigkeit von Schweißverbindungen aus der Sicht einer örtlichen Beanspruchungsmessung, Fraunhofer-Ges. z. Förderung d. Angewandten Forschung eV, 1968.
[32] E. Niemi, Stress determination for fatigue analysis of welded components, Woodhead Publishing, 1995.
[33] Z.-G. Xiao, K. Yamada, A method of determining geometric stress for fatigue strength evaluation of steel welded joints, International Journal of Fatigue, 26 (2004) 1277–1293.
[34] P. Lazzarin, T. Lassen, P. Livieri, A notch stress intensity approach applied to fatigue life predictions of welded joints with different local toe geometry, Fatigue & Fracture of Engineering Materials & Structures, 26 (2003) 49–58.
[35] H. Neuber, Über die Berücksichtigung der Spannungskonzentration bei Festigkeits berechnungen, Konstruktion, 20 (1968) 245–251.
[36] L. Feng, X. Qian, A modified Neuber's rule for welded cruciform joints under low-cycle actions, The Journal of Strain Analysis for Engineering Design, 52 (2017) 488–505.
[37] D.W. Hoeppner, Fatigue testing of weldments, ASTM International, 1978.
[38] N. Nemati, B. Metrovich, A. Nanni, Acoustic emission assessment of fatigue crack growth from a transverse weld toe, Journal of Materials in Civil Engineering, 28 (2015) 04015103.
[39] J. Otegui, U. Mohaupt, D. Burns, Effect of weld process on early growth of fatigue cracks in steel T joints, International Journal of Fatigue, 13 (1991) 45–58.

[40] J. Schijve, Fatigue of structures and materials, Springer, 2001.
[41] X. Liao, Y. Wang, Z. Wang, L. Feng, Y. Shi, Effect of low temperatures on constant amplitude fatigue properties of Q345qD steel butt-welded joints, Engineering Failure Analysis, 105 (2019) 597–609.
[42] J. Lemaitre, A course on damage mechanics, Springer Science & Business Media, 2012.
[43] P. Lazzarin, R. Zambardi, The equivalent strain energy density approach re-formulated and applied to sharp V-shaped notches under localized and generalized plasticity, Fatigue & Fracture of Engineering Materials & Structures, 25 (2002) 917–928.
[44] C. Lim, W. Choi, E.A. Sumner, Parametric study using finite element simulation for low cycle fatigue behavior of end plate moment connection, Steel and Composite Structures, 14 (2013) 57–71.
[45] A.M. de Jesus, R. Matos, B.F. Fontoura, C. Rebelo, L.S. da Silva, M. Veljkovic, A comparison of the fatigue behavior between S355 and S690 steel grades, Journal of Constructional Steel Research, 79 (2012) 140–150.
[46] L. Feng, X. Qian, Size effect and life estimation for welded plate joints under low cycle actions at room and low ambient temperatures, Thin-Walled Structures, 132 (2018) 195–207.
[47] L. Feng, X. Qian, R. Jiang, A uniform volume-based fatigue indicator for high-and low-cycle assessment of notched components, International Journal of Fatigue, 144 (2021) 106048.
[48] K. Wallin, Statistical modeling of fracture in the ductile to brittle transition region, Defect Assessment in Components-Fundamentals and Applications (1991) 415–445.
[49] A. Abdul-Latif, D. Razafindramary, J. Rakotoarisoa, New hybrid cycle jump approach for predicting low-cycle fatigue behavior by a micromechanical model with the damage induced anisotropy concept, International Journal of Mechanical Sciences, 160 (2019) 397–411.
[50] R. Moslemian, A.M. Karlsson, C. Berggreen, Application of a cycle jump technique for acceleration of fatigue crack growth simulation, in: 9th international conference on sandwich structures, 2010.
[51] L. Feng, X. Qian, Low cycle fatigue test and enhanced lifetime estimation of high-strength steel S550 under different strain ratios, Marine Structures, 61 (2018) 343–360.
[52] G. Abaqus, Abaqus 6.11, Dassault systemes simulia corporation, Providence, RI, USA, 2011.
[53] C. Zhiping, Fatigue initiation life assessment for offshore structural details, PhD Thesis, National University of Singapore, 2015.
[54] R. Darveaux, Effect of simulation methodology on solder joint crack growth correlation and fatigue life prediction, The Journal of Electronic Packaging, 124 (2002) 147–154.
[55] J.H. Lau, S.H. Pan, C. Chang, A new thermal-fatigue life prediction model for wafer level chip scale package (WLCSP) solder joints, The Journal of Electronic Packaging, 124 (2002) 212–220.
[56] L. Wang, X. Qian, Welding residual stresses and their relaxation under cyclic loading in welded S550 steel plates, International Journal of Fatigue, 162 (2022) 106992.
[57] A. Hobbacher, Recommendations for fatigue design of welded joints and components, Springer, 2016.
[58] A. Nussbaumer, L. Borges, L. Davaine, Fatigue design of steel and composite structures: Eurocode 3: Design of steel structures, part 1–9 fatigue; Eurocode 4: Design of composite steel and concrete structures, John Wiley & Sons, 2012.
[59] J.A. Goldak, M. Akhlaghi, Computational welding mechanics, Springer, 2005.
[60] L. Wang, X. Qian, Optimization-improved thermal–mechanical simulation of welding residual stresses in welded connections, Computer-Aided Civil and Infrastructure Engineering, 39 (2024) 1275–1293.
[61] A.S.U.s. Manual, Abaqus 6.11, http://130.149, 89 (2012) v6.
[62] L. Wang, X. Qian, Effects of pre-tension and fatigue loadings on the evolution of welding residual stresses in welded plates, Engineering Structures, 301 (2024) 117272.

7 Digital Twin-Based Fatigue Assessment of Welded Plate Joints

7.1 REVIEW OF DIGITAL TWIN STUDY

7.1.1 Application of Digital Twin

Originally proposed by Michael Grieves in 2003, the concept of the digital twin has become an indispensable technology across different sectors (as shown in Figure 7.1). A digital twin serves as a real-time virtual replica of a physical entity, which can be an asset or a system, that is continuously updated with data from IoT sensors. Integration empowers industries to enhance monitoring, predictive maintenance, and decision-making.

As shown in Figure 7.1, the industry-specific examples of digital twins can involve the aerospace and defense, automotive, healthcare, energy and utilities, manufacturing Industry 4.0, retail and e-commerce, smart cities and urban planning, agriculture, construction and real estate, etc. In the Aerospace and Defence sector, aerospace companies create digital twins for each aircraft, tracking the engine's performance and structural wear. These digital twins assist engineers in planning the maintenance

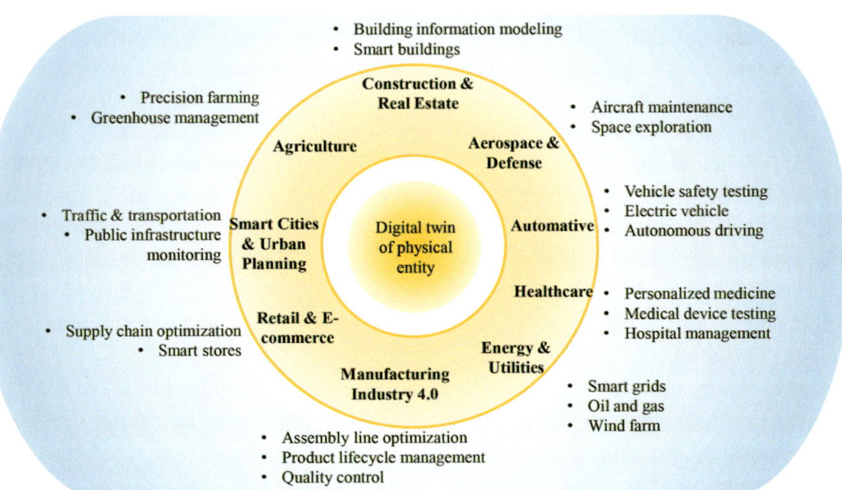

FIGURE 7.1 Application of digital twin concepts in different industry sectors.

DOI: 10.1201/9781003509806-7

schedule and detecting potential issues before the final failures, thus enhancing the safety tolerance and reducing downtime. Bisanti *et al.* [1] comprehensively reviewed the state of the art on digital twin-related technology applied to aircraft operations and maintenance. Lai *et al.* [2] developed the digital twin-based structural health monitoring framework combining the data from the measurement and computational techniques. NASA uses digital twins to simulate space environments and test scenarios before sending spacecraft on missions. For example, the Mars Rover built a digital twin on Earth that helps engineers troubleshoot issues remotely and plan optimal paths on Mars's surface. Liu *et al.* [3] explored the advanced application of digital twins in space environments. Ambalkar *et al.* [4] addressed the impact of digital twin technology on space exploration.

In the automotive industry, car manufacturers employ digital twins to simulate crash tests and refine vehicle safety features without physical prototypes. Shi *et al.* [5] optimized the thermal performance of the vehicle through the digital twin management system using various testing sensors. Digital twins of EV batteries monitor charge levels, temperature, and wear patterns to optimize battery life and safety, and predict the service and replacement of the battery. Rajesh *et al.* [6] highlighted the digital twin technology in enhancing the battery design, increasing longevity, ensuring safety, etc. Digital twins of roads and traffic patterns allow for testing autonomous driving systems in simulated real-world environments, enhancing the algorithms' learning performance before real-world trials. Shoukat *et al.* [7] presented a digital twin to speed up the development and testing of autonomous driving technology.

In the healthcare industry, digital twins of individual patients are created using health data, genetics, and medical history. The digital twin technique simulates the treatment effects of new medicines and the response of patients under different medications and enables promoting personalized care plans. Yingngam *et al.* [8] utilized the digital twin to mirror, predict, and optimize an individual's health in real time. Dziopa *et al.* [9] combined the digital twinning approach with generalist AI to accelerate the implementation of precision medicine, recognition, and prevention of cardiovascular disease. Medical equipment manufacturers adopt digital twins to ensure quality and reliability. For instance, a digital twin of a surgical robot can simulate different surgical scenarios, ensuring it performs safely and effectively before use in actual procedures [10]. Rowan [11] reviewed the safe and sustainable opportunities for medical devices using the digital twin and extended reality applications. Hospitals use digital twins of their facilities to optimize workflows, manage patient flow, and ensure the availability of resources like beds, operating rooms, and staff based on real-time data. Han *et al.* [12] proposed a conceptual framework of digital twining for smart hospital management to promote the automation, intelligence in hospital operations.

In the energy and utilities industry, utility providers use digital twins of the electrical grid to monitor power demand and supply, further optimize energy distribution, reduce outages, and integrate renewable sources like solar and wind more effectively. Das *et al.* [13] summarized the functions of the digital twin in forecasting, anomaly detection, and security. Digital twins of offshore drilling rigs or refineries enable remote monitoring, predictive maintenance, and safety analysis, reducing the demand for on-site personnel in hazardous environments [14, 15]. The digital twin of the wind turbine aims to monitor wind conditions, blade wear, and energy output through optimizing turbine angles and predicting maintenance timings. Kandemir *et al.* [16] developed a novel approach to optimize the wind farm layout during operation based on digital twin and dynamic repositioning. Mousavi *et al.* [17] generated a digital twin for a compromised structure, utilizing a physics-based model to analyze damage scenarios. Fahim *et al.* [18] utilized the machine learning approach to predict the wind turbine's energy production.

In the Manufacturing and Industry 4.0 sector, digital twins of manufacturing plants help to identify bottlenecks and inefficiencies in real-time through tracking machinery and product flow. Sensors in production lines feed data into digital twins of products and machines. Manufacturers can address the defects and improve the product reliability. Soori *et al.* [19] evaluated the previous papers relating to the digital twin techniques for continuous improvement of manufacturing systems, part production monitoring and modification, process and product performance optimization, downtime reduction, hazards and risks identification, and predictive maintenance.

In the Retail and E-commerce sector, digital twins of the supply chain record the inventory levels, transportation, and logistics in real-time, which can be used to forecast demand, reduce delays, and manage stock levels across warehouses. Bakhshi *et al.* [20] identified the existing applications of DT in dealing with current construction supply chain challenges. Retailers can create digital twins of physical stores to monitor customer traffic, optimize shelf layouts, and improve product placements. Data from customer movement and purchase behavior help adjust layouts for better sales and customer satisfaction. Rizqi *et al.* [21] proposed a 3D-based computer simulation technology based on an asset administration shell proposed by platform Industry 4.0.

In smart cities and urban planning sector, digital twins of road networks monitor the traffic flow, optimize signals, and manage congestion [22–24]. Digital twins of bridges, buildings, and other infrastructures can monitor structural integrity, detect wear, and predict maintenance, and prevent potential hazards. Torzoni *et al.* [25] developed a predictive digital twin approach for health monitoring, maintenance, and management planning of civil engineering structures. In the Agriculture sector, digital twins of farmland integrate data from soil sensors, weather stations, and crop health monitoring to optimize irrigation, fertilization, and pest control. Ubina *et al.* [26] designed a digital twin infrastructure for intelligent fish farming in aquaculture, including fish feeding, environmental monitoring, health monitoring, etc. Digital twins of greenhouses help to optimize environmental controls (light, temperature, humidity) to improve crop growth and determine the optimal settings for various plant species. Bousnina and Guerassimoff [27] introduced a novel approach to energy management in smart energy systems using deep reinforcement learning to optimize the management of flexible energy systems, including heating, cooling, and electricity storage systems.

In the construction and real estate sector, digital twins of construction projects track project timelines, budgets, and resources, which provide virtual walkthroughs, allow stakeholders to visualize the completed project and address design issues before construction. Wang *et al.* [28] developed the interactive closed-loop digital twin framework that integrates a BIM into human-robot collaborative construction workflows. Xu *et al.* [29] proposed a digital twin health monitoring information model to simultaneously locate sensors in space and monitor data. Building management systems use digital twins to monitor and control HVAC, lighting, and security systems, enhancing energy efficiency and occupant comfort. Eneyew *et al.* [30] introduced a calibration framework for physics-based models, addressing the challenges of continuous model calibration in smart building digital twins while considering the uncertain environment of physical buildings.

7.1.2 Features and Purposes of Digital Twin

Overall, digital twin technologies have been adopted in multiple industries. Digital twin aims to build connections with physical assets and virtual twin through sensing technology for human decision-making, optimization, and predictions. As shown in Figure 7.2, through reviewing different existing research work, the critical components in digital twin include 1) physical entity, which can be a real-world asset, system, process, or environment that the digital twin is replicating; 2) sensors – The physical assets should be equipped with IoT sensors, to collect real-time monitoring data, which reflects the operational states and environmental conditions of the physical entity; 3) data collection and processing – Database or cloud-based systems store and process the captured data from the IoT sensors and further analyzes the data to build the relationship between monitoring data and targeted performance-related information of the physical asset, through machine learning models, edge computing, and other simulation techniques; 4) virtual replicas – Information obtained from the sensors is imported into the virtual replica, thereafter, provides further feedback, optimization, or prediction, updating or decision-making systems. According to Sharma *et al.* [31], there are three types of digital-related models. Digital Model only involves the manual information interaction between physical and virtual systems. Digital shadow highlights the automatic communication from the physical entity to the virtual replica, yet manual feedback from the virtual replica to the physical entity. Digital twin is the most complicated one, which emphasizes seamless automatic forward and backward communication between the physical entity and virtual replica.

In Figure 7.2, this study summarizes the existing research work from different industries and categorizes the physical entity into updateable and non-updateable. For the updateable physical entity, such as system-based [32, 33] or information-based [34, 35], which can be optimized or updated after receiving the feedback of the virtual replica seamlessly. Some physical entities are non-updatable, such as the built civil engineering structures [36, 37]. The non-updatable physical entities can only be monitored or maintained periodically. To have a better understanding of the physical entities, two groups of sensors are utilized, such as global and local sensors. Global sensors reflect the summative or representative features of the whole physical entity, such as vibration-related sensors [38], temperature sensors [39], and unmanned aerial vehicle (UAV) sensors [40]. The local sensors mounted at the physical entity can only reflect the local behavior, such as the strain-gauge sensors [41] and crack-sized sensors [42].

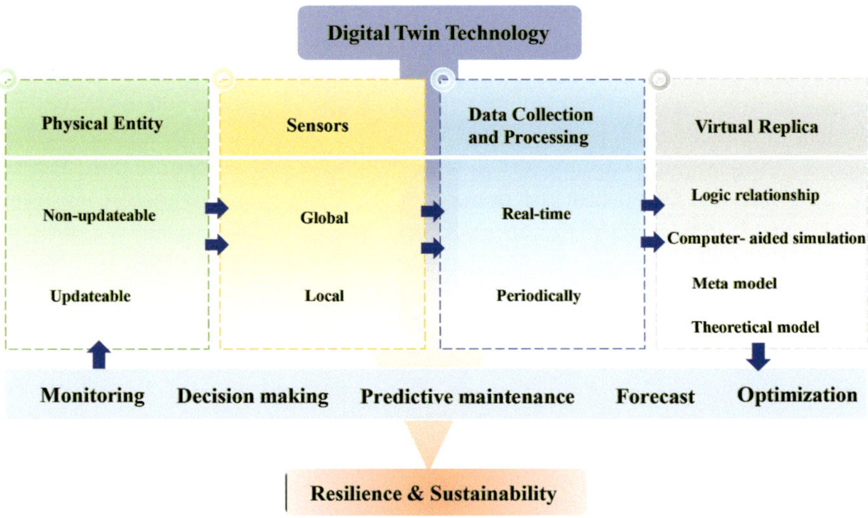

FIGURE 7.2 Features and purposes of digital twin technology.

The data collection and processing approach can be real-time or periodical, which depends on the features and demand of the monitored physical entity and operation cost. For example, the digital twin of the transport or traffic [23], which is always changeable, should be real-time to prevent some accidental cases. Similarly for the precision farming digital twin [43], the autonomous driving system [44]. The periodical data collection can be days, months, or even years, such as fatigue crack propagation [36, 45, 46]. The real-time data collection and processing impose critical challenges on the computation and storage capabilities of hardware and software.

The virtual replica maps the whole data from the sensor deployed in the physical entity (independent variables) to the targeted information (dependent variables). The virtual replica of the physical entity can be constructed in different approaches. The logical mathematical relationship in this study represents the targeted variable of the physical entity mainly consists of monitored variables, which are discrete states, such as the valves' on/off in wastewater treatment [47], switches' on/off in electricity management [48]. The computer-aided simulation approaches consist of some advanced numerical simulation techniques, such as FE models, computational fluid dynamics, and thermodynamic modeling. Zhang *et al.* [37] determined the structural behavior through localized monitoring data with FE model. Molinaro *et al.* [49] introduced a paradigm for the exploitation of computational fluid dynamics data into the machine learning-based digital twin. Meta model or surrogate model by data-based approaches (machine learning and deep learning, etc.) are alternative approaches to build virtual replicas. Fahim *et al.* [18] utilized the machine learning approach to predict the wind turbine's energy production. Feng and Qian [50] adopt the classification ANN model to predict the cracks in welded joints using the measured strain data. Kreuzer *et al.* [51] reviewed the artificial intelligence in the digital twin research work. These data-based approaches are "black box," which can be fast and capture the main relationship, but lack some reliability and explainability for fast scaling and deploying to other systems. In addition, the surrogate model by physical or theoretical approaches are "white box," which depicts the physical correlation of the targeted variable and independent variables. Peng *et al.* [36] incorporated the physical Paris' law into the digital twin framework to predict the fatigue life. Fu *et al.* [52] updated the complex boundary conditions using the structural dynamic models in the digital twin framework. However, this physical model can be limited due to simplified hypothesis and assumptions. The above four models of virtual replica are not solely used. Hybrid approaches integrating the above virtual replica models are employed usually.

Summarizing the existing literature, the critical purpose of the digital twin includes are as follows:

Real-time monitoring – Digital twins enable continuous monitoring of assets, systems, or processes in real time. For example, a digital twin of a wind turbine can provide live data on performance metrics, environmental conditions, and maintenance needs.

Decision making – Digital twins simulate the test scenarios in a virtual environment without interrupting actual operations, even assess how a process will respond to changes, such as different loads or settings, and experiment with innovations, designs, or workflows safely.

Predictive maintenance – Digital twins can predict potential failures by analyzing data patterns, furthermore, reduce downtime and optimize maintenance schedules, and save costs and improve the asset's lifespan.

Forecast – Digital twin can provide the forecast function, such as the weather forecast, environmental loadings, and evolution of the autonomous control to prevent failure of the manufacturing system.

Optimization – Digital twin helps in optimizing processes by analyzing data and providing insights on adjustments to improve efficiency and reduce waste.

7.1.3 Challenges of Digital Twin

The challenges of digital twin technology still exist. This study categorizes the challenges as follows:

- *Data integration and interoperability:* Integrating data from diverse sources (IoT devices, cloud platforms, etc.) is complex. Different systems entail different data formats and protocols, which lack some standardization and universal formats.
- *Scalability:* Managing and scaling digital twins for large, complex systems (e.g., smart cities or large manufacturing plants) requires significant computational power and sophisticated architecture.
- *Cybersecurity and privacy:* Some digital twins rely on real-time data exchange, which is vulnerable to cyberattacks. Sensitive data relating to the healthcare and energy sectors requires a strong protection strategy.

- *Data accuracy and realism:* The effectiveness of a digital twin depends on the accuracy of its underlying model and the real-time data feeding it. Discrepancies between the digital twin and the physical counterpart always exist due to noise, simplification, etc.
- *Cost and resource intensity:* Developing and maintaining digital twins involves high costs, requiring advanced sensors, computing infrastructure, and domain expertise.
- *Complexity in simulation and modeling:* Creating a high-fidelity model that captures all relevant physical and operational aspects of a system is challenging. Models need to balance detail and computational feasibility.
- *Skill gap:* Digital twin is a multidisciplinary approach which requires professionals skilled in different technologies, such as AI, machine learning, and advanced simulation.

7.2 INTELLIGENT APPROACH TO SIZE SURFACE CRACKS

7.2.1 Strain Relaxation Phenomenon under Crack's Presence

The cracks emerging from geometrically discontinuous locations under cyclic environmental loadings [53, 54] are critical concerns for the safety of the existing structural components. The fatigue assessment is essential for the evolving digital twin of sustainable infrastructures, including bridges, ships, and offshore platforms, to optimize the lifetime cost of these structures. The capability to forecast small surface cracks has clear advantages in reducing the maintenance cost of structures and in preventing cascading failures in large-scale infrastructure. The determination of the crack initiation and propagation relies on advanced crack-measurement technologies, such as the ACPD approach [55, 56], acoustic emission approach [57–59], UPA approach [42, 60], and eddy current. While offering engineering approaches to non-destructively size the cracks in civil engineering infrastructures, deploying these techniques is often costly and requires special technical skills. For example, the ACPD typically applies only to electrically conductive materials, and fixing the measurement probes remains challenging for topologically complex structural components. Acoustic emission approaches utilize energy dissipation-induced acoustic signals to detect the cracking events without quantifying the crack size. Ultrasonic-based sensors require detailed fabrication, and the capability to detect the crack size depends on the frequency and wavelength of the ultrasonic signals. The inhomogeneity of the material near the welds further compromises the accuracy and robustness of the ultrasonic sensors. In contrast, strain gauges [61–65] provide cheap and accurate strain measurement in concrete, steel, and composite structures. In addition, strain data via strain gauges or other strain sensors [66] provide real-time fast-speed data acquisition and long-term durability for the entire life span of the structures. However, the research on quantitative crack sizing in both length and depth directions of welded plate joints based on strain measurement is limited.

Instead of costly experimental tests, the numerical simulation provides an affordable tool to reveal the material and structural response for different crack profiles. Figure 7.3 presents the full FE model of welded plate joints (BT) specimens under

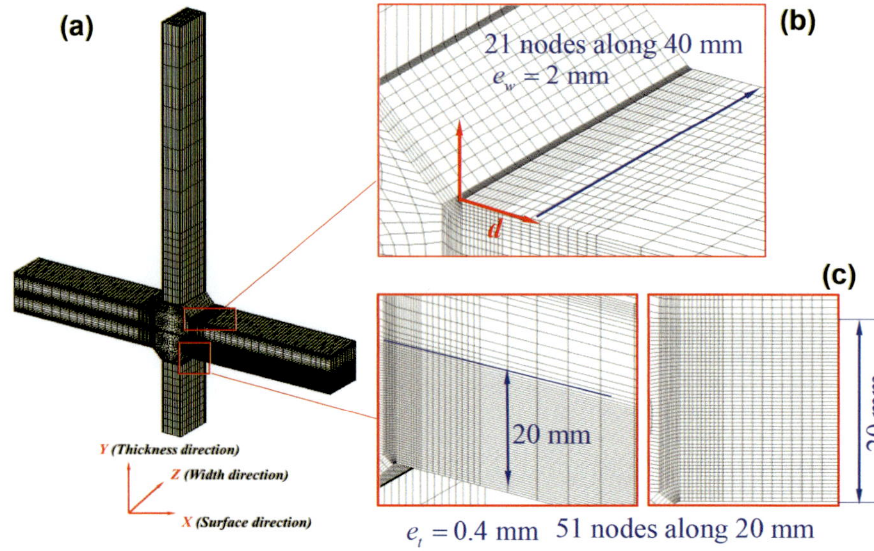

FIGURE 7.3 FE model of the full model of welded plate joints and detailed mesh size in different directions.

three-point bending (maximum load of 50 kN), the same as those in Chapter 6. The unsymmetric crack geometry is created on one side of the welded plate joints (Figure 7.4). The element size (e_w) and (e_t) is 2 mm and 0.4 mm, respectively. Therefore, there are 50 elements (51 nodes) in the bottom half of the main plate and 20 elements (21 nodes) in the width direction of the main plate. The element type is C3D8 element in ABAQUS [67]. The investigated material is S550 with a yield strength of 650 MPa. Without the presence of cracks, the stress and strain concentrate at the weld toe region in the welded plate joints under linear-elastic analysis.

Figure 7.4 demonstrates that the presence of a through-width crack redistributes stress and strain concentration near the crack tip. The yield strength of the S550 is 650 MPa. Similarly, different crack depths lead to different degrees of stress and strain relaxation along the bottom surface of the welded plate joints, which are easily tractable by the sensor. Figure 7.5 plots the strain variation along with the normalized distance (d/t) from the weld toe at the bottom surface for different crack depths (c) for the welded plate joints with a through-width crack. The strain relaxation manifests as the crack depth increases, as compared to the intact joint. For the welded plate joints under bending loads. The strain relaxation at $0.4t$ away from the weld toe portrays the difference in the through-width crack depths as well. Therefore, this study adopts the strain at $0.4t$ away from the weld toe in the bottom surface to size the crack profile in welded plate joints.

Figure 7.6 presents the strategy to vectorized the crack profiles. Through dividing the crack profile into multiple layers in the thickness direction and width direction, the crack front can be approximated by connecting a series of discrete nodes, which represent the crack-front positions. The crack sizing is mathematically equivalent to

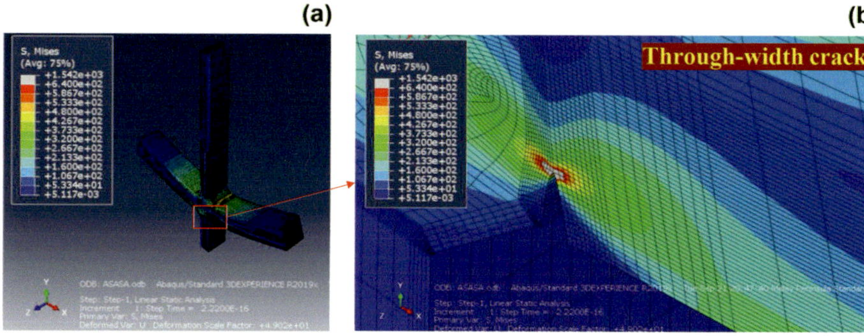

FIGURE 7.4 The representative numerical results of the Mises stress under the through-width crack.

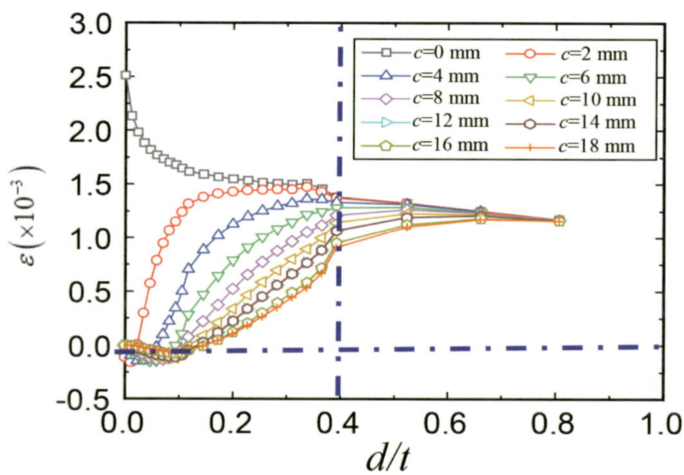

FIGURE 7.5 The distribution of the strain concentration under different through-width crack sizes along the normalized distance (d / t) in the welded plate joints.

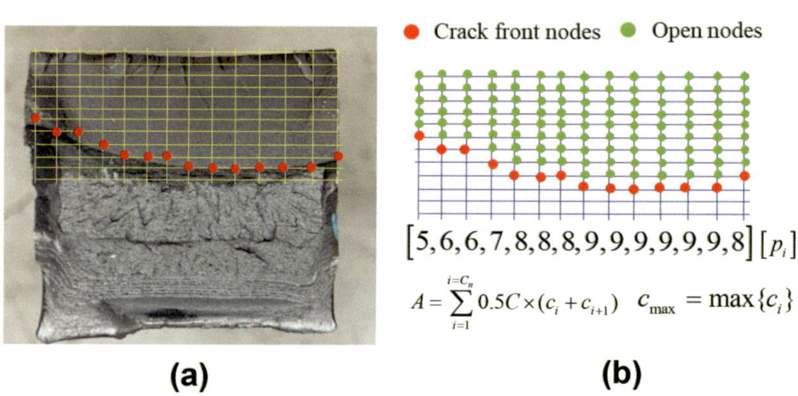

FIGURE 7.6 The concept to divide the failure surface into discreet FE nodes.

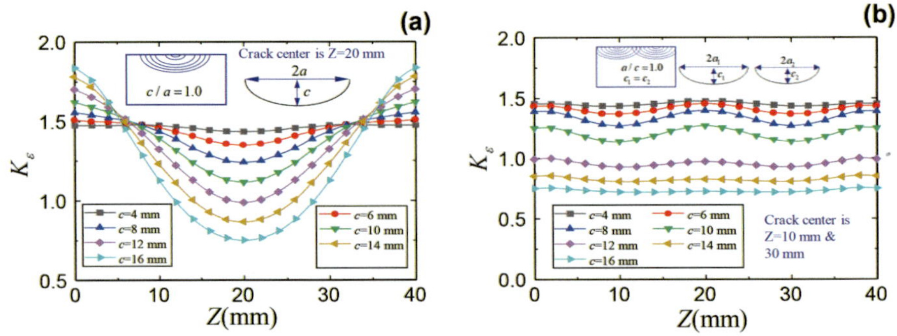

FIGURE 7.7 The variation of the strain concentration factors along the width direction under different crack profiles.

determining the number of open nodes along the width of the main plate $\lfloor p_i \rfloor$. Based on the above assumption and approximation, Figures 7.7a and b present the strain concentration variations ($K_\varepsilon = \varepsilon_{0.4t} / \varepsilon_{nom}$) at $0.4t$ away from the weld toe for a plate with one semi-elliptical crack and a plate with two semi-elliptical cracks, respectively. The strain concentration variation can reflect the crack depth and length.

7.2.2 Metamodel in Sizing the Cracks

Feng and Qian [50] proposed multiple classification neural network models to determine the number of open nodes along the width direction of the main plate. However, the number of the classification NN models depends on the node numbers along the width direction, which significantly increases the number of weighted parameters in all NN models due to the fine mesh size required to describe the crack-front profile with sufficient accuracy. In addition, separate calibration of these neural network models for each number of open nodes along the width direction tends to neglect the critical overall features of the crack front, including the total areas enveloped by the crack front profile and the maximum depth along the crack front. Therefore, the proposed approach in this study constructs a single regression neural network to describe all the crack front points along the width of the main plate. Following the procedure in constructing the full FE in the previous chapter, this study trains the metamodel utilizing a numerical database. For the FE model in Figure 7.3, there are 50 elements in the bottom half of the main plate thickness. Along the width direction of the main plate, there are 20 elements. Therefore, this study adopts the regression NN as indicated by the blue-dot block in Figure 7.8.

The input variables are the normalized strain relaxation values along the width direction. As the plate contains 20 elements across the width, there are 21 columns of nodal strain variables across the plate width. The 21 input variables (normalized strain relaxation indicator $\overline{\varepsilon}_i$) are:

$$\overline{\varepsilon}_i = \frac{\varepsilon_i^{\text{o}} - \varepsilon_i}{\varepsilon_i^{\text{o}}}, i = 1,2,3...,21 \tag{7.1}$$

where ε_i^{o} represents the initial strain value at ith column at $0.4t$ away from the weld toe without any crack; ε_i denotes the actual strain at the ith column at $0.4t$ away from the weld toe with a crack; $\overline{\varepsilon}_i$ is the normalized nondimensional strain relaxation value at the ith column.

To enhance the training efficiency, this study conducts the feature scaling approach to prep-process the 21 normalized strain relaxation indicators and map them to [–1,1];

$$x_i = \frac{2\times\left(\overline{\varepsilon}_i - \left[\overline{\varepsilon}_i\right]_{\min}\right)}{\left[\overline{\varepsilon}_i\right]_{\max} - \left[\overline{\varepsilon}_i\right]_{\min}} - 1,\ i = 1,2,...,21 \tag{7.2}$$

where $\left[\overline{\varepsilon}_i\right]_{\min}$ and $\left[\overline{\varepsilon}_i\right]_{\max}$ represent the minimum and maximum value of the normalized strain relaxation variables $\overline{\varepsilon}_i$; Therefore, the input layer entails 21 neurons, representing 21 feature-scaled normalized strain relaxation indicators along the width direction. There are 21 neurons in the output layer, denoting 21 crack front depths along the width direction (b_k, k = 1,2 . . .,21). By balancing the effectiveness and minimizing the parameters in NN, this study sets 10 neurons in the one hidden layer. The transfer functions among these neurons in different layers follow,

$$Q_j = \sum_{i=1}^{n} \omega_{j,i} x_i + \xi_j \quad a_j = \frac{1}{1+e^{-Q_j}} \tag{7.3a}$$

$$P_k = \sum_{j=1}^{m} \omega_{k,j} a_j + \xi_k \quad b_k = \left(P_k + 1\right)\frac{\left[y_k\right]_{\max} - \left[y_k\right]_{\min}}{2} + \left[y_k\right]_{\min} \tag{7.3b}$$

where ω represents the weights and ξ is the bias parameters; x_i, a_j and b_k are the neurons in the input layer, hidden layer, and output layer, respectively; $\left[y_k\right]_{\min}$ and $\left[y_k\right]_{\max}$ denote the minimum and maximum value of the real crack depth y_k along the width direction. To utilize the backpropagation algorithm, the cost function root mean square error (RMSE) in the NN model in this study follows

$$RMSE = \sqrt{\frac{1}{M \bullet 21} \sum_{j=1}^{M} \sum_{k}^{21} \left(b_k^j - y_k^j\right)^2} \tag{7.4}$$

where y_k^j and b_k^j represent the real crack front depths and the predicted crack front depths along the width direction, respectively; M denotes the size of the database.

This study adopts the Levenberg-Marquardt backpropagation approach [68] for training the NN model. The LM approach combines the Gauss-Newton algorithm and the steepest descent method, as shown in Eq. (7.5).

$$\alpha_{k+1} = \alpha_k - (L_k^T L_k + \mu I)^{-1} L_k e_k \tag{7.5}$$

where α_{k+1} and α_k represent the current and previous weighted parameters, respectively. L_k denotes the Jacobian matrix of the error vector e_k with respect to α_k . I denotes the identity matrix. μ represents the combination coefficient between the Gauss-Newton algorithm and the steepest gradient algorithm.

However, the crack length in the FE model is discrete. The smallest value of Eq. (7.4) does not represent the optimal NN model in considering the discrete output variables in the FE model. Therefore, this study rounds the output variables b_k to c_k correspondingly. Another performance function is proposed as

$$J = \sum_{j=1}^{M} \sum_{k=1}^{21} \left(c_k^j - y_k^j \right)^2 \tag{7.6}$$

The strategy adopted in this study is to conduct 500 times of the NN model training and selects the optimal NN model with the smallest J values. To generate a sufficient database for training the NN model, this study selects four types of cracks in the FE model: 1) through width crack; 2) one semi-elliptical crack; 3) two semi-elliptical cracks, and 4) three semi-elliptical cracks, as shown in Figure 7.8. In total, 181 sets of crack profiles and corresponding normalized strain relaxation indicators at 0.4t away from the weld toe are utilized for training the NN model. The study adopts randomly 141 sets of the total database to train the NN model, while the remaining databases to examine the NN model's accuracy.

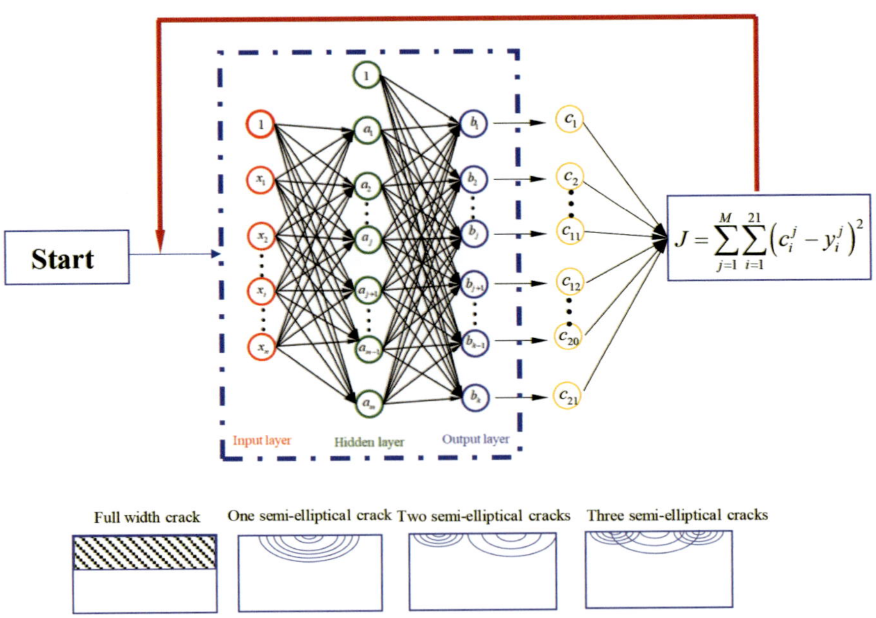

FIGURE 7.8 The proposed regression NN model to determine the metamodel of the crack profiles and normalized strain relaxation indicators.

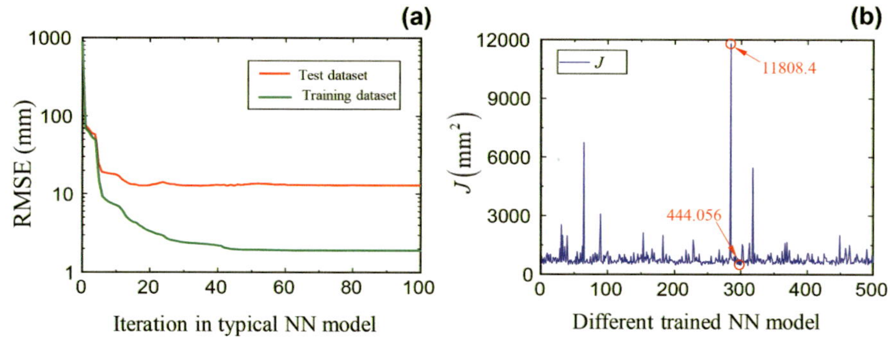

FIGURE 7.9 (a) The variation of the RMSE with increasing iteration numbers for train and test dataset; (b) the CWC values of the 500 times of NN trainings.

Figure 7.9a presents the representative evolution of the cost function for the training dataset and test dataset. The cost function RMSE quickly approaches the convergent values within 50 iterations. Figure 7.9b presents the final J values for the repeated 500 times of NN model training. The maximum value of the training is 11808.4 mm, while the minimum value is 444.056 mm at the 325th NN model training. Therefore, the weighted parameters from the 325th NN model are utilized for the NN model for the subsequent sections. Figures 7.10a and b illustrate the capability of the final optimal NN model for the representative crack in the training dataset and test dataset, respectively. The predicted crack shapes achieve close agreement with the numerical crack shapes. The area enveloped by the crack profile (or the crack area) and the maximum crack depth are two critical indicators. Figures 7.10c and d compare the area enveloped by the crack profile (A) and the maximum crack depth ($c_{\max}$). For both the training dataset and test dataset, the area enveloped by the crack profile achieves excellent agreement between the numerical crack (A_{num}) and the predicted crack (A_{pre}). The maximum depth of the multiple cracks by the predicted cracks ($c_{\max}^{pre}$) reaches good agreement with the numerical values ($c_{\max}^{num}$), yet with some reasonable scatters. Such scatters are attributed to the discrete numerical value and the rounding strategy.

Figures 7.11a and b present the statistical distribution of the absolute difference in the crack depth for the training dataset and test dataset, respectively. The vertical axis represents the number of data points located within the corresponding crack error bins. The fitted probability density function (PDF) is indicated by the red curve. The mean value for the training dataset is 0.468, while for the test dataset is 0.636. More than 90% of the absolute error in the crack depth is less than 2 mm for both the training and test dataset. Specifically, as shown in Figures 7.11c and d, for the training dataset, the absolute error in the crack depth is less than 0.5 mm, which accounts for 71.46%, while it accounts for 62.86% for the test dataset. The absolute error in crack depth less than 1.0 mm accounts for 87% of the training dataset and 78% of the test dataset. The comparison in Figures 7.10 and 7.11 validates the effectiveness of the optimal NN model.

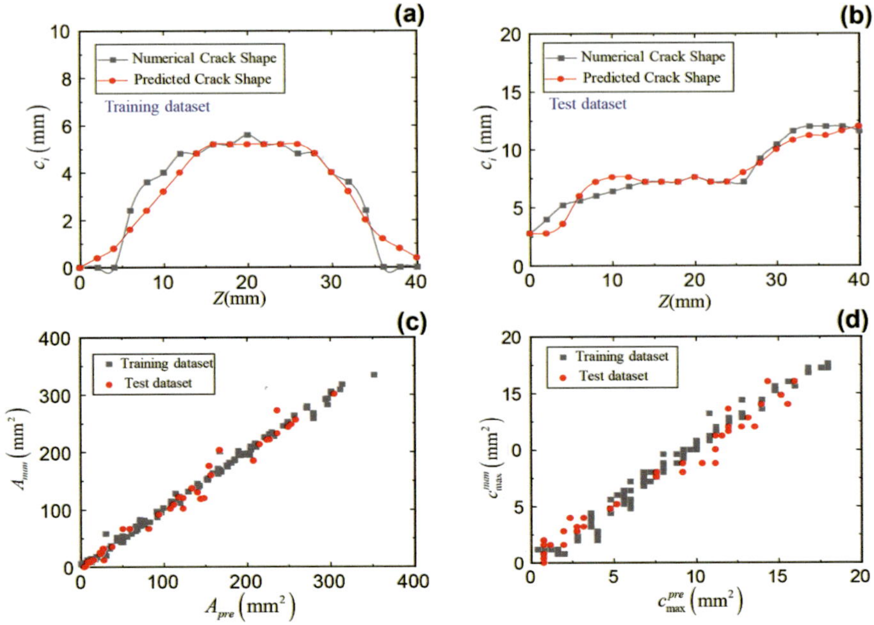

FIGURE 7.10 (a, b) Comparison between the numerical crack profile and the predicted crack by the NN model for train dataset and test dataset, respectively; (c) comparison in the area enveloped by the numerical crack and predicted crack for the train and test dataset; (d) comparison in the maximum crack depth by the numerical crack and predicted crack for the train and test dataset.

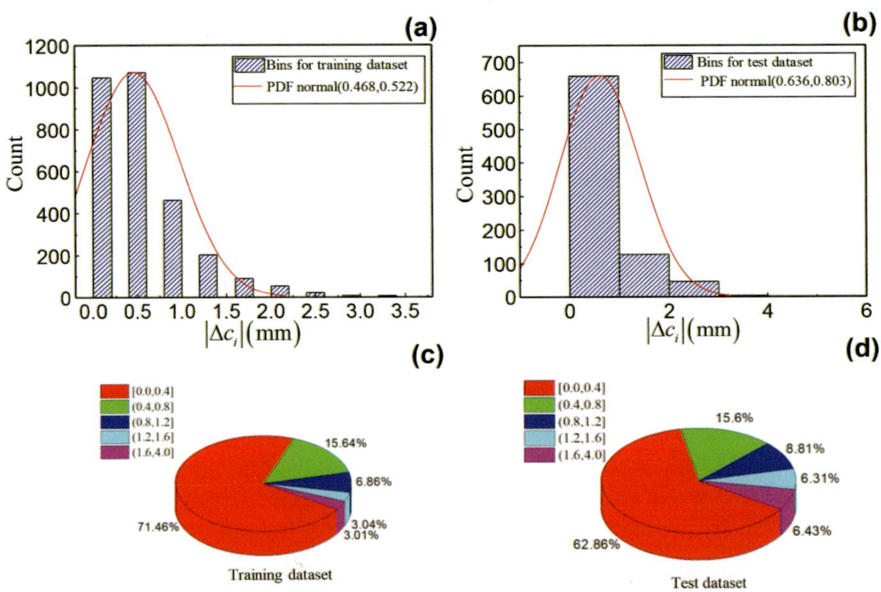

FIGURE 7.11 The statistical distribution of errors in the crack depths for the (a, c) training dataset and (b, d) test dataset.

7.3 UPDATING OF CRACK SIZING ANCHORED TO SENSING DATA

7.3.1 Experimental Crack Measurement Setup

Numerical simulation provides a powerful and cost-effective tool to obtain reasonable results instead of expensive experimental tests. However, the numerical simulation depends heavily on simplifying assumptions on the material property, loading conditions, and perfect workmanship [69, 70]. Therefore, the mathematical metamodel solely based on numerical results lacks confidence in application in engineering practices due to the uncertainty in geometrical configurations, material properties, and measurement techniques. The inconsistencies caused by such uncertainties have impacted the structural health monitoring of civil engineering structures over the last few decades. Therefore, an enhanced updating approach to improve the forecast accuracy and reduce the inaccuracy of the metamodel is essential.

Engineering structural health monitoring, however, does not entail the luxury of abundant data. Instead, the sensor locations for the health monitoring and damage measurement remain typically limited to some discrete positions in a structure [71–74]. Such limited information in the engineering structure or experimental study adversely affects the effectiveness of local damage quantification if dealt with inappropriately. For the structural health monitoring of structural components, the accuracy and predictability of the intelligent algorithm depend significantly on the number of measured data points and the different measurement techniques. Consequently, integrating and updating the crack sizing algorithms, adapting to limited experimental strain and crack data, to enhance their accuracy and robustness, requires further study and comprehensive experimental validation.

Aligning with the engineering fatigue assessment protocols, this study extends the capability of strain data to quantify the crack size, validated by the high-cycle fatigue of welded plate connections (Figure 7.12). This study utilizes the high-strength steel

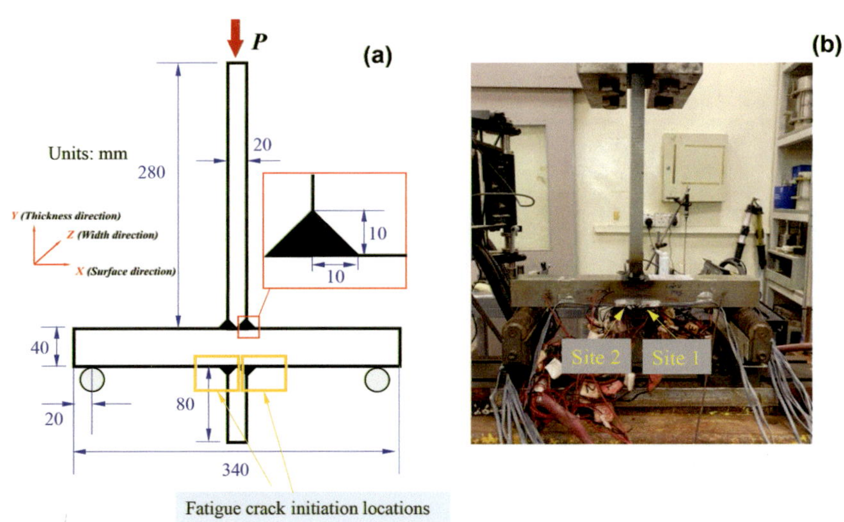

FIGURE 7.12 (a) Geometrical configuration and (b) experimental setup of welded plate joints under high-cycle fatigue bending.

S550 to fabricate the welded plate joints. The detailed material information of S550 can be found in [75]. The welded plate joint is under three-point bending, with the main plate simply supported at the ends, as shown in Figure 7.12. The supported span of the welded plate joint is 300 mm. The thickness of the main plate is 40 mm with a width of 40 mm. The full penetration welding used for the welded plate joint has a welding leg of 10 mm in length. The total height of the specimen is 400 mm, with the thickness of the vertical attachment equal to 20 mm. The downward loading creates tensile bending stress at the bottom of the welded plate joints. Therefore, fatigue-induced cracks tend to occur in these two weld toes (yellow block in Figure 7.12a). The load-controlled high-cycle fatigue test has a maximum loading value of 50 kN and a minimum value of 5 kN. Correspondingly, the estimated fatigue life of 120,000 cycles.

This study employs two nondestructive techniques: the ACPD (Figure 7.13a) and the UPA (Figure 7.13b) to measure the crack size near two weld toes (Figure 7.12a). The fatigue cracks' initiation and propagation affect the path of the alternating current due to the skin effect in the steel. As shown in Figure 7.13a, the ACPD directly determines the crack depth through the potential drop between the non-cracked areas (reference points) and the cracked areas (crack-crossing points). The ACPD has been validated widely in welded connections [76] and entails a relatively small scatter (0.5 mm). Figure 7.13c presents the close-up view of the ACPD at the bottom areas of the specimen and the UPA at the top areas. Figure 7.12d indicates multiple strain gauges of 1 mm gauge length along the width of the welded plate joints (Figure 7.13d).

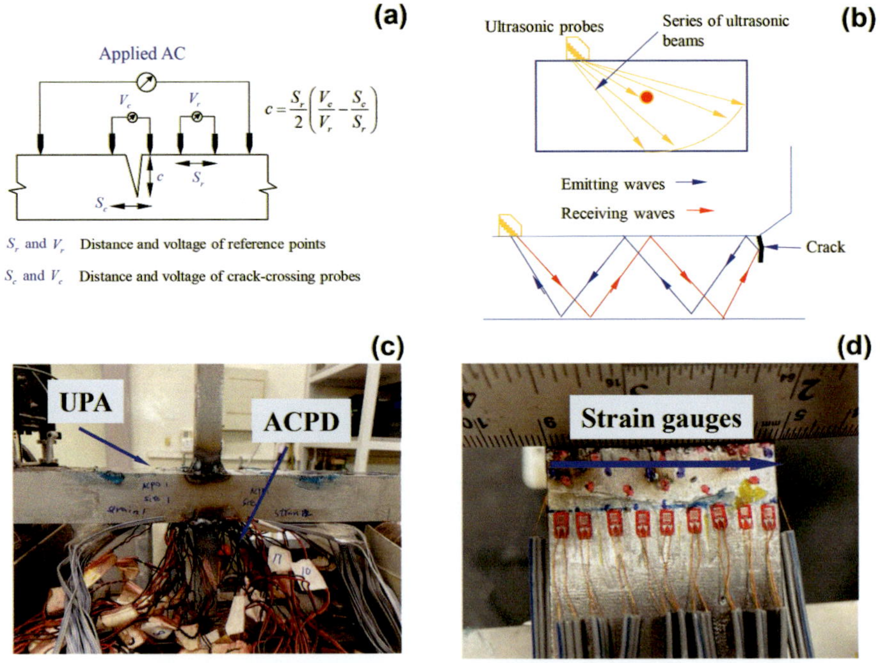

FIGURE 7.13 (a) Alternating Current Potential Drop (ACPD); (b) ultrasonic phased array (UPA); (c) location of the UPA and ACPD; (d) strain gauges along the width direction.

Figure 7.14 presents the detailed arrangement of the measurement points by the ACPD, UPA, and strain gauges. To describe the crack profile (crack length in the width direction and crack depth in the thickness direction), there are 7 measurement points by the ACPD and 3 measurement points by the UPA, located along the width of the main plate. Table 7.1 summarizes the detailed location of the seven crack measurement points and nine strain measurement locations. The nine strain gauges along the width of the plate are located 0.4t away from the weld toe, where t denotes the thickness of the main plate. UPA measures the crack depth at three locations in the width direction (10 mm, 20 mm, and 30 mm). At each crack length measurement location, there are three measurement points in the surface direction at 10 mm, 20 mm, and 30 mm away from the weld toe, as shown in Figure 7.14. The crack depth at the edge surface of the main plate is visible to the naked eye and is thus determined by rulers periodically.

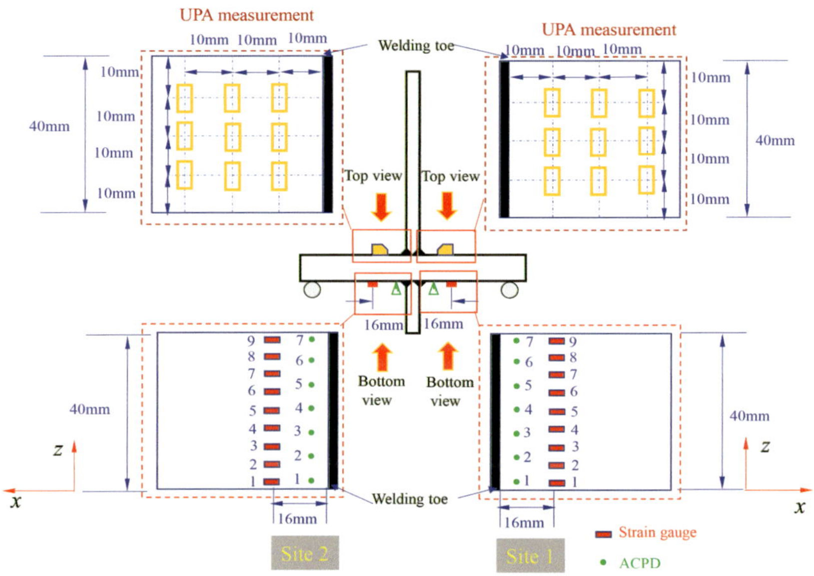

FIGURE 7.14 Detailed arrangement of the ACPD, UPA, and strain gauges at different locations in the welded plate joints.

TABLE 7.1
The Measured Locations of the ACPD Points and Strain Gauges.

		Distance measured from the specimen edge in the width direction (Z)								
		Point1 (mm)	Point2 (mm)	Point3 (mm)	Point 4 (mm)	Point5 (mm)	Point6 (mm)	Point7 (mm)	Point 8 (mm)	Point 9 (mm)
Site 1	ACPD	2	9	15	20	25	30	38		
	Strain gauge	2	7	11.5	15.5	20	24.5	29	34	38
Site 2	ACPD	1	7	16	21	29	35	39		
	Strain gauge	2	5	10.5	14.5	19.5	24	28.5	33.5	38.5

The loading frequency for the high-cycle fatigue test is 5 Hz. The data collection using ACPD to capture the crack depth requires one minute for each measurement. The strain data measurement by the DL 850 oscilloscope has a frequency of 100 samples per second. The manual crack measurement by UPA is performed every 5000 to 10000 fatigue cycles.

7.3.2 Experimental Crack Sizes and Strain Relaxation

Figure 7.15a presents the final failure of the welded plate joints. Figures 7.15b and c present the cracks on the two sides of the weld toe in the bottom surface. Observable cracks (Figure 7.15c) emerge only on one side of the weld toe (Site 1), while there are no noticeable cracks (Figure 7.15b) on the opposite weld toe (Site 2). Figure 7.15d presents the final failure surface of the welded plate joint, which reveals clearly the final fatigue crack front profile. Figures 7.16a and b show the evolution of crack depth (c) with increasing fatigue cycles (N) on two sides of the weld toe in the bottom areas, measured from the ACPD. As shown in Figure 7.16a and b, the crack initiates and grows mainly at Site 1, while Site 2 entails a small crack with negligible propagation till the specimen's failure. The legends indicate the location of the strain gauges or ACPD in

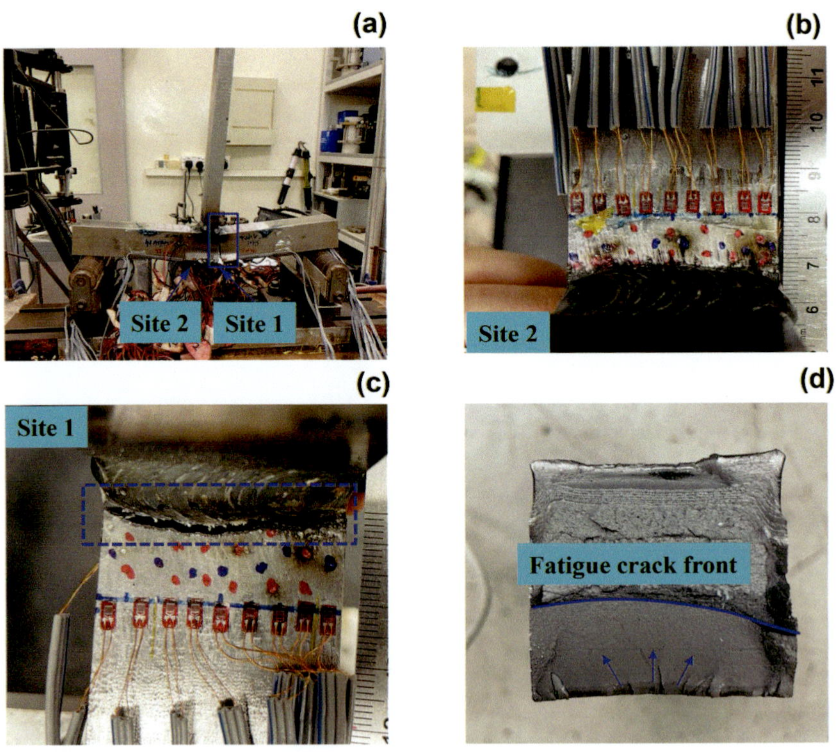

FIGURE 7.15 (a) The experimental failure of the welded plate joints; (b, c) the crack on either side of the welded plate joints; (d) the final failure surface of the welded plate joints.

the width direction (Z direction). Correspondingly, the peak strain of each fatigue cycle in Site 1 (Figure 7.16c) varies significantly with increasing crack depth and length. The peak strain in Site 2 (Figure 7.16d) remains stable in most fatigue cycles and only fluctuates at the end phase of the fatigue life. Therefore, the strain variation can reflect the crack initiation and propagation in the thickness direction (depth direction). The experimental welded plate connections are subjected to a constant loading amplitude. Therefore, this study adopts the normalized strain relaxation indicator $\overline{\varepsilon}$ [Eq. (7.1)] to represent the change of strain values under different crack profiles.

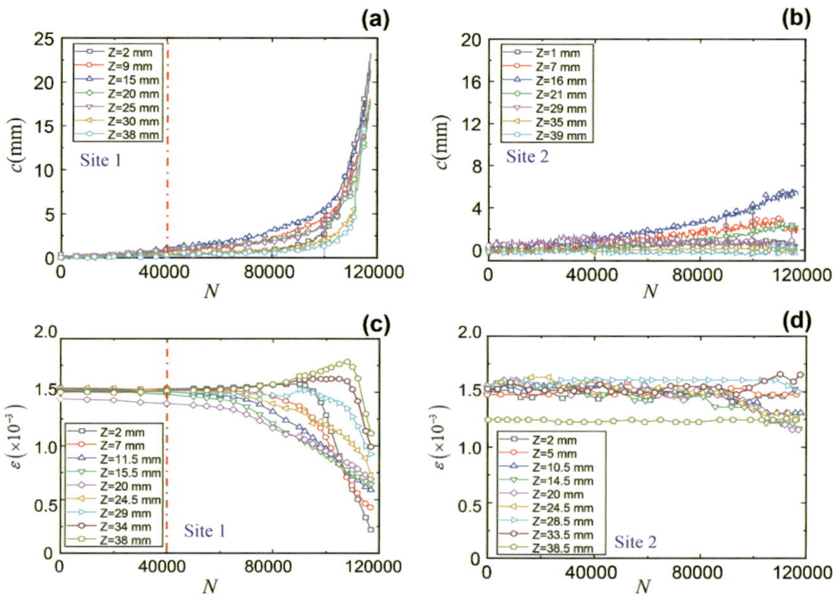

FIGURE 7.16 (a, b) The crack initiation and propagation measured by the ACPD at Site 1 and Site 2, respectively; (c, d) the peak strain variation measured by the strain gauges at Site 1 and Site 2, respectively.

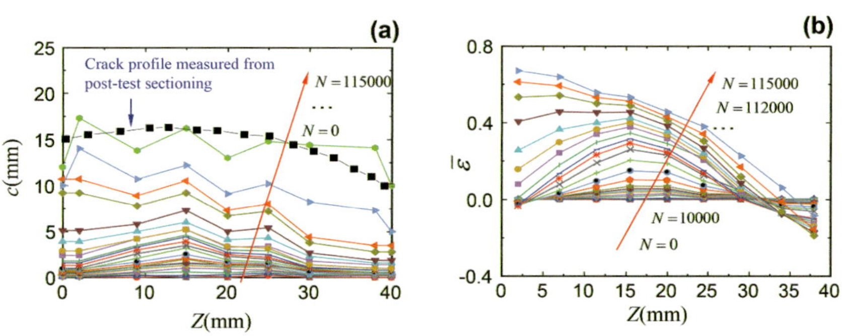

FIGURE 7.17 (a) The evolution of the crack profile measured by the ACPD under different fatigue cycles; (b) the evolution of the normalized strain relaxation indicators under different fatigue cycles.

To examine the effectiveness of the strain variation in characterizing the depth of the crack front along the width of the main plate, Figure 7.17a presents the crack front profile along the width of the main plate under different measured fatigue cycles (*N*) from ACPD. Figure 7.17b elucidates the normalized strain relaxation indicators [Eq (7.1)] along the width of the main plate under different fatigue cycles (*N*) measured from strain gauges. The apparent correlation shown in Figure 7.17 lays the foundation for utilizing the normalized strain relaxation indicators to depict the crack front profile.

7.3.3 Prediction of Crack Size Based on Metamodel Using Measured Strains

Due to the uncertainty caused by the limited strain measurement points, uncertainties in strain data and nondestructive techniques, the ability of the metamodel (optimal NN model) becomes compromised and requires further investigation through the experimental study. The experimental setup in Section 7.3.2 records the strain data 100 times every second. Figure 7.18a presents the variation of the peak strain measured in the test from the 9950th to 10050th fatigue cycle. The peak strain fluctuates and does not maintain a constant value. To determine the peak strain value used for the following sections, this study plots the statistical distribution of the 100 peak strain values around the 10000th fatigue cycle in Figure 7.18b. This study utilizes the normal distribution to fit the statistical distribution of the 100 peak strain values and adopts the mean value as the peak strain value at the 10000th fatigue cycle. Figure 7.18c presents the nine values of the measured peak strain (black dot) and the 21 peak strain data (red dot) from the FE model without any cracks at the initial 1000th fatigue cycle along the width of the main plate. The comparison in Figure 7.18c demonstrates the discrepancy between the strain measurement and numerical simulation. The required input variables in the trained NN model are 21 normalized strain relaxation indicators $\left[\bar{\varepsilon}_i\right]$. However, there are only nine strain measurement points along the width direction of the main plate in the experimental tests (Figure 7.18). Therefore, this study adopts the interpolation approach to obtain the 21 normalized strain relaxation indicators at the specified locations. The interpolation approach follows

$$\begin{gathered}\bar{\varepsilon}_i = \bar{\varepsilon}_{j-1}^{mea} + \frac{\left(\bar{\varepsilon}_j^{mea} - \bar{\varepsilon}_{j-1}^{mea}\right)\times\left(Z_i - Z_{j-1}^{mea}\right)}{Z_j^{mea} - Z_{j-1}^{mea}}, \\ i = 1,2,\ldots 21; j = 1,2,\ldots 9 \;\; \text{if } Z_i \in \left[Z_{j-1}^{mea}, Z_j^{mea}\right]\end{gathered} \tag{7.7}$$

where $\bar{\varepsilon}_{j-1}^{mea}$ and $\bar{\varepsilon}_j^{mea}$ denote the measured normalized strain relaxation indicator at Point *j*-1 and *j*; Z_{j-1}^{mea} and Z_j^{mea} represent the location at Point *j*-1 and *j*; Z_i and $\bar{\varepsilon}_i$ are the *i*th location and normalized strain relaxation indicators corresponding to the FE arrangement (Figure 7.17). Figure 7.18d shows the 21 interpolated normalized strain relaxation indicators based on Eq. (7.7).

Figure 7.19 compares the predicted cracks based on the trained NN model and 21 interpolated normalized strain relaxation indicators and the measured cracks by

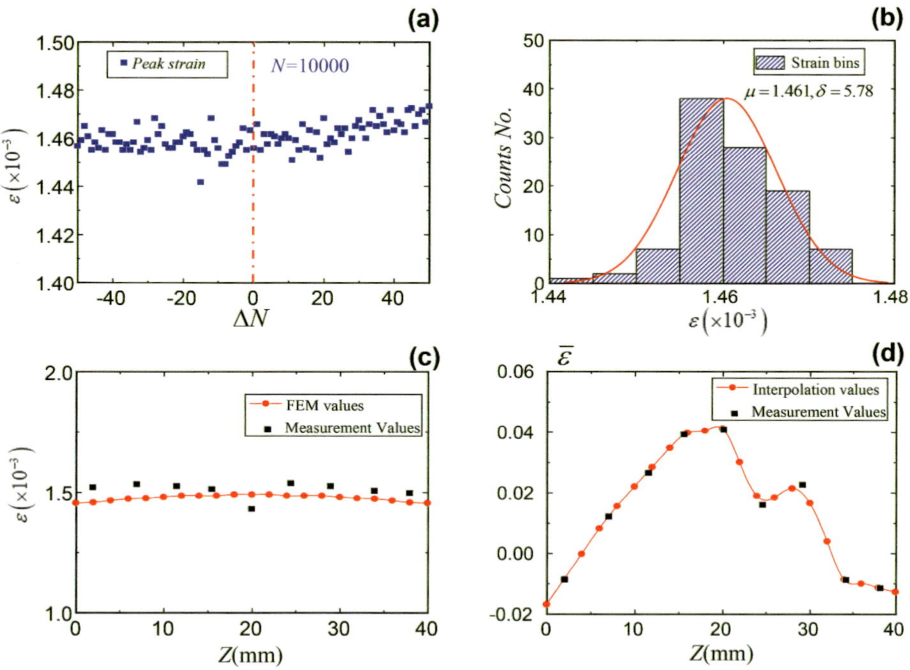

FIGURE 7.18 (a) The variation of the peak strain values from the 9950th to 10050th fatigue cycles; (b) the statistical distribution of the 100 number of peak strain values; (c) the comparison between the measured strain at the initial fatigue cycles and the FE values; (d) the distribution of the normalized strain relaxation indicators based on the extrapolation scheme.

the ACPD at different fatigue cycles. The blue curve represents the predicted crack based on the 21 interpolated, normalized strain indicators [Eq. (7.7)] and the trained regression NN model. The red curve indicates the nine crack measurement points (seven by the ACPD and two physically measured by the side surface). The discrepancy between the predicted crack profile and the measured crack profile increases with increasing fatigue cycles (N). Therefore, solely based on the measured normalization strain relaxation indicators and trained regression NN model, the predicted crack fails to capture the actual crack accurately at different fatigue cycles. The discrepancy increases with larger actual crack sizes. Such discrepancies raise challenges and potential issues in further fatigue and risk assessments.

7.3.4 Metamodel Updating Anchoring to Physical Measurements

To address the discrepancy caused by a numerical-based metamodel, this study adopts the Bootstrap Particle Filtering approach to update the weighted parameters in the trained regression NN model, as

$$\begin{aligned} \theta_t &= \theta_{t-1} + \mathrm{N}_1(\mu_1, \delta_1) \\ C_t &= f(\theta_t, \bar{\varepsilon}_t) + \mathrm{N}_2(\mu_2, \delta_2) \end{aligned} \tag{7.8}$$

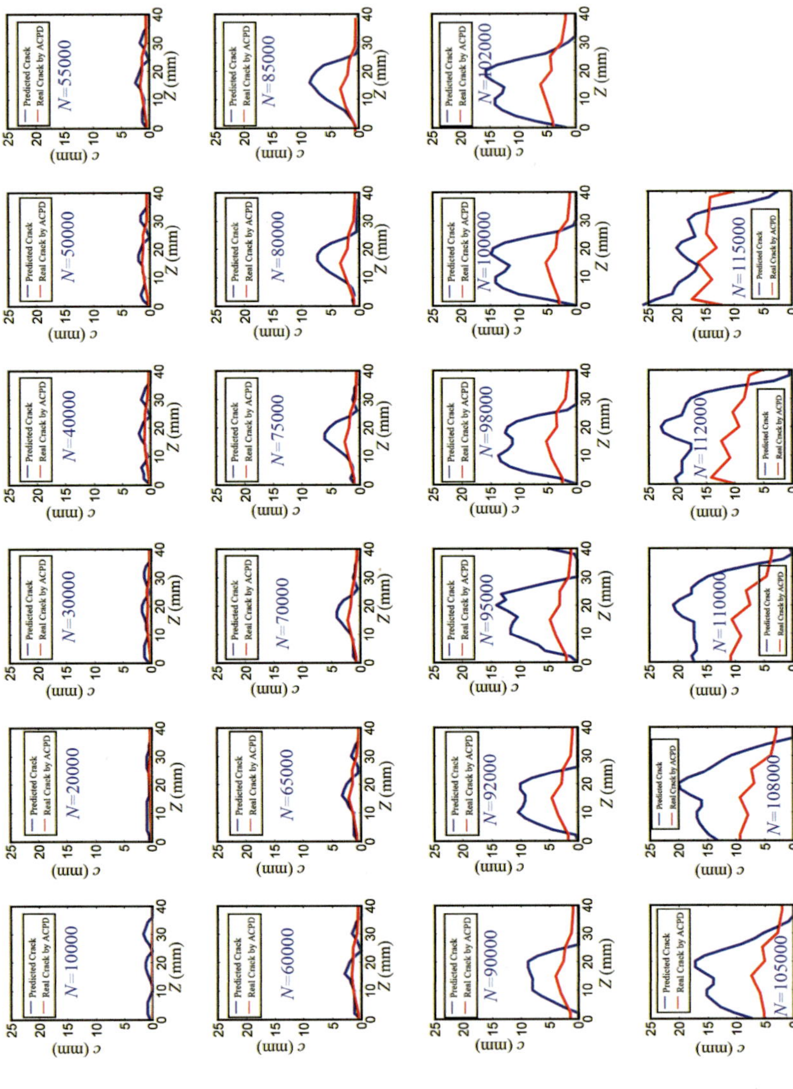

FIGURE 7.19 The comparison between the measured crack profiles and the predicted crack profiles based on the NN model under different fatigue cycles.

where θ_t and θ_{t-1} represent the parameters (ω and ξ) in the trained NN models at time t and t-1; $\mathrm{N}_1(\mu_1,\delta_1)$ and $\mathrm{N}_2(\mu_1,\delta_2)$ denote the normal distribution with the mean value of μ_1,μ_2 and the standard deviation of δ_1,δ_2; $f(\theta_t,\overline{\varepsilon}_t)$ represents the previously trained NN models; C_t is the predicted crack depth based on different sets of θ_t. The predicted crack depth C_t entails an uncertainty distribution N_2. Based on the crack measurement techniques (ACPD and UPA nondestructive approaches), the uncertainty distribution tends to have a normal distribution with a mean value of zero. The candidate parameters leading to the best-predicted crack depth closest to the crack measurement data have the highest assigned weight in the BPF algorithm. Based on the BPF procedure, this study selects the uncertainty characterized by

$$\theta_t = \theta_{t-1} + \varsigma_t \tag{7.9a}$$

$$\varsigma_t \sim N(0,0.1) \tag{7.9b}$$

After examination of different distribution parameters, the bias parameter (ς_t) assumes a normal distribution with the mean value of 0.0 and a standard deviation of 0.1 for illustration purposes. For the ANN model, this study assigns the noise terms in the bias parameters (ξ). Figure 7.20 presents the whole framework of the NN-BPF algorithm in this study.

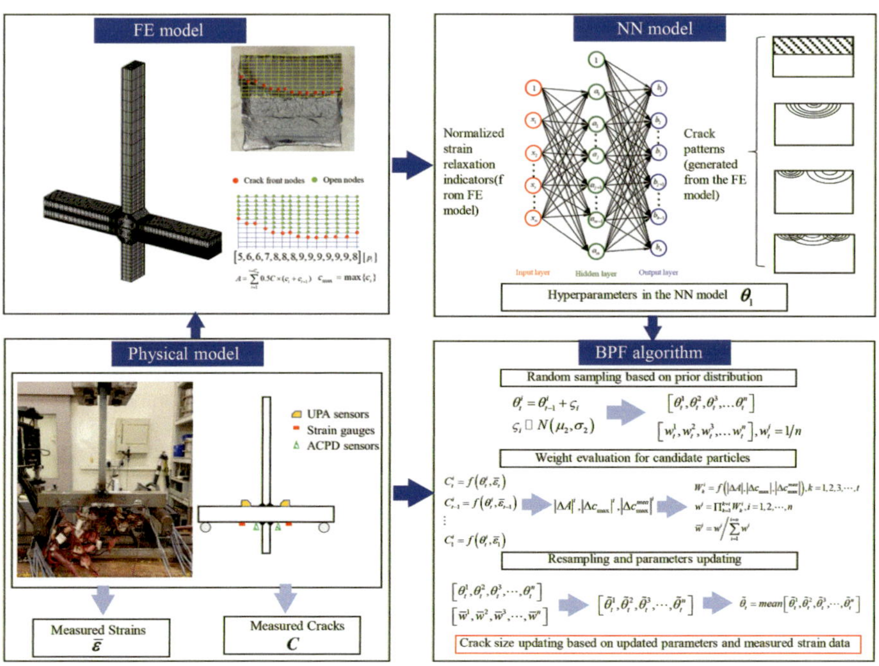

FIGURE 7.20 The framework to size and update the crack information.

- Generation of Random Particles Based on Normal Distribution
  Relying on Eq. (7.9), this study generates 10000 candidates of $\left[\theta_t^1, \theta_t^2, \theta_t^3, \ldots \theta_t^n\right]$. Initially, the weight W_t^i of each candidate θ_t^i has a constant value of $1/n$.
- Calculate the Weights of Each Candidates
  Considering all the crack measurement data from 1 to *t*, the summative updated weights w_t^i of each candidate θ_t^i equals the product of a series of weights W_n^i using Eq. (7.10).

$$w^i = \prod_{n=1}^{t} W_n^i \tag{7.10}$$

$$\begin{aligned} W_t^i &= 1/\left(W_t^i\right)_1 \times \left(W_t^i\right)_2 \times \left(W_t^i\right)_3 = f\left(|\Delta A|, |\Delta c_{\max}|, \left|\Delta c_{\max}^{mea}\right|\right) \\ &= 1/\left(|\Delta A| \times |\Delta c_{\max}| \times \left|\Delta c_{\max}^{mea}\right|\right) \end{aligned} \tag{7.11}$$

$$\begin{aligned} &\left(W_t^i\right)_1 = \left|\Delta c_{\max}^{mea}\right| + 0.01 = \max\left\{\left|c_k^{mea} - y_k^{mea}\right|\right\} + 0.01 \\ &\left(W_t^i\right)_2 = |\Delta A| + 0.01 = \max\left\{\left|A^{NN} - A^{mea}\right|\right\} + 0.01 \\ &\left(W_t^i\right)_3 = |\Delta c_{\max}| + 0.01 = \max\left\{\left|\left[c_m\right]_{\max} - \left[y_k^{mea}\right]_{\max}\right|\right\} + 0.01 \\ &(m = 1, 2, \ldots, 21;\ k = 1, 2 \ldots, 9) \end{aligned} \tag{7.12}$$

where A^{NN} denotes the predicted crack area based on the trained NN model; A^{mea} represents the measured crack area in the test; c_k^{mea} denotes the crack at *k*th measured location extrapolated by the 21 predicted crack front points; y_k^{mea} is the *k*th measured crack front depth. $\left[c_m\right]_{\max}$ is the maximum value of the 21 predicted crack front depths; $\left[y_k^{mea}\right]_{\max}$ denotes the maximum value of the nine measured crack front depths (seven points by the ACPD and two points at the side surface by rulers). The detailed calculation of the A^{NN}, A^{mea}, and c_k^{mea} follows

$$A^{NN} = \sum_{m=2}^{m=21} 0.5\left|Z_m - Z_{m-1}\right| \times \left(c_m + c_{m-1}\right) \tag{7.13}$$

$$A^{mea} = \sum_{k=2}^{k=9} 0.5\left|Z_k - Z_{k-1}\right| \times \left(c_k^{mea} + c_{k-1}^{mea}\right) \tag{7.14}$$

$$\begin{aligned} &c_k^{mea} = c_{m-1} + \frac{\left(c_m - c_{m-1}\right) \times \left(Z_k^{mea} - Z_{m-1}\right)}{Z_m - Z_{m-1}}, \\ &m = 1, 2, \ldots 21; k = 1, 2, \ldots 9 \ \text{ if } Z_k^{mea} \in \left[Z_{m-1}, Z_m\right] \end{aligned} \tag{7.15}$$

- Importance cation

After obtaining the summative weighted values w^i of different sets of θ_t^i data, this study derives the normalized weighted values $\bar{w}^i$ (Eq. (7.16))

$$\bar{w}^i = w^i \Big/ \sum_{i=1}^{i=n} w^i \tag{7.16}$$

- Resampling Based on Normalized Weights and Update the Parameters

Based on the normalized weights (Eq. (7.16)), this study generates n numbers of candidates $\theta^i, i = 1, 2 \cdots n$. Therefore, the updated parameters of the bias parameters in the NN model derive from

$$\theta_t^f = mean\{\theta_t^1, \theta_t^2, \ldots \theta_t^n\} \tag{7.17}$$

Therefore, based on the updated parameters, the fatigue crack at time $t + 1$ can be determined based on the measured strain information at time $t + 1$.

This study utilizes the parameters updated in a previous step to predict the crack profile for subsequent steps until the parameters are updated again. For example, the predicted crack profile at the 70000th fatigue cycle employs the updated parameters θ_t^f in the 65000th fatigue cycle. Figure 7.21 compares the predicted crack profile based on the NN-BPF algorithm and measured crack profile. The blue curve represents the predicted crack profile at time t based on the updated parameters at time t-1.

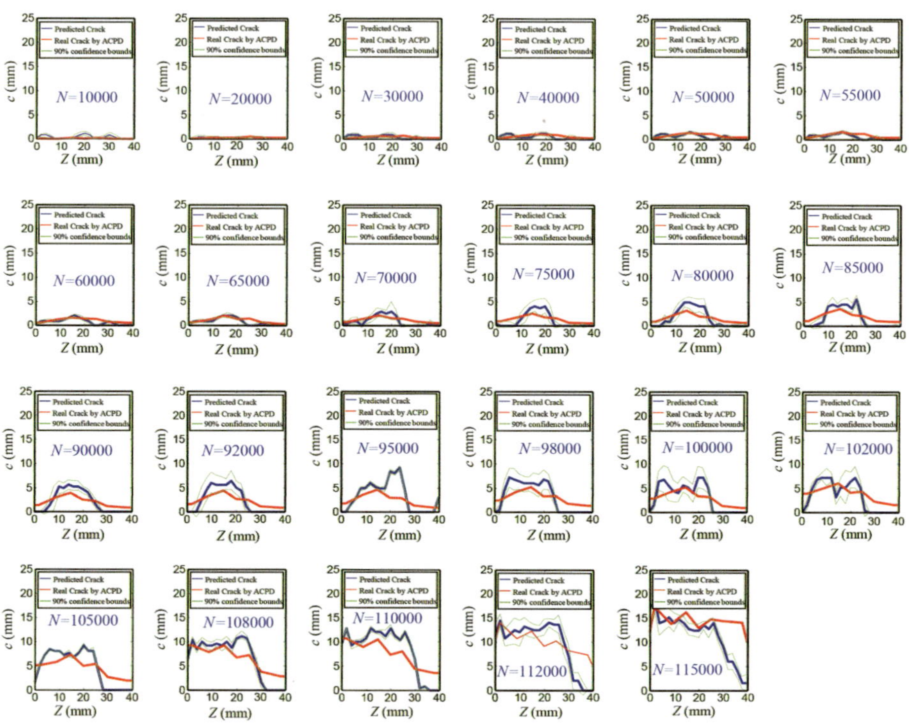

FIGURE 7.21 The comparison between the measured crack profiles and the updating predicted crack profiles based on the NN-BPF algorithm under different fatigue cycles.

The red curve indicates the measured crack profile. In comparison with Figure 7.19, the updated parameters lead to pronounced improvement in the crack sizing.

7.4 DIGITAL TWIN OF WELDED JOINTS UNDER FATIGUE LOADINGS

7.4.1 Experimental Setup

As demonstrated in Section 7.1, the digital twin concept has been applied across various industry sectors. This section integrates the FEM analysis in Chapters 6 and 7 with intelligent methodologies to construct a digital twin of welded plate joints. The system achieves key functionalities, including real-time monitoring, fatigue damage prediction, crack detection, and estimation of remaining fatigue life. As illustrated in Figure 7.22, the hardware system for the digital twin comprises a webcam and an oscilloscope. The digital twin targets both global and local behaviors of cruciform welded plate joints. Global behavior includes parameters such as displacements and external loads, while the local behavior focuses on aspects such as strain distribution, crack initiation, and propagation. The virtual replica of the welded plate joints is modeled using the ABAQUS FEM software, enabling accurate simulation and analysis.

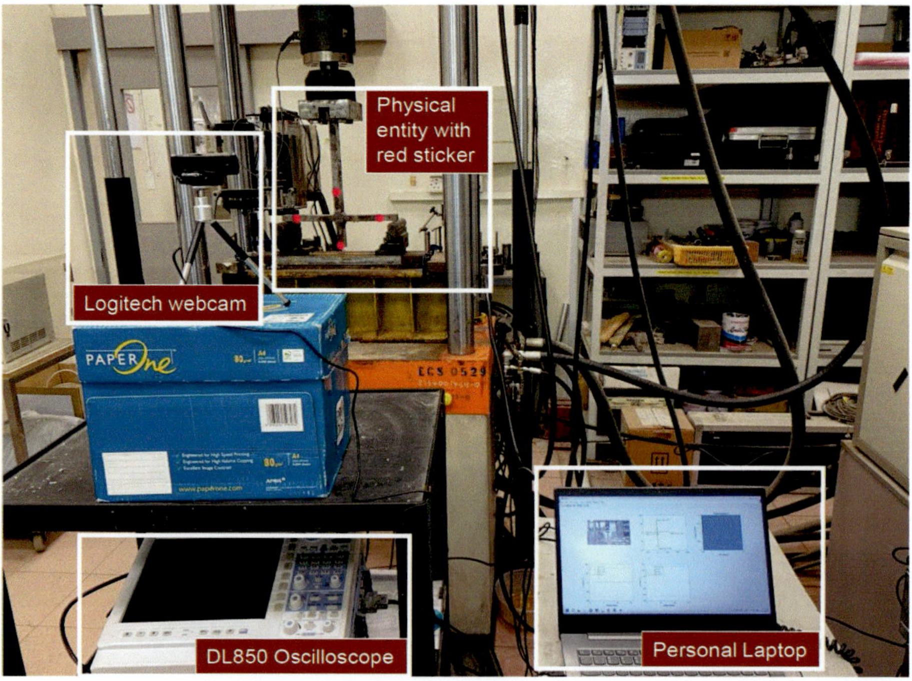

FIGURE 7.22 Hardware and software setup for the digital twin of the welded plate joint.

As illustrated in Figure 7.23, the digital twin comprises several key components, including the physical entity (welded plate joint) marked with four red adhesive stickers. These red stickers are affixed to both horizontal and vertical plates. The experimental setup is demonstrated in Section 7.3.1. The welded cruciform joints are subjected to three-point bending. A Logitech webcam positioned in front of the welded joint captures real-time movements of the red stickers. Strain gauges adhesive to the welded plate joints follow the same location and arrangements in Section 7.3. The strain gauge data is captured through the DL850 oscilloscope at a high frequency of 100Hz. The hardware (webcam and oscilloscope) are integrated via the in-house developed algorithm in Matlab, as shown in Figure 7.23. The digital twin operates as follows:

- Video Capture and Processing: The webcam captures videos, which are automatically uploaded to the Zoom video cloud. The custom Matlab algorithm downloads and processes these videos, tracking the positions of the red stickers.
- Data Acquisition and Processing: Strain, load, and displacement data recorded by the DL850 oscilloscope are transferred to Matlab platform using the Matlab-Yokogawa module. The custom algorithm processes the strain data and monitors cracks using the strain-based crack monitoring method described in Sections 7.2 and 7.3.
- FEA: Crack information derived from the processed data is input for updating the FE model in ABAQUS. The updated model simulates crack propagation and predicts the remaining fatigue life using the damage constitutive model detailed in Chapter 6.

The subsequent sections provide a comprehensive explanation of the methodologies and algorithms employed in developing the digital twin for welded joints.

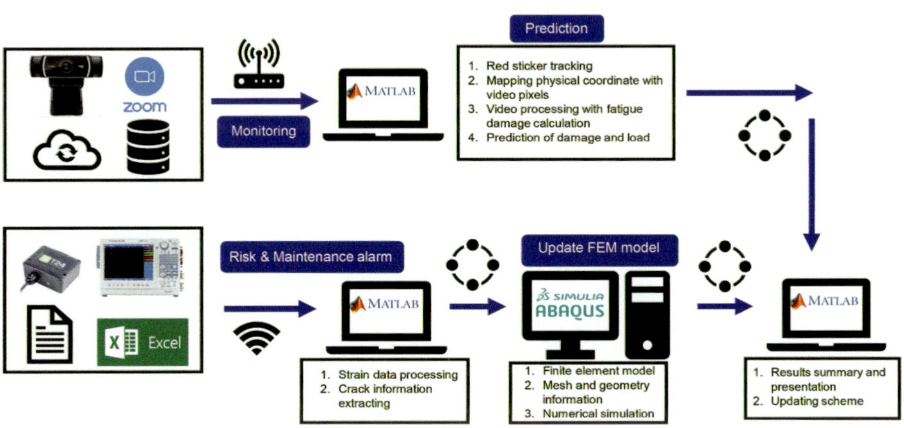

FIGURE 7.23 Functional features of digital twins of the welded plate joint.

7.4.2 Calibration of Hardware and Software

Integrating multiple sensors necessitates calibration and connection testing. The calibration process involves applying monotonic loading to the welded plate, as illustrated in Figure 7.24. For the webcam calibration, the physical coordinates of four red stickers are mapped to their corresponding virtual pixel coordinates as captured in the webcam video. This study proposes the transfer matrix *[T]* to correlate these two coordinate systems, as expressed in Eq. (7.18):

$$\begin{bmatrix} X \\ Y \\ Z \end{bmatrix} = \begin{bmatrix} T_{11} & T_{12} & T_{13} \\ T_{21} & T_{22} & T_{23} \\ T_{31} & T_{32} & T_{33} \end{bmatrix} \begin{bmatrix} PX \\ PY \\ PZ \end{bmatrix} \tag{7.18}$$

where $\begin{bmatrix} X & Y & Z \end{bmatrix}'$ represents the physical coordinates of the four red stickers, while $\begin{bmatrix} PX & PY & PZ \end{bmatrix}'$ denotes the pixel or virtual coordinates of the four red stickers captured by the webcam. In this study, since Z and PZ remain constant, Eq. (7.18) can be simplified to:

$$\begin{bmatrix} X \\ Y \end{bmatrix} = \begin{bmatrix} T_{11} & T_{12} \\ T_{21} & T_{22} \end{bmatrix} \begin{bmatrix} PX \\ PY \end{bmatrix} \tag{7.19}$$

To determine the transfer matrix, this study adopts the genetic optimization algorithm [42]. The second calibration procedure is to optimize the virtual replica FE model. From the DL 850 oscilloscope, the external load on the top attachment plate of the welded joints, and the strain gauge data can be obtained. Simultaneously,

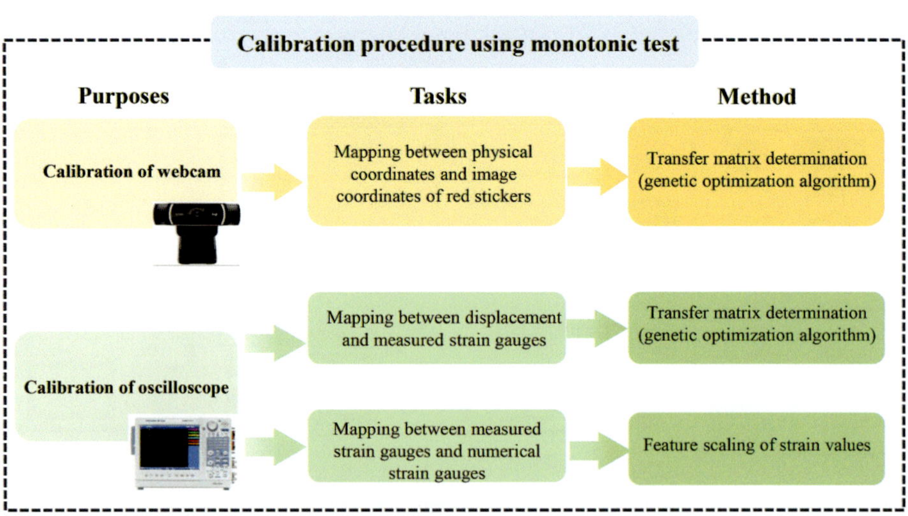

FIGURE 7.24 Calibration procedure for webcam and oscilloscope.

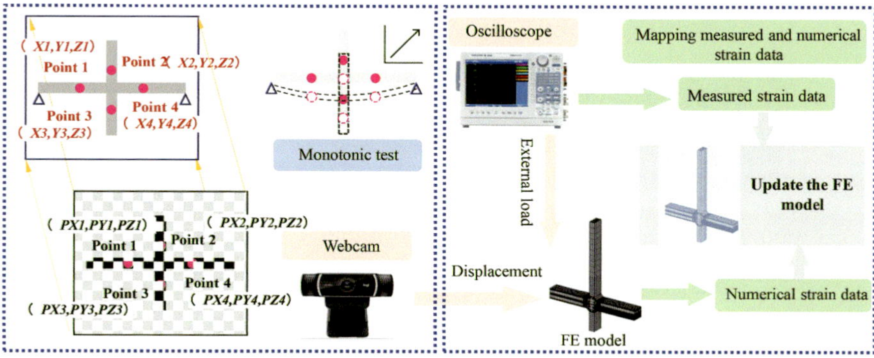

FIGURE 7.25 Detailed calibration procedure integrating different sensors and FE model.

the displacements of the red stickers are tracked. The correlation of the displacement and load reflects the overall stiffness of the welded joint and servers as validation data for the boundary conditions of the welded joint's FE model. Based on the updated boundary condition, numerical strain data can be extracted from the FE model. By comparing this numerical strain data with the strain gauge measurements, the local weld geometry of the joint can be iteratively refined to align the numerical and experimental results, as shown in Figure 7.25. Performing the calibration test illustrated in Figure 7.25 is essential to achieving high accuracy in the digital twin model.

7.4.3 Monitoring Function of the Digital Twin

As shown in Figure 7.26, the monitoring functions include: 1) real-time tracking the displacement of four red stickers $\begin{bmatrix} PX \\ PY \end{bmatrix}$ and $\begin{bmatrix} X \\ Y \end{bmatrix}$ (Figures 7.26a, b, and e); 2) fatigue cycles counting (Figure 7.26c) through the rain flow counting method [77]; 3) the overall accumulative fatigue damage D (Figure 7.26d); 4) the hot spot stress interpreted from the global displacement (Figure 7.26f).

Figure 7.26a shows the real-time video with the virtual coordinates of four red stickers, while Figure 7.26b represents the physical coordinates of the four red stickers based on Eq. (7.18). Figure 7.26e presents the historical data of the Y physical coordinates of the four red stickers. To count the total fatigue cycles, this study adopts the rain flow counting method, as shown in Figure 7.27. Details of the rain flow counting method can be found in [77]. To evaluate the accumulative fatigue damages, this study proposes the relationship between the displacement of red stickers and the hot spot stress on the FE model, as shown in Figure 7.27. The relationship [Eq. (7.20)] is linear with the assumption of the linear elastic analysis.

$$\Delta\sigma_{hs} = \begin{bmatrix} a & b \end{bmatrix} \begin{bmatrix} \Delta X \\ \Delta Y \end{bmatrix} \tag{7.20}$$

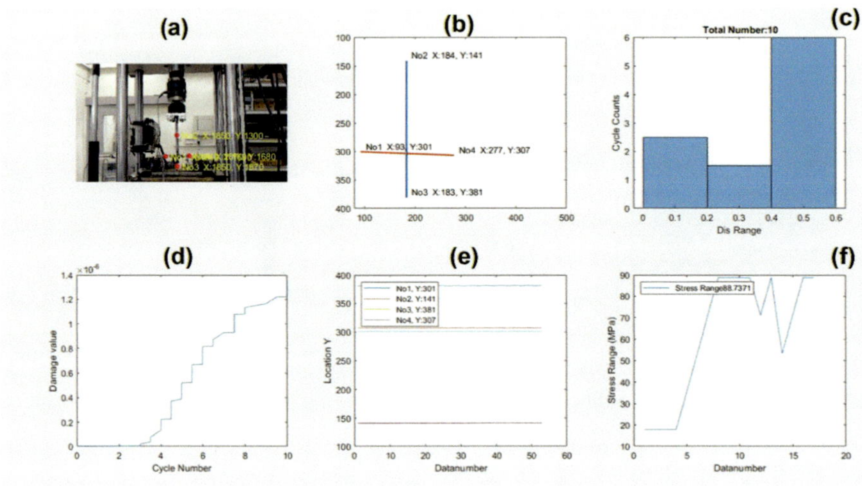

FIGURE 7.26 Visualization of the monitoring function.

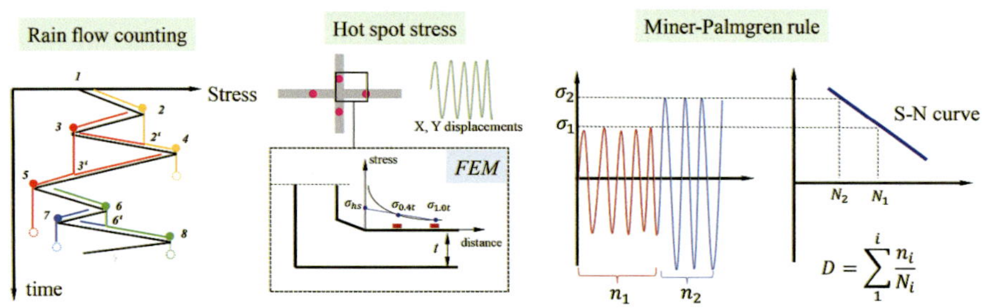

FIGURE 7.27 Rain flow counting, hot-spot stress determination, and Miner-Palmgren damage accumulation rule.

where $\Delta\sigma_{hs}$ represent the range of the hot spot stress defined based on IIW [78], $\begin{bmatrix}\Delta X\\ \Delta Y\end{bmatrix}$ denotes the range of the displacement of the red stickers, and $\begin{bmatrix}a & b\end{bmatrix}$ are the fitted constants to reflect the linear relationship between the displacement and the hot spot stress. Therefore, through the monitored displacement values and Eq. (7.20), the hot spot stress range can be monitored in real-time.

The accumulative fatigue damage indicator D is determined through the linear damage accumulation law (Miner-Palmgren damage accumulation rule) [7.101], as shown in Figure 7.32. Eq. (7.21) presents the general form of the linear damage accumulation rule.

$$D = \sum_{i=1}^{i=n} \frac{n_i}{N_i} \tag{7.21}$$

where D denotes the cumulative total damage, n_i represents the fatigue cycles under ith stress range and N_i denotes the total fatigue life subjected to the ith hot-spot stress range. The total fatigue life N_i can be determined using the following S-N curve FAT 100 [79]

$$LogN_f = 12.30 - 3LogS \tag{7.22}$$

Therefore, through the real-time video captured by the webcam, the hot-spot stress, the fatigue cycles, and the total fatigue damage can be determined using the above methods.

7.4.4 Prediction of the Fatigue Displacements and Damage

As shown in Figure 7.28, this study develops the algorithm to record the historical load information from the oscilloscope (red hollow circles) and predict the load information using the LSTM algorithm [80]. The LSTM network is a type of recurrent neural network (RNN) that processes sequential data by iterating over time steps and updating its internal state. This state retains information from all prior time steps, enabling the network to effectively capture temporal dependencies. As shown in Figure 7.29, the LSTM architecture comprises four primary components responsible for managing the cell state and hidden state of the layer. The trainable parameters of an LSTM layer include the input weights $\boldsymbol{W}$ (InputWeights), the recurrent weights $\boldsymbol{R}$ (RecurrentWeights), and the bias $\boldsymbol{b}$ (Bias). The input gate, forget gate, cell candidate, and output gate are as follows:

$$i_t = \sigma_g\left(W_i x_t + R_i h_{t-1} + b_i\right) \tag{7.23a}$$

$$f_t = \sigma_g\left(W_f x_t + R_f h_{t-1} + b_f\right) \tag{7.23b}$$

$$g_t = \sigma_g\left(W_g x_t + R_g h_{t-1} + b_g\right) \tag{7.23c}$$

$$o_t = \sigma_g\left(W_o x_t + R_o h_{t-1} + b_o\right) \tag{7.23d}$$

The cell state at time step t is given by

$$c_t = f_t \odot c_{t-1} + i_t \odot g_t \tag{7.24}$$

where $\odot$ denotes the Hadamard product, the hidden state at time t is given by

$$h_t = o_t \odot \sigma_c\left(c_t\right) \tag{7.25}$$

where σ_c denotes the state activation function. Therefore, based on the predicted displacement (Figure 7.28a) and the framework in Section 7.4.2, the predicted fatigue damage can be derived (Figure 7.28b).

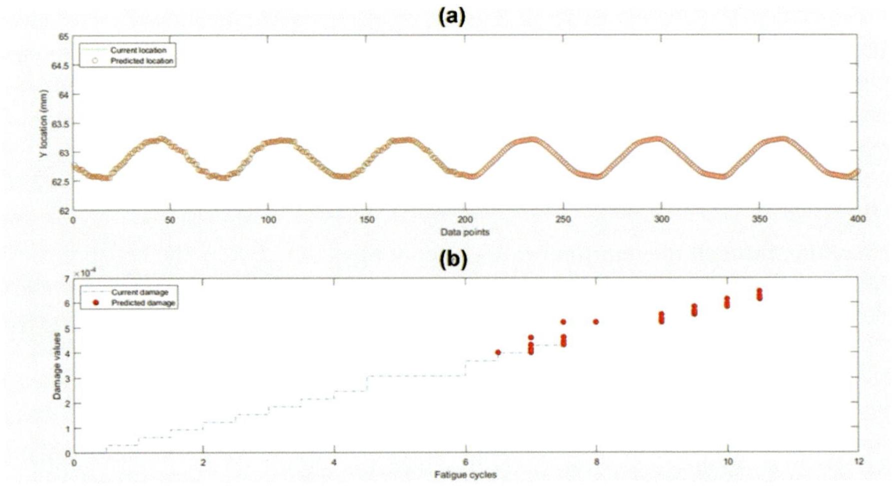

FIGURE 7.28 Prediction of the displacement of red stickers and fatigue damage.

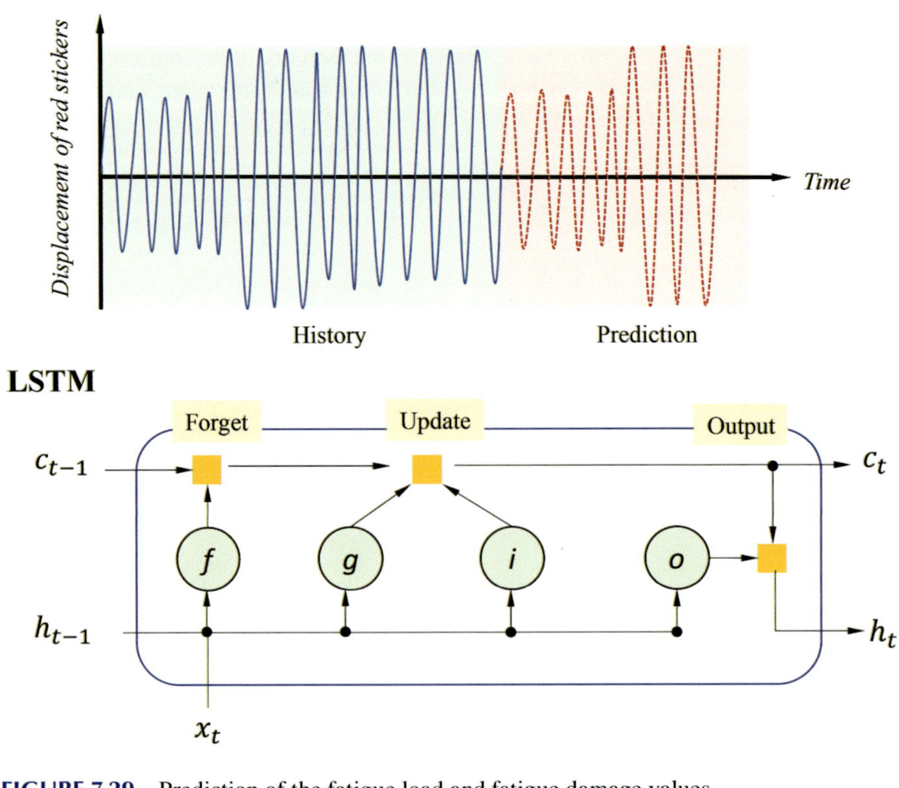

FIGURE 7.29 Prediction of the fatigue load and fatigue damage values.

7.4.5 Crack Size Alarm and FE Model Update

By monitoring the normalized strain relaxation indicator (Eq. (7.1)) in Figure 7.30c, a crack size alarm is established based on its value. In this study, two crack size alarms are configured, corresponding to a 10% (Figure 7.30a) and 20% (Figure 7.30b) reduction in the normalized strain relaxation indicator. Incorporating the intelligent approach described in Section 7.3, the crack profile of the welded plate joint, along with the width direction (Figure 7.30d) is also effectively monitored.

Based on the derived crack information, the FE model will be updated periodically. However, the numerical calculation for fatigue life is computationally intensive. As illustrated in Figure 7.31, this study tracks fatigue crack information at 50000th, 80000th, 100000th fatigue cycle. Based on these observations, a pre-crack model is generated in ABAQUS FE software, reflecting the measured crack profiles shown in Figure 7.32. Figures 7.32a–d present the damage failure under non-existing crack model, and pre-existing cracked model under 50000th, 80000th and 100000th fatigue cycle, respectively. The final failure indicated by the damage distribution shows some difference along the side surface in the FE model. Figure 7.33 presents the crack evolution under different fatigue cycles along the width direction based on different pre-existing crack profiles. The crack evolves differently with different pre-existing crack profiles. Without any existing crack profile, the crack depth along the width of the welded plate joints is almost uniform. With pre-existing crack profile, the evolution of the crack maintains a similar shape to the pre-existing crack profiles. Figure 7.34a compares the final crack profiles by the numerical simulation and the silicon replica from the experimental tests. With the information from the third crack measurement, the final crack profile aligns closely with the experimental results,

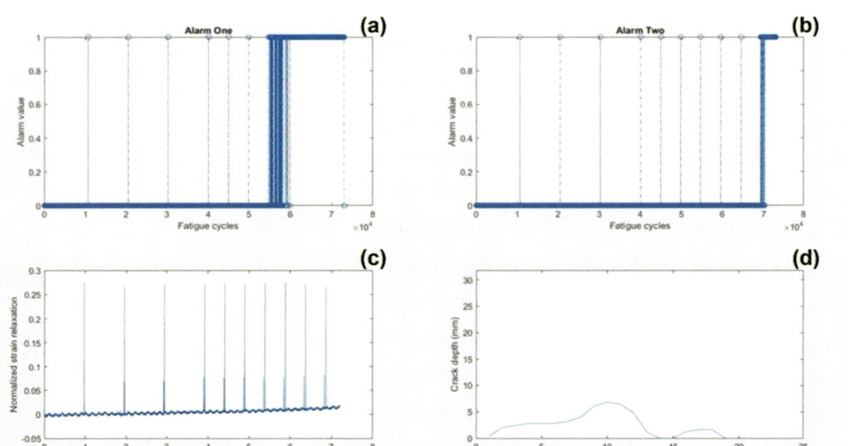

FIGURE 7.30 (a, b) Alarm based on the 10% and 20% reduction of the normalized strain relaxation indicator; (c) the historical data of the normalized strain relaxation indicator; (d) derived crack profile in the width direction based on the intelligent approach.

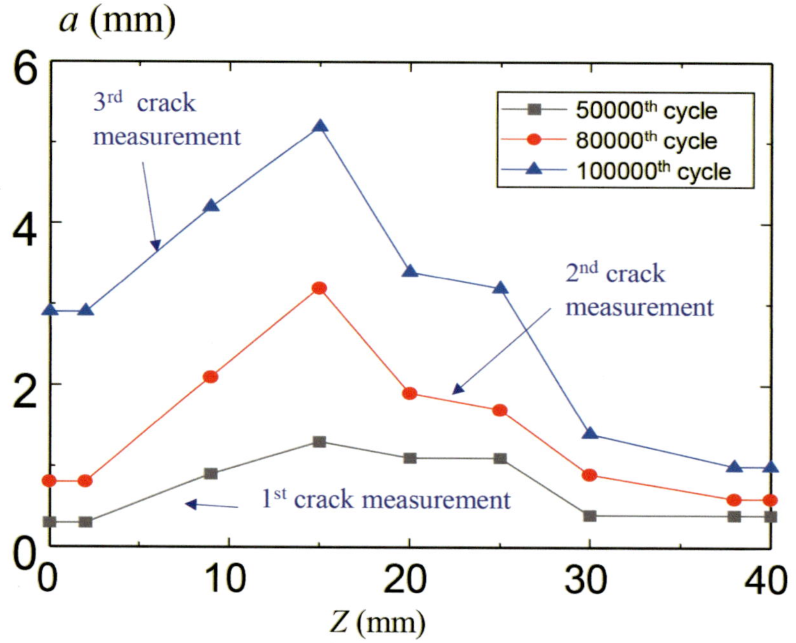

FIGURE 7.31 The measured crack information and corresponding fatigue cycles.

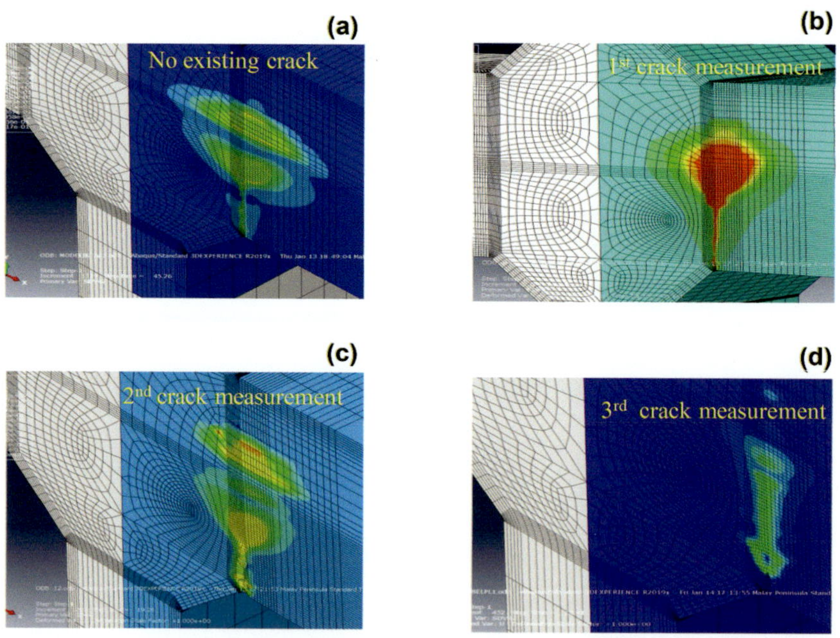

FIGURE 7.32 (a, b) Alarm based on the 10% and 20% reduction of the normalized strain relaxation indicator; (c) the historical data of the normalized strain relaxation indicator; (d) derived crack profile in the width direction based on the intelligent approach.

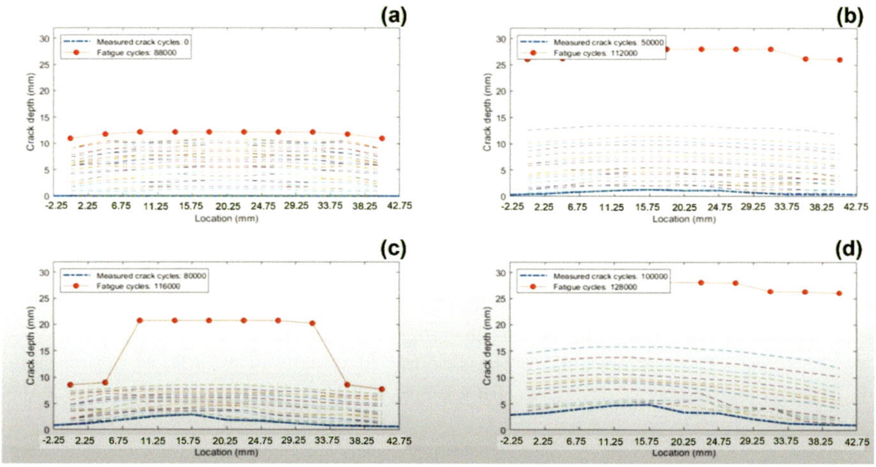

FIGURE 7.33 The fatigue crack propagation behavior under different pre-crack profile.

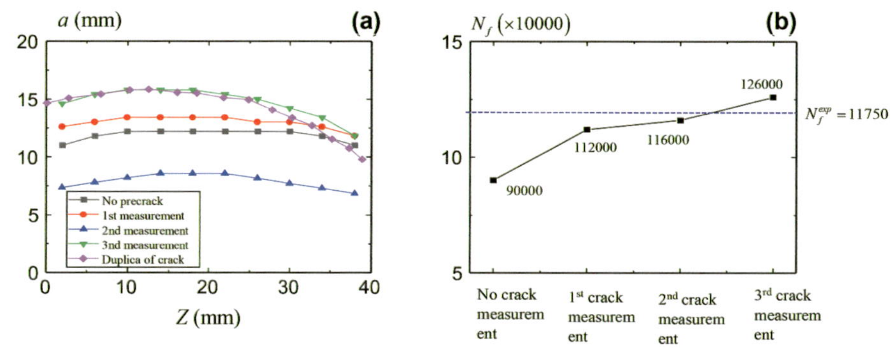

FIGURE 7.34 The final failure based on the damage model with different pre-existing measured cracks.

validating that this crack data can significantly improve the prediction of fatigue crack propagation. Figure 7.34b summarizes the numerical simulation of the fatigue life for this BT40 specimen. Without any pre-existing crack, the predicted fatigue life is 90000, in comparison with the fatigue life value of 117500. In contrast, with any of the three crack measurements, the predicted fatigue life can be significantly enhanced, with a much smaller gap between the numerical simulation fatigue life and experimental fatigue life.

REFERENCES

[1] G.M. Bisanti, L. Mainetti, T. Montanaro, L. Patrono, I. Sergi, Digital twins for aircraft maintenance and operation: A systematic literature review and an IoT-enabled modular architecture, Internet of Things, 24 (2023) 100991.

[2] X. Lai, L. Yang, X. He, Y. Pang, X. Song, W. Sun, Digital twin-based structural health monitoring by combining measurement and computational data: An aircraft wing example, Journal of Manufacturing Systems, 69 (2023) 76–90.
[3] W. Liu, M. Wu, G. Wan, M. Xu, Digital twin of space environment: Development, challenges, applications, and future outlook, Remote Sensing, 16 (2024) 3023.
[4] P.S. Ambalkar, P.M. Mandhare, P.P. Ninawe, S.R. Shingne, R.R. Singh, P. Jaronde, Digital twins and ai, the cosmic revolution in space technology: Revolutionizing space operations, digital twins, and ai in action, in: Harnessing AI and digital twin technologies in businesses, IGI Global, 2024, pp. 132–142.
[5] J. Shi, X. Wang, Z. Zhang, C. Zhang, J. Chen, Optimization of energy flow in thermal management of electric vehicles based on real vehicle testing and digital twin simulation, Case Studies in Thermal Engineering, 60 (2024) 104607.
[6] P. Rajesh, T. Soundarya, K. Jithin, Driving sustainability-the role of digital twin in enhancing battery performance for electric vehicles, Journal of Power Sources, 604 (2024) 234464.
[7] M.U. Shoukat, L. Yan, Y. Yan, F. Zhang, Y. Zhai, P. Han, S.A. Nawaz, M.A. Raza, M.W. Akbar, A. Hussain, Autonomous driving test system under hybrid reality: The role of digital twin technology, Internet of Things, 27 (2024) 101301.
[8] B. Yingngam, J. Khumsikiew, R. Netthong, Digital twins for telemedicine and personalized medicine, in: Digital twins for smart cities and billages, Elsevier, 2025, pp. 295–325.
[9] K. Dziopa, K. Lekadir, P. van der Harst, F.W. Asselbergs, Digital twins: Reimagining the future of cardiovascular risk prediction and personalised care, Hellenic Journal of Cardiology, 81 (2025) 4–8.
[10] A. Trabelsi, J. Sandoval, A. Mlika, S. Lahouar, M.A. Laribi, Design and implementation of a digital twin for a multi-robot surgical platform, in: IFToMM international symposium on robotics and mechatronics, Springer, 2024, pp. 207–216.
[11] N.J. Rowan, Digital technologies to unlock safe and sustainable opportunities for medical device and healthcare sectors with a focus on the combined use of digital twin and extended reality applications: A review, Science of the Total Environment, 926 (2024) 171672.
[12] Y. Han, Y. Li, Y. Li, B. Yang, L. Cao, Digital twinning for smart hospital operations: Framework and proof of concept, Technology in Society, 74 (2023) 102317.
[13] O. Das, M.H. Zafar, F. Sanfilippo, S. Rudra, M.L. Kolhe, Advancements in digital twin technology and machine learning for energy systems: A comprehensive review of applications in smart grids, renewable energy, and electric vehicle optimisation, Energy Conversion and Management: X, 24 (2024) 100715.
[14] M. Samnejad, M. Gharib Shirangi, R. Ettehadi, A digital twin of drilling fluids rheology for real-time rig operations, in: Offshore technology conference, OTC, 2020, pp. D011S010R006.
[15] T.R. Wanasinghe, L. Wroblewski, B.K. Petersen, R.G. Gosine, L.A. James, O. De Silva, G.K. Mann, P.J. Warrian, Digital twin for the oil and gas industry: Overview, research trends, opportunities, and challenges, IEEE Access, 8 (2020) 104175–104197.
[16] E. Kandemir, J. Liu, A. Hasan, Digital twin-driven dynamic repositioning of floating offshore wind farms, Energy Reports, 9 (2023) 208–214.
[17] Z. Mousavi, S. Varahram, M.M. Ettefagh, M.H. Sadeghi, W.-Q. Feng, M. Bayat, A digital twin-based framework for damage detection of a floating wind turbine structure under various loading conditions based on deep learning approach, Ocean Engineering, 292 (2024) 116563.
[18] M. Fahim, V. Sharma, T.-V. Cao, B. Canberk, T.Q. Duong, Machine learning-based digital twin for predictive modeling in wind turbines, IEEE Access, 10 (2022) 14184–14194.

[19] M. Soori, B. Arezoo, R. Dastres, Digital twin for smart manufacturing, a review, Sustainable Manufacturing and Service Economics, 2 (2023) 100017.
[20] S. Bakhshi, A. Ghaffarianhoseini, A. Ghaffarianhoseini, M. Najafi, F. Rahimian, C. Park, D. Lee, Digital twin applications for overcoming construction supply chain challenges, Automation in Construction, 167 (2024) 105679.
[21] Z.U. Rizqi, S.-Y. Chou, W.N. Cahyo, A simulation-based digital twin for smart warehouse: Towards standardization, Decision Analytics Journal, 12 (2024) 100509.
[22] F. Jiang, L. Ma, T. Broyd, W. Chen, H. Luo, Digital twin enabled sustainable urban road planning, Sustainable Cities and Society, 78 (2022) 103645.
[23] K. Kušić, R. Schumann, E. Ivanjko, A digital twin in transportation: Real-time synergy of traffic data streams and simulation for virtualizing motorway dynamics, Advanced Engineering Informatics, 55 (2023) 101858.
[24] W.J. Steyn, A. Broekman, Development of a digital twin of a local road network: A case study, Journal of Testing and Evaluation, 50 (2022) 2901–2915.
[25] M. Torzoni, M. Tezzele, S. Mariani, A. Manzoni, K.E. Willcox, A digital twin framework for civil engineering structures, Computer Methods in Applied Mechanics and Engineering, 418 (2024) 116584.
[26] N.A. Ubina, H.-Y. Lan, S.-C. Cheng, C.-C. Chang, S.-S. Lin, K.-X. Zhang, H.-Y. Lu, C.-Y. Cheng, Y.-Z. Hsieh, Digital twin-based intelligent fish farming with Artificial Intelligence Internet of Things (AIoT), Smart Agricultural Technology, 5 (2023) 100285.
[27] D. Bousnina, G. Guerassimoff, Optimal energy management in smart energy systems: A deep reinforcement learning approach and a digital twin case-study, Smart Energy, 16 (2024) 100163.
[28] X. Wang, H. Yu, W. McGee, C.C. Menassa, V.R. Kamat, Enabling building information model-driven human-robot collaborative construction workflows with closed-loop digital twins, Computers in Industry, 161 (2024) 104112.
[29] J. Xu, X. Shu, P. Qiao, S. Li, J. Xu, Developing a digital twin model for monitoring building structural health by combining a building information model and a real-scene 3D model, Measurement, 217 (2023) 112955.
[30] D.D. Eneyew, M.A. Capretz, G.T. Bitsuamlak, Continuous model calibration framework for smart-building digital twin: A generative model-based approach, Applied Energy, 375 (2024) 124080.
[31] A. Sharma, E. Kosasih, J. Zhang, A. Brintrup, A. Calinescu, Digital twins: State of the art theory and practice, challenges, and open research questions, Journal of Industrial Information Integration, 30 (2022) 100383.
[32] K. Chen, X. Zhu, B. Anduv, X. Jin, Z. Du, Digital twins model and its updating method for heating, ventilation and air conditioning system using broad learning system algorithm, Energy, 251 (2022) 124040.
[33] C. Gehrmann, M. Gunnarsson, A digital twin based industrial automation and control system security architecture, IEEE Transactions on Industrial Informatics, 16 (2019) 669–680.
[34] W. Hofmann, F. Branding, Implementation of an IoT-and cloud-based digital twin for real-time decision support in port operations, IFAC-PapersOnLine, 52 (2019) 2104–2109.
[35] A.N. Nasaruddin, T. Ito, T.B. Tuan, Digital twin approach to building information management, in: The proceedings of manufacturing systems division conference 2018, The Japan Society of Mechanical Engineers, 2018, p. 304.
[36] A. Peng, Y. Ma, K. Huang, L. Wang, Digital twin-driven framework for fatigue life prediction of welded structures considering residual stress, International Journal of Fatigue, 181 (2024) 108144.

[37] Y. Zhang, R. Hao, J. Niiranen, Y. Yang, E. Brühwiler, D. Su, T. Nagayama, Computational engine for finite element digital twins of structural dynamics via motion data, Engineering Structures, 316 (2024) 118630.
[38] Z. Zhu, J. Zhang, S. Zhu, J. Yang, Digital twin technology for wind turbine towers based on joint load–response estimation: A laboratory experimental study, Applied Energy, 352 (2023) 121953.
[39] Z. ElArwady, A. Kandil, M. Afiffy, M. Marzouk, Modeling indoor thermal comfort in buildings using digital twin and machine learning, Developments in the Built Environment, 19 (2024) 100480.
[40] S. Chen, G. Fan, J. Li, Improving completeness and accuracy of 3D point clouds by using deep learning for applications of digital twins to civil structures, Advanced Engineering Informatics, 58 (2023) 102196.
[41] L. Feng, T. Liu, X. Qian, C. Chen, A complete integrity assessment of welded connections under high and low cycle fatigue followed by fracture failure, Steel and Composite Structures, an International Journal, 43 (2022) 465–481.
[42] L. Feng, X. Qian, Enhanced sizing for surface cracks in welded tubular joints using ultrasonic phased array and image processing, NDT & E International, 116 (2020) 102334.
[43] D. Shamia, S. Suganyadevi, V. Satheeswaran, K. Balasamy, Digital twins in precision agriculture monitoring using artificial intelligence, in: Digital twin for smart manufacturing, Elsevier, 2023, pp. 243–265.
[44] S. Almeaibed, S. Al-Rubaye, A. Tsourdos, N.P. Avdelidis, Digital twin analysis to promote safety and security in autonomous vehicles, IEEE Communications Standards Magazine, 5 (2021) 40–46.
[45] E.W.W. Cheok, X. Qian, C. Chen, S.T. Quek, M.B.I. Si, A local digital twin approach for identifying, locating and sizing cracks in CHS X-joints subjected to brace axial loading, Engineering Structures, 299 (2024) 117085.
[46] T. Wang, K. Feng, J. Ling, M. Liao, C. Yang, R. Neubeck, Z. Liu, Pipeline condition monitoring towards digital twin system: A case study, Journal of Manufacturing Systems, 73 (2024) 256–274.
[47] S.-W. Zhou, S.-S. Guo, W.-X. Xu, B.-G. Du, J.-Y. Liang, L. Wang, Y.-B. Li, Digital twin-based pump station dynamic scheduling for energy-saving optimization in water supply system, Water Resources Management, 38 (2024) 2773–2789.
[48] B. Arsecularatne, N. Rodrigo, R. Chang, Digital twins for reducing energy consumption in buildings: A review, Sustainability, 16 (2024) 9275.
[49] R. Molinaro, J.-S. Singh, S. Catsoulis, C. Narayanan, D. Lakehal, Embedding data analytics and CFD into the digital twin concept, Computers & Fluids, 214 (2021) 104759.
[50] L. Feng, X. Qian, An adaptive learning approach to determine and update crack sizes from strain relaxation data for welded plate joints, Engineering Fracture Mechanics, 259 (2022) 108165.
[51] T. Kreuzer, P. Papapetrou, J. Zdravkovic, Artificial intelligence in digital twins – A systematic literature review, Data & Knowledge Engineering, 151 (2024) 102304.
[52] Y. Fu, J. Liang, Y. Wang, J. Ou, Updating complex boundary conditions using latent system internal forces towards structural digital twin models, Mechanical Systems and Signal Processing, 224 (2025) 112088.
[53] C.V. Dung, H. Sekiya, S. Hirano, T. Okatani, C. Miki, A vision-based method for crack detection in gusset plate welded joints of steel bridges using deep convolutional neural networks, Automation in Construction, 102 (2019) 217–229.
[54] B. Masserey, P. Fromme, Fatigue crack growth monitoring using high-frequency guided waves, Structural Health Monitoring, 12 (2013) 484–493.

[55] L. Feng, X. Qian, Size effect and life estimation for welded plate joints under low cycle actions at room and low ambient temperatures, Thin-Walled Structures, 132 (2018) 195–207.
[56] M. Ho, S. El-Borgi, D. Patil, G. Song, Inspection and monitoring systems subsea pipelines: A review paper, Structural Health Monitoring, 19 (2020) 606–645.
[57] A. Keshtgar, M. Modarres, Detecting crack initiation based on acoustic emission, Chemical Engineering Transactions, 33 (2013) 547–552.
[58] D. Li, K.S.C. Kuang, C.G. Koh, Rail crack monitoring based on Tsallis synchrosqueezed wavelet entropy of acoustic emission signals: A field study, Structural Health Monitoring, 17 (2018) 1410–1424.
[59] T.B. Quy, J.-M. Kim, Crack detection and localization in a fluid pipeline based on acoustic emission signals, Mechanical Systems and Signal Processing, 150 (2021) 107254.
[60] Z. Bai, S. Chen, Q. Xiao, L. Jia, Y. Zhao, Z. Zeng, Compressive sensing of phased array ultrasonic signal in defect detection: Simulation study and experimental verification, Structural Health Monitoring, 17 (2018) 434–449.
[61] I. Farreras-Alcover, M.K. Chryssanthopoulos, J.E. Andersen, Regression models for structural health monitoring of welded bridge joints based on temperature, traffic and strain measurements, Structural Health Monitoring, 14 (2015) 648–662.
[62] L. Feng, X. Qian, A hot-spot energy indicator for welded plate connections under cyclic axial loading and bending, Engineering Structures, 147 (2017) 598–612.
[63] L. Groo, J. Nasser, D.J. Inman, H.A. Sodano, Transfer printed laser induced graphene strain gauges for embedded sensing in fiberglass composites, Composites Part B: Engineering, 219 (2021) 108932.
[64] J.M. Romera, I. Marcos, M. Skaf, V. Ortega-López, An alternative experimental methodology to determine the diagonal cracking resistance of steel-reinforced concrete beams, Engineering Structures, 244 (2021) 112741.
[65] A. Valeev, Condition monitoring based on locating defects via strain gauge analysis, Mechanical Systems and Signal Processing, 182 (2023) 109587.
[66] D. Cristiani, C. Sbarufatti, F. Cadini, M. Giglio, Fatigue damage diagnosis and prognosis of an aeronautical structure based on surrogate modelling and particle filter, Structural Health Monitoring, 20 (2021) 2726–2746.
[67] G. Abaqus, Abaqus 6.11, Dassault systemes simulia corporation, Providence, RI, USA, 2011.
[68] S. Sapna, A. Tamilarasi, M.P. Kumar, Backpropagation learning algorithm based on Levenberg Marquardt Algorithm, Computer Science and Information Technology (CS and IT), 2 (2012) 393–398.
[69] I. Singh, B. Mishra, S. Bhattacharya, R. Patil, The numerical simulation of fatigue crack growth using extended finite element method, International Journal of Fatigue, 36 (2012) 109–119.
[70] P. Tajdary, L. Morin, C. Braham, G. Gonzalez, A reduced single-pattern model for the numerical simulation of multi-pattern metal forming, International Journal of Material Forming (2021) 1–14.
[71] J. Gordon, E.H. Shortliffe, The dempster-shafer theory of evidence, Rule-Based Expert Systems: The MYCIN Experiments of the Stanford Heuristic Programming Project, 3 (1984) 832–838.
[72] F.P. Kopsaftopoulos, S.D. Fassois, Vibration based health monitoring for a lightweight truss structure: Experimental assessment of several statistical time series methods, Mechanical Systems and Signal Processing, 24 (2010) 1977–1997.
[73] A.G. Poulimenos, S.D. Fassois, Output-only stochastic identification of a time-varying structure via functional series TARMA models, Mechanical Systems and Signal Processing, 23 (2009) 1180–1204.

[74] J. Prawin, K. Lakshmi, A.R.M. Rao, A novel vibration based breathing crack localization technique using a single sensor measurement, Mechanical Systems and Signal Processing, 122 (2019) 117–138.
[75] L. Feng, X. Qian, Low cycle fatigue test and enhanced lifetime estimation of high-strength steel S550 under different strain ratios, Marine Structures, 61 (2018) 343–360.
[76] L. Feng, X. Qian, Enhanced crack sizing and life estimation for welded tubular joints under low cycle actions, International Journal of Fatigue, 137 (2020) 105670.
[77] C. Amzallag, J. Gerey, J.L. Robert, J. Bahuaud, Standardization of the rainflow counting method for fatigue analysis, International Journal of Fatigue, 16 (1994) 287–293.
[78] A. Hobbacher, Recommendations for fatigue design of welded joints and components, Springer, 2016.
[79] L. Wang, X. Qian, L. Feng, Effect of welding residual stresses on the fatigue life assessment of welded connections, International Journal of Fatigue, 189 (2024) 108570.
[80] S. Hochreiter, J. Schmidhuber, Long short-term memory, Neural Computation, 9 (1997) 1735–1780.

8 Conclusions and Summary

8.1 SUMMARY AND CONCLUSIONS

8.1.1 Chapter 1: Introduction

Fatigue and fracture mechanics are traditional and long-standing scientific topics, originating from historical disasters. The overall framework scope of fatigue and fracture has been built by brilliant pioneers. Advances in experimental techniques and simulation tools promote understanding, monitoring, and prediction of fatigue and fracture behaviors. The research in fatigue and fracture has witnessed the progress from small smooth specimens to huge complicated structural components; from macroscopic description to microscopic, even nanoscopic mechanisms verification; from metal alloys to composite materials; from analytical foundation to data-driven approaches. Specifically, this book aims to provide more accurate and convenient approaches and methodologies proposed by the authors, to deal with some specific but important assessment topics in fatigue and fracture of welded joints, with consideration of loading modes, crack extension, life-extension, residual stress and monitoring techniques.

8.1.2 Chapter 2: Experimental Approaches to Quantify Fracture Resistance

This chapter presents innovative and simple experimental approaches (hybrid numerical and experimental approaches, and DIC-based approaches) to quantify the fracture resistance of non-conventional specimens (surface-cracked plates and welded pipes with circumferential surface cracks). The framework of the hybrid approach is developed and validated for through-thickness SE(B) specimens under mode I and mixed mode loadings, which requires only a single test specimen to generate the load deformation curve with multiple unloading-reloading cycles, and replaces the remaining test specimens with FE models.

For surface-cracked specimens, the challenges relating to 1) the determination of the varying energy release rate along the curved front of a surface crack from experimentally measurable quantities and 2) the evaluation of the extending curved crack front are tackled with a pre-determined relationship [e.g., Eqs. (2.29) and (2.36)] between the J-values and crack-front location parameters using large-deformation, elastic-plastic analysis, and the fitted polynomial relationship between the *CMOD* compliance and the crack size relationship calculated in linear-elastic FEA [e.g., Eqs. (2.28) and (2.33)].

For surface-cracked plates and pipes, the average value of η equals to 1.1. The J-R curves measured at three different crack-front positions demonstrate similar

DOI: 10.1201/9781003509806-8

levels of fracture resistance, which agrees with the fracture resistance obtained from the C(T) specimens, at short crack extensions. As the load increases, the crack front near the free surface experiences more significant crack extensions than the crack-front material near the deepest crack-front point, due to the higher crack driving forces near the free surface. The *J-R* curves derived by assuming that the crack extends only in the thickness direction remain much lower than the plane-strain *J-R* curves determined from the C(T) specimen.

For pipes with circumferential surface cracks, the η values depend on the *CMOD* [Eq. (2.34)]. For small crack extensions, the pipe specimens exhibit fracture resistance similar to the SE(B) and SE(T) specimens. However, as the crack extends further, the fracture resistance of the pipe specimens surpasses that of the SE(B) and SE(T) specimens.

The DIC provides a contactless approach to measure the load, the deformation, and the crack profile. This chapter also includes the framework of DIC in determining the *J-R* curve of through-thickness SE(B) specimens. The challenges in noise, *J-i*ntegral calculation, and rotation center using DIC have been addressed. In comparison to the traditional determination approach, the proposed DIC method for the *J-R* curve measurement of side-grooved specimens shows significant robustness and accuracy.

8.1.3 Chapter 3: Ductile Tearing and Its Engineering Assessment

This chapter addresses the engineering assessment of ductile tearing in steel structures, emphasizing the integration of fracture resistance measurements into structural integrity evaluations. Specifically, the FAC and TAC curves are discussed in detail for welded tubular joints mainly under monotonic loadings and reversed loadings. Furthermore, this chapter introduces the TAD approach to predict the load-crack behavior of fracture specimens and welded joints.

The FAC, as described in BS 7910 [1], varies based on material properties and the required level of conservatism. It includes: (1) Option 1 FAC independent of structural geometry and with minimum material input; (2) the material-specific Option 2 FAC independent of structural geometry; and (3) the material and geometry-specific Option 3 FAC. Different factors, including the crack front locations, tubular joint types, strain hardening, yield strength, crack depth ratio, etc., have been investigated. Overall, Option 3 FACs provide more accurate failure assessments for CHS X- and K-joints under brace axial loading. The authors proposed FACs for tubular joints from nonlinear regression analysis [Eqs. (3.17), (3.18), and (3.19)], which capture the experimental FACs well.

The ductile crack extension before the final brittle failure has been observed in the experimental tests of tubular joints. Option 1 assessment curve does not provide a conservative estimation of the failure load associated with the brittle fracture observed in the experiment. Option 3 assessment curve demonstrates strong agreement with experimental ductile tearing assessment. Therefore, the proposed framework of TAC construction for a fracture specimen or a cracked structural component with an initial crack size consists of two phases: before and after the onset of ductile tearing. A TAD-based approach for predicting the load versus crack extension behavior of cracked structural components, utilizing material fracture tests and linear-elastic numerical analysis, avoiding computationally intensive damage mechanics models. The results demonstrate that the TAD-based assessments closely match experimental data: load resistance predictions differ by a maximum of 6% for fracture specimens, remain within 10% for surface-cracked pipes after initial crack growth, and vary less than 5% for cracked X-joints. These findings validate the TAD method as an accurate and efficient tool for structural integrity evaluations based on material fracture resistance.

Fracture tests under reversed loading with $R = -1$ show significantly reduced load capacity and ductility in comparison with monotonic loading. The study extends the η-approach ASTM E1820 [2] to estimate fracture resistance for cyclically loaded joints, using load and *CMOD* data. In cyclic fracture tests, smaller displacement increments between load cycles produce lower *J-R* curves. This effect is attributed to an increased number of load cycles that facilitate crack propagation, as well as higher tensile residual stresses at the crack tip. However, under both monotonic and reversed loading conditions, TAD predictions of the joint's P-Δa curve closely match experimental results. Moreover, using fracture resistance curves measured from smaller-scale specimens (SE(T)) also shows good agreement, indicating that specimen-derived data can be reliably used to assess full-scale CHS joints.

8.1.4 Chapter 4: Assessment of Brittle Fracture Failure

This chapter delivers a low-cost yet robust calibration framework for the Weibull-stress parameters, originally developed for through-thickness fracture specimens, to assess the probability of fracture in surface-cracked specimens. The proposed calibration procedures utilize both the extensive fracture toughness data in the European Union project [3], standard specimens made of S690, which show similar magnitudes for cleavage fracture probability compared with the existing calibration method. In addition, A Monte-Carlo subsampling study shows that (i) a constraint contrast of $\geq 15\%$ between high- and low-constraint specimen sets and (ii) $\gtrsim 20$ valid tests per set are required to attain $\approx 80\%$ confidence in the modulus m and scale parameter K_0; inadequate contrast or smaller samples bias m downward and erode safety margins.

This chapter examined the cleavage initiation angle by combining the amount of crack plane rotation with the microscopic measurement of the fracture surface orientation near the crack tip. The measured cleavage initiation angles exhibit significant scatter under mixed-mode I and II loadings. The cumulative distribution of cleavage angle is thickness-independent, load-invariant, and Weibull exponent (m)-insensitive. The coplanar energy release rate criterion demonstrates the closest agreement between the calculated probability of cleavage initiation angle and the experimental measurement.

Ductile tearing prior to the final cleavage failure affects the calculation of the history-dependent Weibull stress values. A small ductile tearing length ($\Delta a \leq 0.2$ mm) leads to very small changes in the fracture toughness data when converted to the 1T plain-strain condition, compared to the fracture toughness converted using the conventional $g(M)$ function. However, the contribution by the prior ductile tearing to the total Weibull stress can be very substantial at relatively large Δa values.

This chapter defines a new Weibull stress σ_{cw} to incorporate the effect of ductile tearing along the crack extension angle under mixed-mode I and II loadings. The value of cumulative Weibull stress σ_{cw} demonstrates strong sensitivity to the cumulative volume of the highly stressed fracture process zone with an advancing crack tip. A new non-dimensional correction function $h(M)$ is proposed and validated to assess the effects of the plasticity-induced and geometry-induced constraint loss, as well as the ductile tearing.

8.1.5 Chapter 5: High-Cycle Fatigue of Welded Joints

This chapter addresses advanced methodologies for the high-cycle fatigue assessment of welded tubular and plate joints, introducing an extrapolation-based approach for determining the ENS of CHS X-joint under various loading conditions, which simplifies FE modeling by avoiding the need for finely meshed rounded local notches. Additionally, a volume-based hot-spot energy indicator is proposed to account for stress gradient effects in both surface and thickness directions, offering improved fatigue life predictions (R^2 = 0.84) compared to traditional point-based Hot-Spot Stress methods (R^2 around 0.6) against 110 fatigue test data of welded plate joints under four-point bending, three-point bending, and axial loading. The hot-spot volume is at 1 mm from the weld toe with 1 mm length in the surface direction and 1 mm depth in the thickness direction, based on the numerically computed stress distribution near the weld toe. This volume incorporates the highly stressed material near the weld toe, over which both the magnitude and gradient of the stresses reflect the size effect and loading effect on the fatigue driving force for the welded plate joints.

Experimental and numerical investigations on concrete-filled tubular X-joints with full-capacity welds confirm the applicability of existing *S-N* curves. The fatigue crack initiation life for all specimens remains lower than that estimated using all three types of the design *S-N* curve (FAT 225 SSN curve in IIW [4], the hot-spot stress-based *S-N* curve in API [5], and the nominal stress-based *S-N* curve for tubular joints in AWS [6]). Compared to the nominal stress *S-N* curve, the fatigue life corresponding to the end of the fatigue test of the concrete-filled joints with weld-toe burr-grinding exceeds the life estimation by the nominal stress *S-N* curve. The fatigue life of the concrete-filled joints without weld-toe burr grinding matches the fatigue life estimated using the nominal stress *S-N* curve. All fatigue tests indicate a final fatigue life well exceeding the fatigue life calculated based on the hot-spot stress *S-N* curve. The notch stress *S-N* curve also conservatively predicts the fatigue life for most of the fatigue tests.

The effect of UIT has been discussed in this chapter on the welded plate joints with pre-fatigued cycles. UIT greatly extends the fatigue life of welded plate joints, with a life extension factor ranging from 1.59 to 8.32, depending on the number of prefatigued cycles. To quantify the effect, this chapter combines the nonlinear damage accumulation las and Paris' law to predict the total fatigue life of pre-fatigued welded joints with *UIT*. A high coefficient correlation (R^2 = 0.98) was achieved.

8.1.6 Chapter 6: Low-Cycle Fatigue of Welded Joints

This chapter investigated the environmental temperatures, size effects, and welding residual stresses in the low-cycle fatigue assessment of welded joints. Different from the high-cycle fatigue of welded plate joints, the size effect in low-cycle fatigue life is negligible based on nominal fatigue prediction indicators. Based on the local energy fatigue driving force, the size effect in the total low-cycle fatigue life is notable, implying that the larger the main plate thickness, the longer the total low-cycle fatigue life. The size effect exists in the low-cycle fatigue initiation life assessment based on the nominal fatigue driving force. The local energy fatigue driving force at the weld toe can integrate the fatigue initiation life regardless of different geometries. Under the same nominal amplitude, the area of the cross-section of the welded plate joints determines the life of the fatigue crack propagation to a large extent. However, the difference in the total fatigue life of the welded plate joints made of S550, caused by the effect of the low temperature, is negligible.

The proposed energy-based fatigue driving force is incorporated into the continuum damage mechanics framework. Therefore, the analysis framework of fatigue assessment in this chapter unifies the treatment for the high- and low-cycle fatigue of welded plate connections. The damage-related material parameters calibrated from the smooth coupon specimens are verified through good agreement with the experimental fatigue life. The threshold value of damage imposes a more substantial effect on joints with a long fatigue life than on those with a short fatigue life. The fatigue life based on the proposed numerical approach aligns with both the high- and low-cycle fatigue data from this study and existing literature.

Furthermore, the above damage simulation incorporates the thermal-mechanical techniques to simulate the welding stress and its evolution of welded joints (S550 welded plate joints and S355 CHS X-joints). The predicted residual stresses effectively capture the relaxation of residual stress before fatigue crack initiation and predict the rapid decrease in residual stresses towards the end of fatigue loading cycles, indicating the accumulation of fatigue damage. The presence of welding residual stresses does not significantly affect the fatigue life prediction for low-cycle fatigue specimens but causes conservative estimations in the high-cycle fatigue life predictions. Before the fatigue crack initiates, the most obvious residual stress relaxation occurs within the first loading cycle, after which the welding residual stress remains relatively stable over subsequent cycles until the initiation of the fatigue crack. Thereafter, a significant decrease in residual stress magnitude manifests due to stress releases caused by the creation of new fatigue crack surfaces.

8.1.7 Chapter 7: Digital Twin-Based Fatigue Assessment of Welded Plate Joints

Digital twin techniques have been reviewed in multiple industries, with summarized critical purposes and challenges. A neural-network (NN) bootstrap particle filtering (BPF) for crack monitoring in welded plate connections has been built. First, an improved NN model is used to build the metamodel of the normalized strain indicators and the crack front profile. More than 90% of the absolute difference in the crack depth is less than 2 mm for both the training and test datasets. The effectiveness of the BPF in this study is verified based on the measured strain values and the crack measured by the ACPD. With limited strain measurement, this study adopts an interpolation method to construct the 21 strain input variables. The performance of the BPF in reducing these three critical indicators is not significant in the initial stage of the fatigue crack propagation, where the crack size is small. With increasing crack size, the BPF can significantly reduce the discrepancy between the predicted crack profile and the measured crack profile.

A comprehensive and illustrative DT framework integrates prior FEM simulation with intelligent techniques, using webcams for displacement and oscilloscopes for strain measurements, which enables fatigue damage prediction, crack detection, alarm triggering and remaining life estimation for cost-effective infrastructure maintenance.

REFERENCES

[1] BS 7910: Guide to methods for assessing the acceptability of flaws in metallic structures: An engineering approach, in, The British Standards Institution, 2019.
[2] ASTME1820, ASTM E1820–22, Standard test method for measurement of fracture toughness, in, ASTM International, West Conshohocken, PA, 2022.
[3] J. Heerens, D. Hellmann, R. Ainsworth, Fracture toughness determination in the ductile-to-brittle transition regime-pre-cracked Charpy specimens compared with standard compact specimens, in: European Structural Integrity Society, Elsevier, 2002, pp. 297–305.
[4] A. Hobbacher, Recommendations for fatigue design of welded joints and components, Springer, 2016.
[5] A.P.I. (API), Recommended practice for planning, designing and constructing fixed offshore platform, in, API RP 2A WSD, 2000.
[6] A.S.W. Committee, structural welding code-steel: ANSI/AWS D1. 1–90, an American National Standard Approved by American National Standards Institute, American Welding Society, January 5, 1990.

Index

Note: Page numbers in *italics* indicate a figure, and page numbers in **bold** indicate a table on the corresponding page.

G

H

I

J

L

M

N

Q

R

S

T

U

For Product Safety Concerns and Information please contact our EU representative GPSR@taylorandfrancis.com Taylor & Francis Verlag GmbH, Kaufingerstraße 24, 80331 München, Germany

Batch number: 10733633

Printed by Printforce, the Netherlands